IEC 61850

Digitizing the Electric Power Grid

For a listing of recent titles in the
Artech House Power Engineering Library,
turn to the back of this book.

IEC 61850

Digitizing the Electric Power Grid

Alexander Apostolov

ARTECH HOUSE

BOSTON | LONDON
artechhouse.com

Library of Congress Cataloging-in-Publication Data
A catalog record for this book is available from the U.S. Library of Congress.

British Library Cataloguing in Publication Data
A catalog record for this book is available from the British Library.

ISBN-13: 978-1-63081-884-5

Cover design by Mark Bergeron

685 Canton Street
Norwood, MA 02062

10 9 8 7 6 5 4 3 2 1

Contents

CHAPTER 6

Preface

Since the beginning of my career as a protection engineer almost half a century ago, I have dreamed of replacing all the hardwired interfaces in substations with something much more flexible and controlled by a computer code that will give us the flexibility to adjust the protection scheme to the changed power system topology or system conditions. It was quite frustrating to make all the drawings by hand and to enter the system model for short-circuit studies or dynamic stability analysis using punched cards that did not even have the printed values of the data in the card on top of it.

Software-based protection was just a dream, but there were some signs that this might be possible. It was George Rockefeller's paper, "Fault Protection with a Digital Computer," *IEEE Transactions on Power Apparatus and Systems,* April 1969, that showed, at least to me, that it is possible to break away from the world of electromechanical relays that were prominent at the time.

Half a century later, we have in our pockets multifunctional devices that integrate a phone, television, radio, music player, Global Positioning System (GPS) receiver with maps, photography and video cameras, and many other things that used to be individual devices in the past. You do not need to go to the library anymore to find information, and you do not need to talk to a travel agent or go to the bank; we live in a digital world that is based on high-performance computers and communications. Most of us have accepted the digitalization of our lives because of all the benefits that it brings. However, it is not exactly the same with the protection, automation, and control (PAC) community. We have been using microprocessor-based devices for several decades, as well as communications for accelerated protection schemes. It has been more than 15 years since the publication of Edition 1 of IEC 61850, which laid the foundation for the development and implementation of digital substations. Although some forward-looking utilities are embracing this new technology and taking advantage of the benefits that it brings, there are still many PAC specialists who are not convinced that this is the way to go. This new technology with which they are not familiar is different from the way they always did it; this is the only explanation that I can find for the resistance to the digitization and digitalization of the electric power grid.

After working for about quarter of a century on the development of Utility Communications Architecture (UCA) 2 and International Electrotechnical Commission (IEC) 61850, after traveling all over the world, and talking about it at conferences, seminars, and workshops, I became more convinced that I needed to write this book. When in 1996 I started learning about object modeling and communications, I realized that many people still do not know about this subject. There is a very complex transition from the world of hardwired PAC systems to digital substations. The goal of this book is to provide a big picture of what digitization and digitalization are and how they can be achieved, and I attempt to answer the three questions that I always ask when I start working on something:

- What are we doing?
- Why are we doing it?
- How are we doing it?

The answer to the first one is quite simple: we are talking about digitizing the grid.

The answer to the second one is a little bit more complicated, but it can be summarized that we are doing it because the electric power grid is not the same anymore and we also have a lot of new technology that is available to us today that we did not have in the last century.

The answer to the third question can be summarized by saying that we are doing it by using IEC 61850.

These short answers are covered in a lot more detail in the book. Each chapter highlights the issues of its particular topic to help the reader to understand what it is and where it fits. The space is very limited, especially if we consider that the material in any chapter can be covered in a dedicated book. The idea is to present the fundamentals on each topic that can help the reader to understand:

- How the electric power grid is changing;
- How the requirements for PAC systems are changing;
- What the available technology is;
- What the IEC 61850 standard is;
- What role the IEC 61850 standard plays in achieving the digitization and digitalization of the grid.

I hope that, after reading the book, anyone who had doubts about needing to change the way that PAC systems work from hardwired to communications-based understand that this is the only way that we will be able to meet the requirements for the reliability, security, and efficiency of the electric power grid and that IEC 61850 is the technology that allows us to do it.

CHAPTER 1

Smart Grid Definition

1.1 Introduction

In the last few years, smart grid has been one of the buzzwords in the electric power industry that draws a lot of attention, both inside and outside our industry. Our industry is going through a major shift in the use of advanced twenty-first century technology in an effort to successfully meet today's challenges and those of the future, which, for the first time, brings together people from different domains who used to work in their own field, knowing their responsibilities, technologies, and processes.

In the past, the structure of the industry and the roles of the different players were defined around a separation of functions, both on the user side and the supplier side. Each group within an electric power utility designed its own system and purchased, installed, commissioned, and maintained its own equipment. The equipment that each group bought was designed and produced by different manufacturers or, in many cases, by different divisions of the same major supplier. This was appropriate, considering that there were dedicated single-function devices for everything: measurements, status indication, control, automation, and protection. There were also a large number of well-trained specialists in each of these domains.

Now we live in a different world in which everything is changing. It is difficult to describe the device being installed in the substation. It provides everything: measurements, status indication, control, automation, and protection. So who is responsible to design, purchase, install, commission, and maintain the device? Where are all the qualified specialists who can refurbish the aging substations all over the world?

The electric power system that needs control and protection is changing as well. In the past, things were relatively simple; we had large power stations connected to the transmission grid that was connected through power transformers with the subtransmission or distribution systems. In the distribution system, everything was radial, making it easy to control and protect. Through more than a century of experience, we have learned what the behavior of synchronous generators is during different system conditions, how to model them for different studies, and how to control and protect them.

With the growing concerns related to global warming and the push for green power, the penetration of distributed energy resources (DERs) of different types is making everything completely different. They can be installed anywhere as individual machines or in large numbers in farms. For many of these DERs, we do not know for sure when they will be available. Their behavior during short-circuit faults or other abnormal system conditions is very different from the synchronous generators that we know and depends on the type of DER. At the same time, connecting them to distribution feeders completely changes the requirements for protection and control (Figure 1.1).

It is clear that something has to change, or maybe it is better to say that everything has to change. The hope is that the smart grid is the solution to all our problems.

The push for a smart grid as a solution to all the challenges that the electric power industry is facing is spreading to all continents and requires that specialists in the protection, automation, and control (PAC) field have a look and think about the most efficient way of achieving the reliable and secure operation of the grid.

In the ongoing evolution of the grid, vendors, utilities, and standard bodies have not been idle. For more than two decades, the PAC community has been implementing advanced, microprocessor-based, multifunctional, intelligent electronic devices (IEDs) that can help speed up and reduce the costs of the deployment of the smart grid.

George Rockefeller's groundbreaking paper, "Fault Protection with a Digital Computer," published in April 1969 [1], started a revolution in our industry that resulted in the development of microprocessor-based protection relays later in the twentieth century, followed by the multifunctional IEDs that are commonly used today. For many years, our industry has been working on innovative solutions that

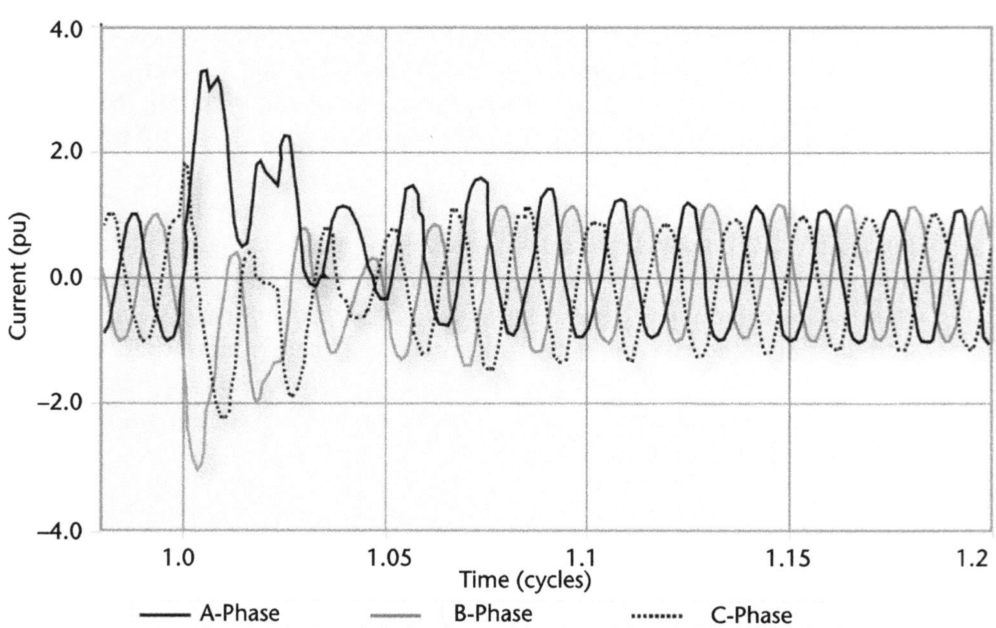

Figure 1.1 Short-circuit current from a double-fed (Type-III) wind generator.

are making PAC systems smarter by allowing them to adapt to changing system and fault conditions, using advanced methods for electric signal processing and continuously monitoring themselves and their environment.

At the same time, the IEDs were adding more nonprotection functions such as measurements, event and fault reporting and recording, disturbance recording, synchrophasor measurements, and user-programmable scheme logic that can meet specific application requirements.

The ability of these multifunctional devices to communicate with their peers and different utility clients supports the development of distributed applications not only within the substation, but also between substations and across different regions of the electric power system.

All of the above support our claim that the world of PAC has been getting smarter for quite some time. Now we can make it even better, by integrating devices through communications and making them the cornerstone of the smart grid.

What is actually the smart grid? In order to avoid any misunderstandings, it is important to have a common definition. Unfortunately, it is not possible to define the smart grid with a few words, unless we say that a "smart" grid is a grid that is "not stupid." This does not help, so we will start with the two visions, the European one and the North American one.

The European SmartGrids Technology Platform: Vision and Strategy for Europe's Electricity Networks of the Future [2] was published in 2006. This vision defines a bold program for the research, development, and implementation of a strategy for modernization of the European grid that will meet Europe's needs in the future. According to this vision, Europe's electric power grid must be:

- *Flexible:* Fulfilling customers' needs while responding to the changes and challenges ahead;
- *Accessible:* Granting connection access to all network users, particularly for renewable power sources and high-efficiency local generation with zero or low carbon emissions;
- *Reliable:* Assuring and improving security and quality of supply, consistent with the demands of the digital age with resilience to hazards and uncertainties;
- *Economic:* Providing best value through innovation, efficient energy management, and level playing field competition and regulation.

A little different definition can be found in the U.S. Energy Independence and Security Act of 2007 [3]. The smart grid is defined as a reliable and secure electricity infrastructure that can meet future demand growth based on the following, which together characterize a smart grid:

- Increased use of digital information and controls technology to improve reliability, security, and efficiency of the electric grid;
- Dynamic optimization of grid operations and resources, with full cybersecurity;

- Deployment and integration of distributed resources and generation, including renewable resources;
- Development and incorporation of demand response, demand-side resources, and energy-efficiency resources;
- Deployment of smart technologies (real-time, automated, adaptive, interactive technologies that optimize the physical operation of appliances, consumer devices, and industrial equipment and processes) for metering, protection, monitoring, control, and communications concerning grid operations and distribution automation;
- Integration of smart power system devices (transformers, breakers);
- Integration of smart appliances and consumer devices;
- Deployment and integration of advanced electricity storage and peak-shaving technologies, including plug-in electric and hybrid electric vehicles, and thermal-storage air conditioning;
- Provision to consumers of timely information and control options;
- Development of standards for communication and interoperability of appliances and equipment connected to the electric grid, including the infrastructure serving the grid;
- Identification and lowering of unreasonable or unnecessary barriers to adoption of smart grid technologies, practices, and services.

What are common and essential in both definitions is that the smart grid is efficient and reliable, based on advanced digital and communications technologies that allow us to integrate DERs of different types anywhere in the electric power system.

In order to achieve this goal, specialists from all different domains in the electric power industry need to continue to work together, to define functional, performance, communications, and tool requirements that can be used in the most efficient way to build the smart grid based on the integration of different functional elements and systems. At the same time, we do not have to start from scratch, but to build the smart grid based on the already available technologies and the lessons that we have learned in the last few decades.

1.2 Power System Hierarchy

As has been already mentioned, the introduction of DERs represents a significant change in the way that electric power systems operate, due to the fact that now we may have energy resources at any level of the system (Figure 1.2). As a result, the principles for PAC that were used in the last century, typically at the transmission level of the system can now be applied at any level of the hierarchy. This requires a proper definition of the hierarchy of the electric power system and the functions that are subject of the efforts to develop the smart grid of the future.

One of the key requirements in the development of any PAC system is the improved reliability and efficiency of operation. Advancements of communications and protection and control technology can lead to significant improvements in

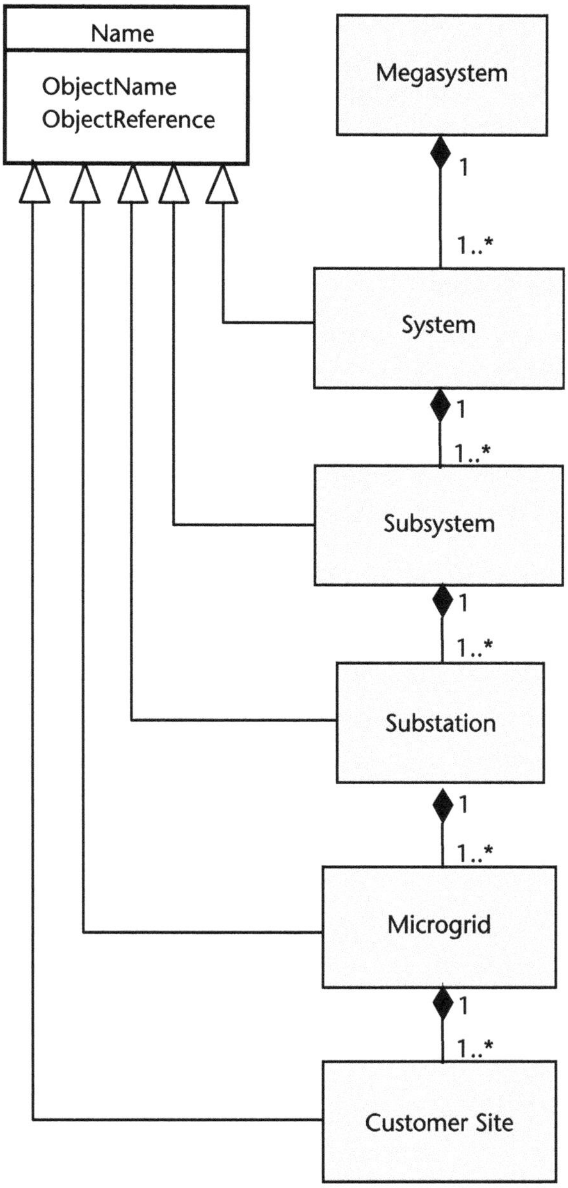

Figure 1.2 Simplified UML model of electric power system.

the efficiency of operation and the reliability of the power system, considering the availability of distributed generators at all levels of the electric power system.

Understanding of the system control hierarchy and development and implementation of new concepts for monitoring and control of the different levels of this hierarchical system are much easier and will bring immediate improvements in the operation of the system.

1.2.1 Megasystem

The megasystem is the collection of all interconnected electric power systems in a large area and contains all individual systems usually controlled by a system operator, as well as the interconnections between them (Figure 1.3). An example of such a system is the United States, which has three electric power systems: East, West, and Texas. At this stage, however, it does not have a global protection, monitoring, and control system. The experience from the North American blackout in August 2003 indicates that there is a need to consider such a system in order to prevent the occurrence of similar events in the future.

1.2.2 Power System

A power system is the collection of all interconnected electric power subsystems in a region and contains all individual subsystems usually owned and controlled by a single utility, as well as the interconnections between them (Figure 1.3).

The development and implementation of PAC systems at the different levels of the system hierarchy require the communications between the different components of the system at the same hierarchical level or between different levels of the hierarchy.

The conventional utility electric power system hierarchy includes the following levels from the transmission and distribution point of view (from the top down):

- *Bulk power system:* Typically 230 kV and above nonradial transmission lines and substations, including interconnections with neighboring systems;
- *Transmission power systems:* Typically 230 kV and below (down to 69 kV) nonradial transmission lines and substations;

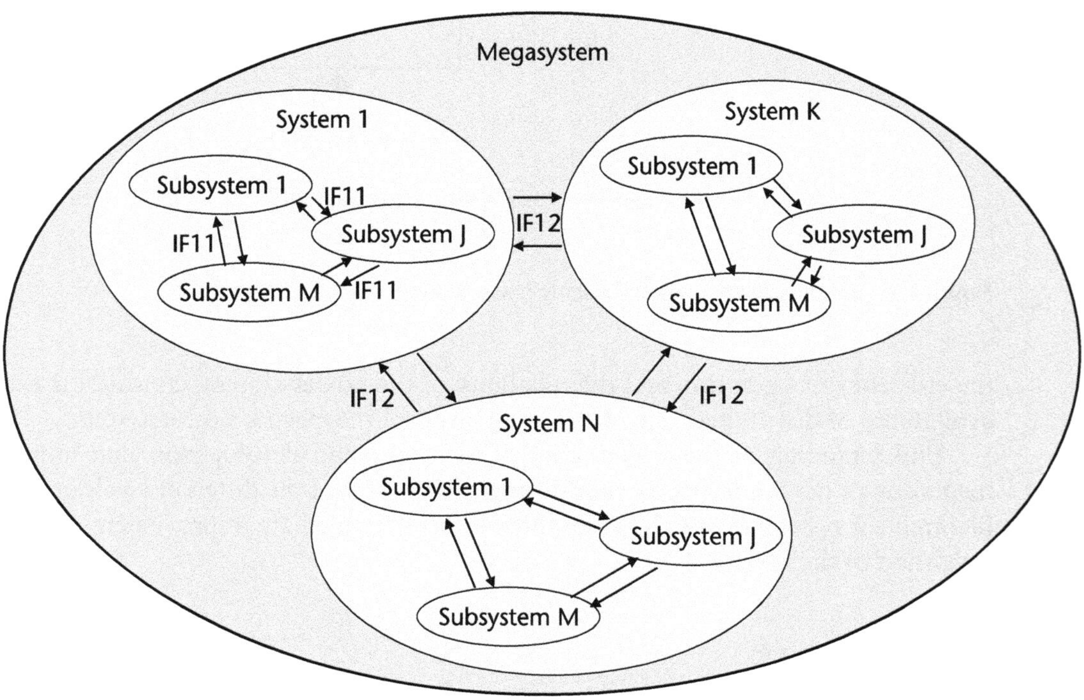

Figure 1.3 Interfaces in the megasystem.

- *Subtransmission:* Typically 69 to 34-kV radial or nonradial lines and substations;
- *Distribution:* Typically radial 34 kV and below distribution circuits.

The voltage levels above are just typical values, but should not be considered as the determining criteria used to classify a system as bulk, transmission, or distribution. The size of the system and the criticality of its components should also play roles in the classification.

Generating stations in the typical utility have been connected to the transmission or bulk system (Figure 1.4). At the same time, some smaller stations might be connected to the subtransmission system.

The control of the power system during abnormal conditions has been typically based on decisions by the system operator or a limited number of underfrequency and undervoltage protection schemes, as well as special protection or remedial action schemes. Some of the characteristics of these solutions are that, in many cases, they operate when it is already too late to execute any optimal control and that they work after the uncontrolled separation of the system and the formation of islands of different sizes and levels of load-generation balance.

The changing utility environment requires the development and implementation of new solutions that will allow the optimization of the protection and control of the electric power system under normal or abnormal system conditions.

The increasing number of DER and their connection to the distribution system lead to a significant change in the characteristics of the distribution system that require the development of new methods for the protection and control of this level of the electric power system hierarchy.

Figure 1.4 Power grid.

Some new concepts have been introduced and discussed in the industry with pilot projects being used to demonstrate the feasibility of these solutions. Some examples are given in the following sections of this chapter.

The success of International Electrotechnical Commission (IEC) 61850 in covering the needs of substation PAC applications provides a good foundation for expanding the models and covering all utility communications. Protection and advanced control of the transmission lines and innovative protection and control of distribution systems, especially the need to cope with DERs on the demand side of the electric power system, impose new requirements for communications between different components of the system.

If we take another look at the power system hierarchy described at the beginning of this section, we can define the following components of the power system hierarchy from the point of view of power system control and communications requirements:

- *The utility power grid (transmission system):* They are the backbone of the utility electric power system that is considered as a subsystem above. They have been the main subject of development for protection, monitoring, control, and energy management for many years.
- *Transmission system islands:* They typically form as a result of the tripping of multiple bulk or transmission lines that isolate a subpart of system from the rest. As this is an uncontrolled action, it is rare that even an approximate load-generation balance is achieved at the time of islanding. After the operation of underfrequency load-shedding schemes, it might be possible to achieve a post-disturbance balance. If, at the moment of separation, the generation is more than the load, generator controls based on overfrequency elements will balance the load and generation. However, this means that valuable generation will be lost that may be used in other areas of the power system outside of the island.
- *The subgrid:* The subgrid is a new concept that can be defined as a controlled transmission island. Based on continuous monitoring of the power flow on the transmission lines, as well as the load and generation in different areas of the system; in case of an emergency, the grid can be split in subgrids with a relative balance between the load and the generation.

  As can be seen from Figure 1.5, under normal conditions all subgrids operate in parallel as part of the grid. They need to be continuously monitored in order for the grid control system to be able to determine the state of each subgrid balance.

  Each subgrid contains one or more substations that can be considered as mini-grids that are increasingly including different types of DERs. Each substation may contain distribution feeders with one or more microgrids (grid in Figure 1.5).

  The subgrid balance can be determined based on different sources of information:

  - System configuration;
  - Generation;

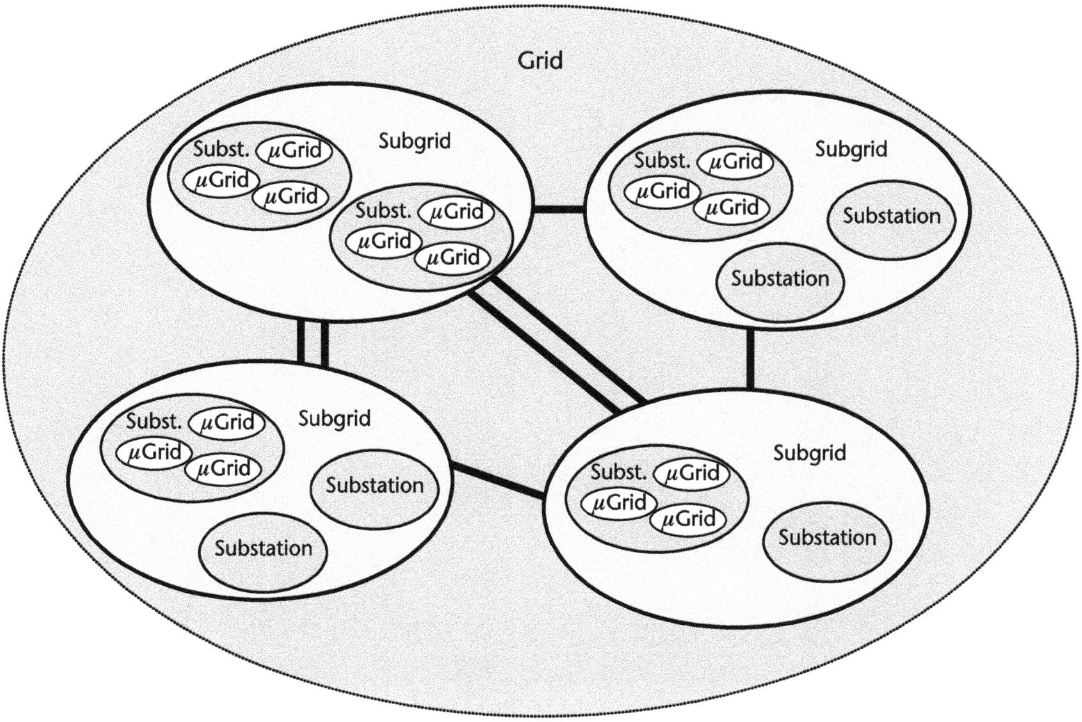

Figure 1.5 Grid hierarchy.

- Loads;
- Power flow between the subgrids.

Based on this real-time data, the control system can make a decision during an emergency to split and put one or more subgrids into an island mode.

The subgrid control system is discussed here in relation to the control strategy of substations connecting the transmission and distribution levels and defined as mini-grids and microgrids. A subgrid will typically include one or more generation sources. It will also include one or more virtual power plants (VPP). A VPP (Figure 1.6) in this case is defined as the total power produced by all DERs in the mini-grid system.

As each substation connected to the subgrid transmission system will be considered as a mini-grid in the case that it has some form of DERs, the control system of the subgrid will know the level of balance of each substation based on the power flow through the power transformers.

- *The substation (mini-grid):* In traditional systems, the distribution system of a substation (Figure 1.7) is supplied through one or more power transformers and does not have any other active source. This has changed with the introduction and increasing number of distributed generators. It might be possible to balance the combined output of all the distributed generators and local storage with at least the critical loads connected to the distribution system. This will require the ability to dynamically shed noncritical load from the system.

Figure 1.6 VPP.

Figure 1.7 Substation: the mini-grid.

This allows the definition of the distribution system of a substation as a mini-grid. It can be isolated from the rest of the grid when all power transformers are taken out of service due to fault, abnormal system condition, or intentionally in order to reduce the effect of a wide area disturbance on the distribution system.

Monitoring and controlling the balance in the substation distribution system will be one of the main tasks of the substation PAC system (SPACS).

A substation's distribution system will typically include one or more distributed generators and wind or other renewable energy parks connected directly to the distribution feeders with the main goal of selling energy to different users in the distribution system.

It will also include one or more micro-VPPs. A micro-VPP in this case is defined as the total power produced by all DERs in a microgrid.

- *The microgrid:* An area or customer facility connected to the substation distribution system will be considered as a microgrid in the case that it has some form of DER (Figure 1.8) and can be disconnected from it. The control system of the substation will know the level of the balance of each microgrid based on the power flow through the distribution transformers at the customer interconnection point.

  Microgrids were described earlier in this chapter. They include one or more different types of DERs and, in conjunction with a load-generation

Figure 1.8 Microgrid.

balancing controller, may support the full load of the microgrid or a subset of critical loads in the case of separation from the mini-grid. If the total output of the VPP is more than the load at any moment in time, the microgrid will export power to the mini-grid.

- *The nano-grid:* With the introduction of very small (a couple hundred watts) solar panels, wind turbines, or other types of DERs that can be directly plugged into a home's outlet to supply part of the energy required by the household, we can now talk about a facility-based nano-grid at the low voltage level of the electric power system (Figure 1.9).

The control system of the smart home will know the load-generation balance of the nano-grid based on the power flow through monitoring of the customer's meter and DER power output. Depending on the available load and generation, it can act at different times both as load or source and in the case of wide area or local disturbance can be isolated from the rest of the distribution system in order to support the critical loads in the nano-grid.

1.3 Grid Control Concept

All of the above-described hierarchical components of the power system have one thing in common: they allow the separation of a balanced load-generation area at each level of the electric power system. This is possible due to the fact that electric power systems in the industrialized world are going through the period of change described above, from being systems with mainly transmission-distribution

Figure 1.9 Nano-grid.

structure and with centralized generation, characterized by requirements for system security, to being consumer-orientated, economically optimized, green energy supply systems with DERs at all levels, characterized by the following features:

- Different types of integrated dynamically changing energy resources are included in the decentralized energy supply concept.
- Improved economic efficiency of all renewable energy resources use available storage and controllable loads at all levels of the system while meeting different consumer requirements.
- There is real-time monitoring of the available energy production and the actual load demand.
- There is optimization of local and inter-grid energy transfers at all levels of the system hierarchy described earlier.

The continuously improving advanced communications and information technologies, especially the development and implementation of cloud computing and services, allow the application of these new systems by combining in the most efficient way new and emerging energy technologies with existing or evolving management functions:

- Energy management of dynamically changing generation, consumption, and storage;
- Load management through DSM, smart metering, and billing contracts;
- System management based on improved information and services, combined with changing organization.

The most important results from these changes will be the improved efficiency, reliability, and quality.

By integrating traditional energy resources and DERs with conventional and smart loads using advanced communications and sophisticated applications based on distributed intelligence, it will be possible in the future to optimize the control, automation, and protection of the overall electric power system and meet future requirements and challenges.

This process can be implemented simultaneously from the top down and from the bottom up based on a changing engineering process using an object-oriented approach to the standardization of PAC systems at all levels of the grid hierarchy discussed above. Standard PAC schemes at the bay, voltage, or substation level can be reused, thus improving the efficiency of the development, installation, and maintenance of PAC systems.

This represents a paradigm shift of the existing approach to the engineering, development, and maintenance of the power systems based on a structure with large power plants connected to the transmission system. With a significant number of DERs being integrated in power systems around the world, in some cases, about 30% to 40%, a fresh look at the different functions in a smart grid is required to support their seamless integration into the networks of the future.

The type of generator, especially whether it is nonvariable or variable (Figure 1.10), has a significant impact on the challenges in its integration and the functions that will be used to support its PAC.

1.4 Reliability

As stated previously, improvement in power system reliability is one of the key characteristics of the smart grid. Reliability is defined generally as the probability that a system performs correctly during a specific time duration. Based on this, electric power reliability can be defined as the degree to which the performance of the elements in an electric power system results in electricity being delivered to customers within accepted standards and in the required amount. The degree of reliability may be measured by the frequency, duration, and magnitude of local, regional, or systemwide disturbances or power interruptions.

The North American Electric Reliability Corporation (NERC) defines the reliability of the interconnected bulk power system in terms of three basic criteria:

- Adequacy;
- Operating reliability;
- Adequate level of reliability.

Adequacy is defined as the ability of the electric power system to supply the required electric power and energy to meet the needs of consumers. This should take into consideration scheduled maintenance outages or probable failures of system components caused by weather conditions or other factors. This means that

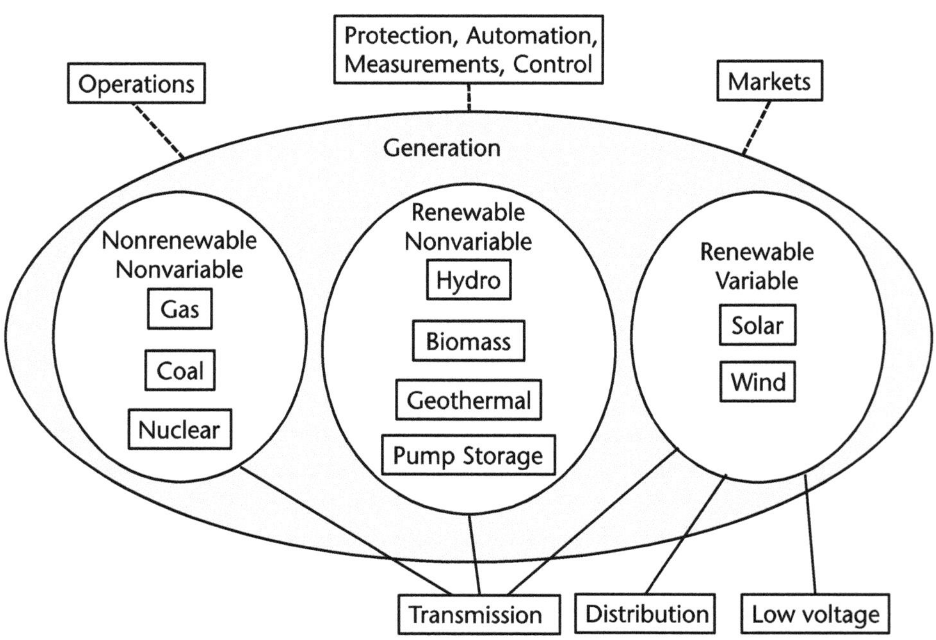

Figure 1.10 Generation integration in electric power grids.

the control systems should continuously monitor the balance between supply and demand in an area of the electric power system at the different levels of the system hierarchy described earlier.

In the conventional synchronous generator-based grid, the balance between generation and load is maintained typically by monitoring the system frequency that depends on it. However, with the continuously increasing penetration of distributed renewable energy resources at all levels of the electric power grid and the fact that they are using inverter-based interfaces to the grid, frequency is no longer a reliable indicator about the level of balance between load and generation. This determination requires the application of robust communications-based continuous monitoring of the output of the DERs and the levels of loading in different areas of the system. This task is complicated by the fact that many of the wind and solar-based resources may have a behavior that is difficult to predict because they depend on the weather conditions and most small solar is unmonitored/measured. This will require the addition of different forms of energy storage at all levels of the system in order to meet the adequacy requirement.

In some cases, the balance between load and generation can be maintained by automatic or system operator controls such as load shedding, voltage reduction, or brownouts.

Operating reliability is described as the ability of the electric power system to withstand the impact of system faults, component failures, protection system misoperation, and human errors during maintenance testing. Traditional grid events such as faults may have an impact on system stability that requires the implementation of protection functions that result in fast fault clearing. The most severe event traditionally is considered a breaker failure during clearing of a short-circuit fault at the transmission level of the electric power system. The use of communications-based accelerated protection schemes is the main tool for reducing the fault clearing times.

In today's electric power systems with DERs, the requirement for fast fault clearing applies for faults at all levels of the electric power system. It is the result of the need for support of the ride-through capability of the DERs. The fact that many DERs have inverter-based interfaces and they do not supply significant short-circuit fault currents makes the traditionally used at the distribution-level overcurrent protection devices unsuitable, which requires the implementation of different methods for detecting short-circuit faults or other abnormal conditions.

Avoiding protection misoperations can be achieved by the use of advanced protection coordination tools and adequate testing of the protection schemes based on power system simulations. Avoiding human errors during maintenance testing can be accomplished by increasing the levels of condition monitoring of the primary electric power system equipment that will allow switching from scheduled maintenance to condition-based maintenance. This will eliminate the need for maintenance testing and the associated power outages.

Meeting the adequacy and operational reliability requirements results in achieving an adequate level of reliability (ALR). This means that the electric power system does not experience the following under normal operating conditions when subject to predefined disturbances:

- Instability;
- Uncontrolled separation;
- Cascading events;
- Voltage collapse;
- Unacceptable frequency variations;
- Unacceptable voltage variations.

At the same time, the electric power system should be able to withstand low-probability major disturbances caused by weather events, cybersecurity attacks, or uncontrolled system component-based outages. Following such events, the system restoration should be executed in the most efficient way based on predefined strategies in a coordinated and controlled manner. A special case of restoration is Black Start, which is the process of bringing the grid back to life after a blackout.

The IEEE 1366-2012 Guide for Electric Power Distribution Reliability [4] introduces a set of terms and definitions that can be used to establish uniformity in the development of distribution service reliability indices, to identify factors that affect them, and to help in defining consistent reporting practices among utilities.

The three most commonly used reliability indices defined in the standard are the System Average Interruption Frequency Index (SAIFI), the System Average Interruption Duration Index (SAIDI), and the Customer Average Interruption Duration Index (CAIDI), where:

- SAIFI is the average frequency of sustained interruptions per customer over a predefined area. It is the total number of customer interruptions divided by the total number of customers served.
- SAIDI is commonly referred to as customer minutes of interruption or customer hours and is designed to provide information for the average time that the customers are interrupted. It is the sum of the restoration time for each interruption event times the number of interrupted customers for each interruption event divided by the total number of customers.
- CAIDI is the average time needed to restore service to the average customer per sustained interruption. It is the sum of customer interruption durations divided by the total number of customer interruptions.

These and many other indices provide valuable information and can be used to establish the impact of the transition to the smart grid. However, there are many challenges related to the inconsistency of their application by different utilities.

1.5 Electric Power System Security

Electric power system security is closely related to reliability and refers to the degree of risk in the ability of the electric power system to survive contingencies without interruption of customer service. It depends on the operating conditions of the system, as well as the probability for occurrence of disturbances.

Power system security is dependent on the design of the electric power grid and the utility's ability to keep pace with the demand for electricity. In the last century, when most of the electric utilities had been national institutions or regulated entities, the requirements for security in the system design were met by building sufficient generation and spinning reserves to ensure the adequacy of supply under all system conditions. Similarly, transmission system performance was continually evaluated and expanded as needed to ensure that stable power can be delivered.

With the transition from centralized generation to systems with large penetration of DERs, many new challenges to power system security emerged. The environmental concerns related to the use of nuclear power are leading to the shutdown of large nuclear power stations that were providing the base of the electric power supply. This resource now is disappearing and is being replaced by large, variable, renewable energy resources, which do not provide the same level of security due to the fact that most of them depend on weather conditions. At the same time, the fact that they are connected to the electric power grid through inverter-based equipment that behaves very differently from the synchronous generators and depends on the control algorithms implemented by its designers makes it difficult to model the impact of electric power system events on the behavior of the grid and its security. In order to at least partially address this challenge, it becomes very important that the state of the electric power grid and all its main components are continuously monitored during operation to ensure that a sufficient security margin exists at all times.

The availability of synchrophasor measurement functions in multifunctional protection IEDs makes it possible to transition from state estimation to state observation and dynamic state estimation, which will significantly help in determining the security margins available for the operation of the grid.

Another challenge has to do with the secure operation of the protection and control systems. This means the degree of risk in the ability of a protection device to correctly detect a short-circuit fault or other abnormal system event under varying electric power system conditions. Protection relays have been developing for more than a century based on our understanding of the transient and dynamic behavior of synchronous generator-based systems characterized by significant amounts of short-circuit current contributions.

The inverter-based interfaces of the DERs do not provide significant levels of short-circuit currents (typically 1.1 per unit (p.u.)) and even if they provide some, it is for a very short period of time after the fault inception, which is very challenging for the traditional protection devices.

The transition of the traditional electric power systems into smart grids based on advanced computer and communications technologies offer significant benefits in improving the reliability, security, and efficiency of grid operations. However, everything has a price and one of the challenges that we are facing is the introduction of cybersecurity vulnerabilities that need to be addressed in order to protect the electric power system and the electricity supply chain.

There is a dedicated chapter to the topic of cybersecurity later in the book, but here we just need to mention briefly that there are two main threats to the electric power utility industry:

- State-sponsored hackers that attempt to gain control of electric power system assets that will allow them to cause electricity supply interruptions or other disturbances in the electric power grid;
- Individual hackers looking for monetary gains by penetrating the secured parameter of a utility and gaining access to sensitive information.

This requires a very serious consideration of the cybersecurity threats and their risk analysis, as well as the identification of critical assets that will have the most significant impact on the security of the grid if subjected to a successful cyberattack.

While cybersecurity is a major concern, the impact of a cyberattack may be limited to a temporary power interruption. However, a physical attack, for example, on a transmission substation, may have a significant impact on the reliability and security of the affected electric power grid. That is why it is important to identify and protect critical transmission stations and substations and their associated primary control centers that, if damaged as a result of a physical attack, could result in widespread instability, uncontrolled separation, or cascading disturbance in the grid.

1.6 Efficiency

Improvement in efficiency is one of the main characteristics in the definition of the smart grid. We need to clearly define what do we understand by efficiency because sometimes there is confusion between the terms effectiveness and efficiency.

Effectiveness is defined as the degree to which objectives are achieved, without consideration of the resources being used. This means to establish a goal and use whatever resources are available to achieve this goal without worrying what is the total cost and the time it takes to accomplish it. Unfortunately, this is the approach that many people take because it requires some investments to analyze the different procedures or processes used in the design, installation, and maintenance of primary and secondary electric power system equipment. There are people that believe only in what they already know, expressed by the principle: That is the way that we always did it. They do not think that, with the development of primary and secondary equipment based on new technologies, everything can be done in a much more efficient way.

Efficiency is defined as the extent to which a resource is used in order to effectively achieve an objective. It is clear that this adds another layer to the analysis of how to approach any task.

Several different resources may be considered in the analysis, such as:

- Equipment purchasing costs;
- Spare equipment costs;
- Engineering costs;
- Installation costs;
- Space;
- Time;

- Human resources;
- Many others.

Different chapters of the book cover many examples of how digitization and digitalization of the electric power grid improve the efficiency of its design, installation, operation, and maintenance. An example is the optimization of the functionality of a protection IED for a specific application by identifying the exact protection, monitoring, and control requirements for a transmission line protection and selecting a device model that meets these requirements as closely as possible. It may save some money in the purchase price. However, as the requirements for each individual application may vary from one to another, doing specific device selection every time will use a time resource in the selection process and may result in a huge variety of different models and a large stock of spares that need to be maintained.

If, instead of taking the above approach, an effort is made to identify a super set of functions that cover all required applications and then purchase equipment that meets this higher level of functional requirements, it may cost more money per individual device, but actually will cost less money in total because of the much smaller number of devices that need to be maintained in stock.

An additional benefit will be that the testing and maintenance specialists will need to familiarize themselves only with a single specific device that will significantly improve the efficiency of their work.

An example of the impact of digitization on the improvement of efficiency is the installation costs for the analog interface between the instrument transformers and the IEDs in the substations. In traditional substations, this interface is based on copper cables located in trenches that have to provide all the three-phase current and voltage circuits. They have to be measured, cut, laid out in the trenches, connected to the terminal blocks at each end, and tested to make sure that everything is working properly. The difference in a digital substation is that many of these cables can be replaced with a single fiber cable that takes much less space, is much easier to install, does not require any terminal block connections, and can be commissioned much faster using the available digital testing tools.

Using smart grid-related routable wide area communications allows us to implement accelerated protection schemes for a fraction of the cost of the traditionally used dedicated communications channels. This results not only in significant cost savings, but also improves the efficiency of operation of the protection system based on the reduced fault clearing time.

1.7 Digitization Versus Digitalization

The transition of the electric power industry from the conventional into a smart grid has been made possible only because of the continuous evolution of computer and communications technology. This evolution enables the development of advanced PAC functions in microprocessor-based IEDs that significantly improve their functionality and also converts a protection device into a measuring device, recorder, and event reporter and a condition monitor into a single physical device.

The switch from electromechanical and solid-state protection devices to numerical relays implementing advanced protection and control algorithms requires the conversion of the secondary analog signals from the instrument transformers into digital format that can be processed within the relay. This was the main shift in the industry that initiated the process of its digitalization.

Here it is important to highlight that we need to distinguish between two terms: digitization and digitalization. They sound quite similar, but actually have very different meanings that need to be clearly understood.

Digitization is basically the process of taking an analog signal and converting into a digital data stream that can be communicated, processed, and stored.

In conventional IEDs, the current and voltage signals are hardwired into the analog inputs of the relay, scaled and filtered to the input level of the analog-to-digital (A/D) converter, and digitized. After digitization by the A/D converter, the digitized values are made available to the various algorithms in the device. Today, in digital substations, this process is performed by process interface devices (merging units) that may be located in the substation yard, which, upon collecting a set (sets) of values, proceed to publish them over the substation local area network.

Another example of digitization is the replacement of the hardwired interfaces between the different protection devices (Figures 1.11 and 1.12) working together in a specific protection scheme with communications messages exchanged over the substation network.

Digitization is also the storage of transient recordings in a digital format for further processing by different analysis tools after a short-circuit fault or other power system event occurs.

We can also see digitization in the engineering process where today we have tools available that will scan an old, printed substation one-line diagram and con-

Figure 1.11 Electromechanical protection relays.

Figure 1.12 Microprocessor multifunctional IEDs.

vert it into a digital Extensible Markup Language (XML-)based file for further processing by different engineering or testing tools.

There are still ongoing discussions around the definition of digitalization, but more people tend to agree that digitalization refers to the conversion of processes or interactions into their digital equivalents. The Covid-19 pandemic and the resulting travel restriction and physical distancing requirement accelerated the reorganization of many business activities around digital technologies. The processes and interactions may not be fully digital, but simply rely more heavily on digital tools than they previously did.

In the past, substation protection and control project documentation was kept as printed documents such as one-line and three-line diagrams and signal lists, in multiple folders that have to be manually processed by the specialists. Today the same documentation is available for digital substations in an XML system configuration file that can be used to perform many different tasks such as testing.

Something similar has happened with protection settings. In the past, they were manually calculated and entered into dedicated setting sheets to be later used to set an electromechanical relay by turning knobs or later manually entering them by pressing keys on the front panel of a numerical relay.

Today the complete process of protective relay setting calculation and coordination, their storage in digital setting files, and their downloading into the protection devices is completely digitalized, which leads to significant improvements in the efficiency of the setting configuration process, but also results in improved quality and reduced numbers of human errors in these files.

The analysis of protection operations or system-wide disturbances is also digitalized. The disturbance records are stored typically as Common Format for Transient Data Exchange (COMTRADE) files, which are automatically collected in centralized systems and further analyzed in order to determine if all protection devices operated correctly and identify the cause of the protection operation or system disturbance.

Finally, we should mention team meetings, working group meetings, or workshops and conferences that have moved from face-to-face events in the past to virtual events handled by videoconferencing tools. This digitalization has benefits and drawbacks. The benefits are that people do not need to travel to participate in a meeting or a conference and, for many specialists who had travel restrictions in the past, it becomes possible to participate. There is a significant cost reduction because of the elimination of the travel expenses; however, everybody is aware that the quality of human interaction is not the same compared to face-to-face meetings.

References

[1] Rockefeller, G., "Fault Protection with a Digital Computer," *IEEE Transactions on Power Apparatus and Systems*, April 1969.

[2] *European Technology Platform for SmartGrids: Vision and Strategy for Europe's Electricity Networks of the Future*, Directorate-General for Research and Innovation (European Commission), Sustainable Energy Systems, EUR 22040, April 12, 2006.

[3] Energy Independence and Security Act of 2007, Public Law 110–140, 110th Congress, 2007.

[4] IEEE 1366-2012 Guide for Electric Power Distribution Reliability, 2012.

CHAPTER 2

Smart Grid Functions

2.1 General Function Categories

Requirements for improving the reliability and security of the electric power system are, to a great extent, related to the need for reduction in the duration of short-circuit faults on transmission lines or distribution feeders and substation power equipment. They are based on the need to maintain system stability, as well as the quality of power required by the significant increase in the numbers of customers with loads sensitive to voltage variations. At the same time, better knowledge of the condition of the equipment, as well as an understanding of the types of loads and their behavior, can help to optimize the performance of the PAC systems under different faults or other abnormal system conditions.

High-speed fault clearing for different faults can be achieved by the use of advanced functions, adaptive protection, and communications-based protection schemes combining traditional and smart grid-related technologies. Significant improvement in the performance of noncommunications-based distribution protection solutions can be achieved by the implementation of adaptive protection at the distribution level of the system. The grounding of transformers and their state, the type and state of distributed generators, the configuration of the system, and its dynamically changing topology and behavior have an impact on the performance of distribution feeders and substation protection relays.

Smart grid functions can be divided into many different groups and the performance and communication requirements are different for different categories of functions. When discussing the power system requirements and how to combine different functions in order to achieve the goals of the smart grid, it is suitable to divide the functions into two main groups:

- Fault-clearing functions;
- Control and monitoring functions.

The different types of functions are defined in a more specific way next.

- *Fault-clearing functions:* In most cases, protection functions are in the center of the fault-clearing functions. To perform successful clearing of power system faults, the fault clearing functions are dependent not only on protection equipment, but also on other devices included in the fault-clearing system as shown in Figure 2.1. The fault-clearing functions include all interfaces with the process required to detect the abnormal condition, as well as the switching devices that clear the fault based on a signal from a protection device.

- *Control and monitoring functions:* Control functions include all other functions needed for operation of the system such as manual and automatic functions for the optimization of operation, voltage and frequency control, and interlocking. The functions for restoration of the operation after a disturbance are also considered control functions.

Functions for acquisition, storage, transmission, and visualization of information to enable the operation of the system and the analysis of network and equipment performance and behavior during both nonsystem and system fault conditions are examples of monitoring functions. Alarm and supervision are other examples of control and monitoring functions. Monitoring and control functions can be used in different domains (Figure 2.2) of the electric power system.

2.2 Types of Functions

Within the PAC system of a substation, several categories of functions can be identified. A generic classification of these functions is proposed below and is intended to help the reader to understand the issues related to the implementation of these functions, more specifically in IEC 61850-based substation PAC systems.

A first distinction can be made between:

- *Primary functions:* Directly acting on the process, such as protection, control, automation, monitoring;
- *Support functions:* Operating in the background or only in specific phases, such as configuration and time synchronization.

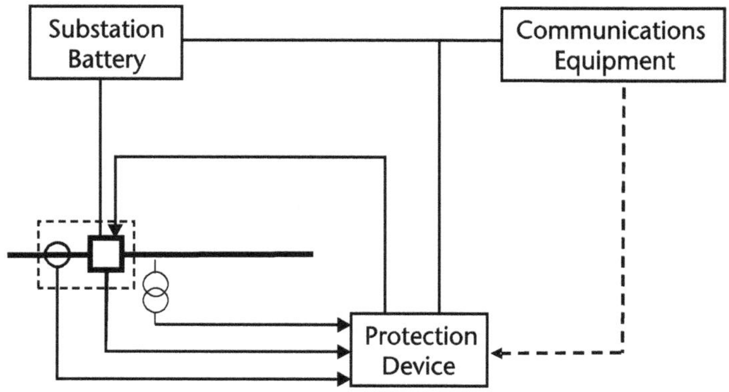

Figure 2.1 Simplified block diagram of fault-clearing system components.

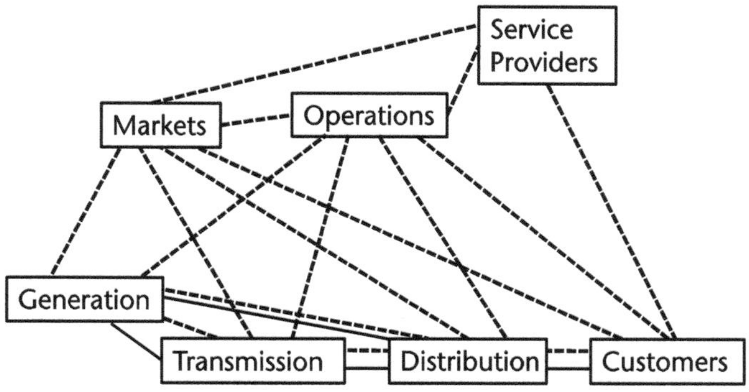

Figure 2.2 Simplified block diagram of electric power system domains.

Another classification can be made considering the level of distribution of a function, such as:

- Functions involving the devices of a single bay (for example, transmission line) and operating based only on locally available process information (for example, overcurrent or distance protection);
- Functions involving devices of more bays in the same substation that require information from multiple process interfaces or device status information within the same substation (for example, busbar protection or breaker failure protection);
- Functions involving devices in different substations that require process or device status information from one or more remote substations (for example, teleprotection, wide-area monitoring systems (WAMS), system integrity protection schemes (SIPS), distribution automation).

2.2.1 Protection Functions

Protection functions operate directly on the substation breakers to isolate faults occurring on the grid or in the plant. The task of any protection function is to measure and monitor the values of different system parameters (voltage, current, frequency, temperature) and issue a trip signal to isolate the faulty object (cable, line, transformer, breaker) when the conditions can damage it or result in a safety situation (e.g., a downed wire). The behavior of any protection function is controlled by a set of parameters known as settings, which may be changed by the protection engineer.

Some examples of protection functions are:

- Distance protection;
- Line differential protection;
- Transformer differential protection;
- Overcurrent protection.

Associated with this category are the protection-related automation functions that are triggered by the protection operation (such as breaker failure protection and autoreclosing) or support the protection function (such as directional elements).

Protection function schemes can involve different installations at the same time. An example of this is the permissive overreaching transfer trip (POTT) function that coordinates the operation of two or more distance protection devices located at the different terminals of a transmission line, using a communication link.

Protection schemes can sometimes involve all the bays of the same voltage level in a substation, for example, bus differential protection. These bays can be protected by a single multifunctional intelligent electronic device (IED) or can be distributed between multiple IEDs communicating over the substation local area network, the ends of the protected line, or a wide area network.

2.2.2 Control Functions

Control functions can be defined as associated with operator actions related to switchgear and process control. They are usually performed by means of a human machine interface (HMI) at different levels (bay, substation, or control center). Control functions allow the user to operate on different parts of the substation high-voltage (HV)/medium-voltage (MV) equipment, auxiliary equipment) through operator control. Today, this control can be visualized through the HMI to analyze events and alarms regarding the equipment or functions in a substation, plant, or network.

2.2.3 Automation Functions

Automation functions are sequences of actions performed automatically, after some trigger criteria have started them. The trigger can be the action of an operator or a process condition.

Automatic functions may have their own safety checks and operate on top of the interlocking and protection functions. They may include sequences of commands, for example, switching sequences that contain a number of switching steps necessary to put a switchyard into an operational state. The operator starts the sequence with a command and all the switching steps are performed automatically in the predesigned sequence.

2.2.4 Monitoring and Recording Functions

Monitoring and recording functions can be implemented as independent or interrelated functions. Monitoring functions collect data in order to identify and alarm for abnormal power system conditions. Such functions are also used to optimize equipment maintenance or verify power system models.

Some examples of monitoring functions are:

- Gas insulated switchgear (GIS) equipment monitoring;
- Transformer or circuit breaker monitoring;

- Power quality monitoring;
- Voltage and current circuit monitoring.

Recording functions capture and record the values of different system parameters related to a specific electric power system event. Which parameters are being recorded, what is the recording rate and duration, and what triggers the recording depend on its purpose. The recordings are typically used for forensic events analysis and testing.

Some examples of recording functions are:

- Transient (waveform) recording;
- Recording of fundamental and/or harmonic signals (power quality, system disturbance);
- Sequence of events recording (including, in some cases, fault location).

2.2.5 Supervision Functions

The supervision functions allow the operator to visualize real-time data on the substation primary and auxiliary equipment. Some examples are:

- Equipment state: positions (open, close);
- Operating condition (normal, faulty);
- Alarms: indication of abnormal condition;
- Measured values: RMS or complex value of electrical parameters;
- Data archiving.

2.2.6 Auxiliary Functions

Auxiliary functions support the PAC and monitoring system operation and management. Some of them are dedicated to the system setup, typical activity occurring during substation commissioning, extension, or modification.

Some are auxiliary functions, fundamental for the system life and operation such as:

- Self-monitoring and diagnostics of the devices;
- Time synchronization;
- Aggregation.

Aggregation is a new type of auxiliary function where data is gathered and expressed in a summary form. The digitization and digitalization of the electric power grid result in the distributed nature of the system. For example, in order to be able to monitor and control large numbers of distributed renewable energy resources, a distribution system operator needs to know the overall state and capabilities of a VPP aggregating numerous DERs.

2.3 Components of Smart Grid Systems

2.3.1 Functions and Function Elements

With today's state-of-the-art IEDs, there is a significant overlapping of the functionality between devices of different types. Typically, in the past, several groups within a utility will install in the substation their own devices, such as:

- Protective relays;
- Measuring devices;
- Metering devices;
- Control devices;
- Communication devices;
- Monitoring devices;
- Disturbance recorders;
- Event recorders;
- Power quality monitoring devices;
- Remote terminal units (RTU).

Each of the above needs to be engineered and installed, wired to the substation equipment with which it interfaces, commissioned, tested, and maintained. Considering the requirements for high availability, many of these devices need a Main 1 and Main 2 or primary and backup, which doubles all of the above costs.

Multifunctional IEDs and now IEC 61850 are significantly changing the way things work in the substation. This is due to the fact that single-function devices are being eliminated as the preference for multifunction and distributed function devices grow in popularity. This advanced functionality results from devices communicating over the substation local area network, the wide area network, or the cloud. All these functions available in protection IEDs should be used as the foundation of any distribution transmission-level smart grid function.

The different smart grid functions can be considered hierarchical and divided into subfunctions and functional elements. The functional elements are the smallest parts of a function that can exchange data with other function elements. They are the basic building blocks of the smart grid, located at the bottom of the function hierarchy. Figure 2.3 shows an example of the implementation of a distributed differential protection function using phasor measuring elements that calculate synchrophasor measurements used by the differential element to determine if there is a fault inside the zone of protection.

In cases when a function requires an exchange of data between two or more function elements located in different physical devices, it is referred to as a distributed function. Distributed functions in the smart grid can be local (within the substation) or remote (including function elements located remote from each other).

The exchange of data is not only between functional elements, but also between different levels of the substation or system functional hierarchy (Figure 2.4).

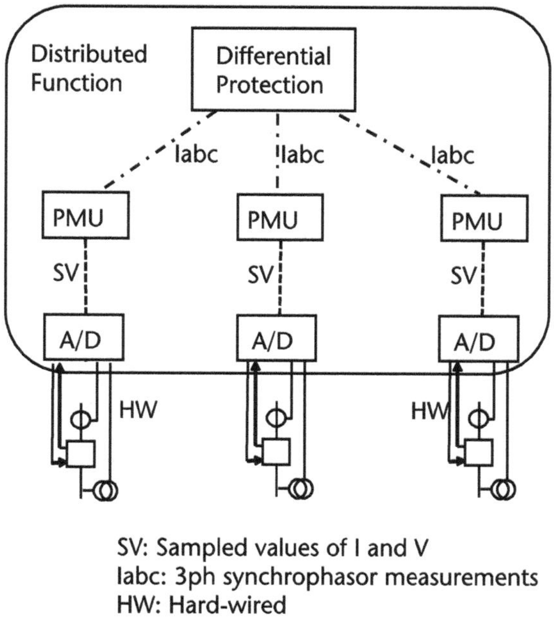

Figure 2.3 Distributed function definition. SF: subfunction, and FE: function element.

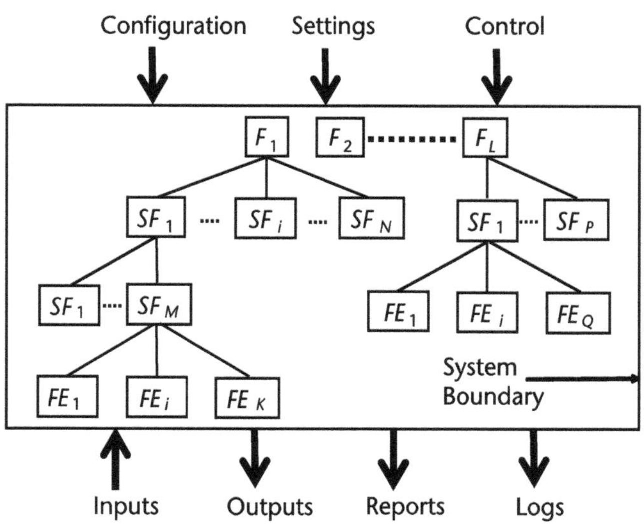

Figure 2.4 System functional hierarchy.

It should be kept in mind that functions at different levels of the functional hierarchy can be located in the same physical device, and, at the same time, different physical devices can be exchanging data at the same functional level.

As can be seen from Figure 2.3, communication interfaces are used between functional elements, in this case:

- Analog-to-digital converters (A/D) performing the digitization;
- Measurement function elements calculating the synchrophasors;
- Protection function element performing the differential protection function.

The communication interfaces may be implemented using dedicated or shared physical interfaces, the communications link between the physical devices.

The allocation of functions and function elements between different physical devices defines the requirements for the physical interfaces and, in some cases, may be implemented into more than one physical LAN.

The functions in the substation can be distributed between IEDs on the same level or on different levels of the substation functional hierarchy, such as:

- Station level;
- Voltage level;
- Bay/unit level;
- Process level.

These levels and the logical interfaces are shown by the logical interpretation of Figure 2.5.

2.3.2 Communication Interfaces in the Smart Grid

Because the digitalization of the electric power system starts with the substation PAC systems, the initial focus is on a subset of interfaces shown in Figure 2.5 and listed next.

The typical logical interfaces shown can be defined as:

- Current transformer (CT) and voltage transformer (VT) data exchange (streaming samples) between the process level and the bay level using peer-to-peer (P2P) connectionless communications;

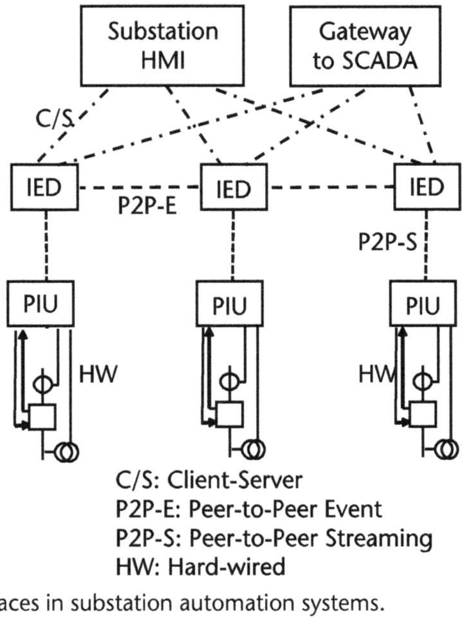

Figure 2.5 Logical interfaces in substation automation systems.

- Switchgear status data exchange between the process level and the bay level using P2P connectionless communications;
- Nonelectrical data exchange (especially samples) between the process level and the bay level using P2P communications;
- Protection-data exchange within the bay level using P2P communications;
- Protection-data exchange between the bay level and the station level using client-server (C/S) communications;
- Protection-data exchange between substations using P2P communications;
- Control-data exchange between the process level and the bay level using C/S communications;
- Control-data exchange between the bay level and the station level using C/S communications;
- Control-data exchange between the substation (devices) and a control center using C/S communications;
- Data exchange between the substation (level) and a remote engineering station using C/S communications or file transfer;
- Direct data exchange between the bays, especially for fast functions such as interlocking using P2P communications;
- Data exchange within the station level using P2P communications.

As can be seen from the analysis of the PAC functions discussed earlier in the chapter, the logical interfaces considered above are not sufficient to cover the needs of all communications-based applications in a smart grid.

Figure 2.6 shows some of the new logical interfaces that need to be standardized as an extension of the interfaces listed above to support different types of smart grid functions.

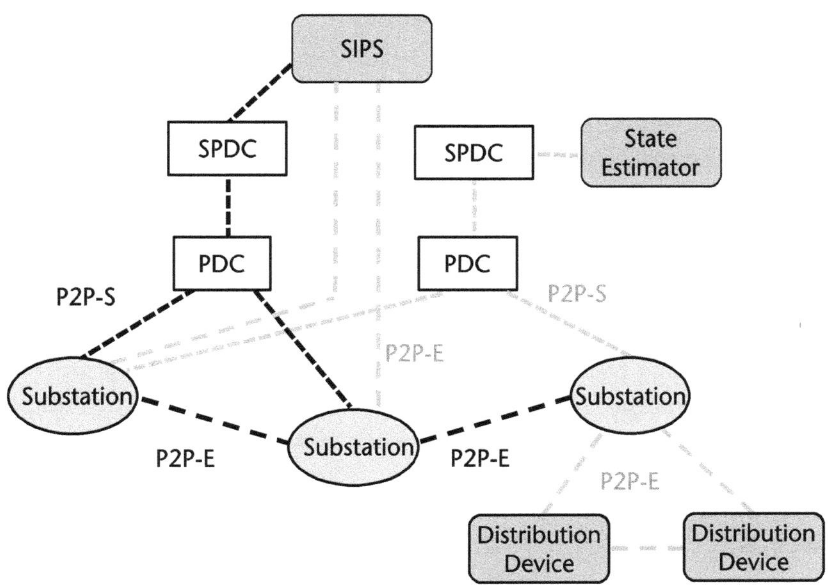

Figure 2.6 Logical interfaces for system-level applications.

The variety of applications covered by the smart grid requires the definition of additional interfaces, such as in the following examples (Figure 2.6):

- Control-data exchange between substations: binary data, for example, for interlocking functions or intersubstation automatics using P2P communications;
- Data exchange between a phasor measurement unit (PMU) and a phasor data concentrator (PDC) using P2P communications;
- Data exchange between PDC and a super PDC (SPDC) using P2P communications;
- Data exchange between an SPDC and a state estimator using P2P communications;
- Data exchange between an SPDC and a system integrity protection scheme (SIPS) using P2P communications;
- Data exchange between the SIPS and a substation using P2P communications;
- Data exchange between the substation level and remote distribution system devices;
- Data exchange between remote distribution system devices.

The above logical interfaces, as well as any other required by applications such as demand side management (DSM) or automatic meter reading (AMR), need to be considered in order to provide complete coverage of all utility communications requirements.

To better understand the communications requirements in digitized system, it is necessary first to define the messages exchanged and their performance and security.

2.4 Messages in Smart Grid Systems

The different distributed functions impose different performance requirements that have to be considered in the design process of substation PAC, monitoring, and recording systems. These performance requirements need to be applied to the various types of messages used in digitalized systems.

To address these issues, it is necessary to define performance requirements for the typical substation functions.

Considering that, in smart grid applications, messages are exchanged over a wide variety of communication links, the transfer time is one of the most significant parameters due to its impact on the overall performance of any specific function. We can define the transfer time as the estimated time for the completion of a data transmission between two function elements in a distributed smart grid function. In most cases, data transfers are not fixed and transfer times may increase and decrease as the data transmission is occurring. This depends on multiple factors, such as:

- The design of the sending and receiving devices;
- The type of communication link used for the message exchange;
- The length of the message being sent;
- The communication speed;
- The amount of traffic on the network;
- The communication architecture;
- The communication services being used.

The measurement of the transfer time (Figure 2.7) starts when the first bit of the message leaves the sending function element until the last bit of the message is received by the receiving function element.

As many smart grid functions are distributed between remote locations, the transfer time in many cases will include the time over a wide area network, which is especially difficult to predict.

In order to define the requirements for implementation of smart grid functions, we need to identify different types of messages and performance classes.

There are two independent groups of performance classes:

- For protection and control applications;
- For metering, monitoring, and power quality applications.

The requirements for control and protection are more stringent because of the effect of the fault-clearing time or control action time on the stability of the system or on sensitive loads. We can define different performance classes for such applications:

- For the distribution level of the substation or in cases where lower performance requirements can be accepted;
- For the transmission level of the system in areas that are not considered critical or if not otherwise specified by the user;
- For the transmission level of the system in areas that are considered critical.

The overall performance requirements also depend on the message type. We can consider several different types that are used for different smart grid applications.

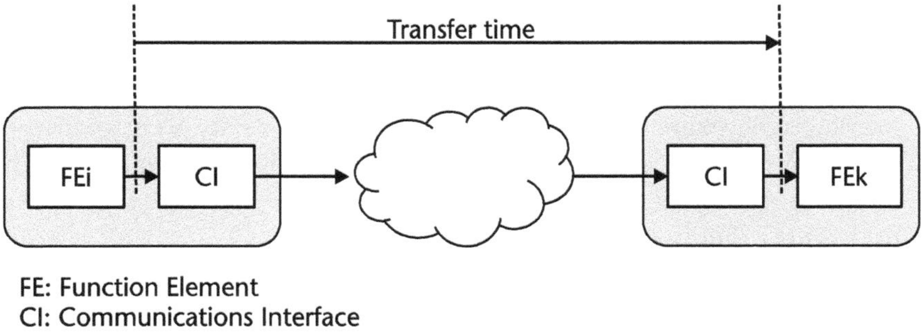

Figure 2.7 Transfer time.

2.4.1 Fast Messages

The most critical are high-speed messages related to fault clearing and have more demanding requirements compared to all other messages. Such messages are typically used to trip the breakers during faults or to initiate breaker failure protection. The expected performance should be comparable to the operating time of electromechanical output contacts of conventional protection relays, which is typically in the range of 3 to 4 ms.

The fast messages are also in a very important category for messaging in the substation and can be used for applications such as interlocking, intertrips, and distributed programmable scheme logic-based distributed protection functions. For this performance class, the total transfer time shall be on the order of half a cycle (i.e., 8 to 10 ms).

Data exchange within the bay level and control-data exchange between the process level and the bay level also require relatively fast message transfer, which is expected to be in the range of one cycle to several cycles.

2.4.2 Medium-Speed Messages

Medium-speed messages are the ones for which the time of the event triggering the message is important but where the transmission time is less critical. The change of state information or nonprotection-related measurement information is classified as this type of message. These messages are typical for some data exchange within the bay level, data exchange between bays or between a bay level and the station level. For such messages, the total transfer time shall be less than 100 ms.

2.4.3 Low-Speed Messages

Time-tagged complex messages that are used for slow-speed automation functions, changes of settings, or other configuration parameters or transmission of event or fault records are considered to be low-speed messages. The total transfer time for such messages shall be less than 500 ms. This type of message is typical for a wide range of interfaces and smart grid applications.

2.4.4 Raw Data Messages

Raw data messages are the output from the different types of analog process interface functions, such as the output data from digitizing transducers and digital instrument transformers independent from the transducer technology (magnetic, optic).

The messages in this case transport time-synchronized data from the process interface digitizing functions that are critical for the performance of protection-related functions. Subsequently, these messages have similar performance requirements to the high-speed messages described earlier in this section: in the range of 3 to 4 ms for the transfer time.

However, we may also be digitizing some nonelectrical signals that are not changing very fast, such as ambient temperature. In this case, we may have to transfer row data with low-speed requirements in the range of 500 ms.

2.4.5 File Transfer

There are different smart grid-related digitalized functions that require the transfer of different types of files, such as:

- Transient recordings;
- Disturbance recordings;
- Harmonic data files;
- Setting files;
- System configuration files.

The impact of the transfer of such files over the communication network that is used for other types of applications with critical transfer time requirements needs to be analyzed, and proper methods for the file transfer should be considered in order to limit their impact on time-critical functions.

Because the file transfer times are not critical, there are no specific time limits on this transfer, which enables higher-speed data to take priority in transmission.

Because the transferred files may contain sensitive information, they may require some form of role-based access control (RBAC) to ensure that only authorized users can access the information.

2.4.6 Time Synchronization

Time synchronization messages are used to synchronize the internal clocks of different devices in the substation PAC system. The requirements for accuracy of the time synchronization depend on the type of applications using time-stamped data. As a result, different communications will be used to ensure that the requirements are met. Typically, there are two main groups of time-synchronized data:

- Event time tags that are used for local or wide area event analysis and typically require accuracy of the time synchronization in the range of 1 ms;
- Sampled analog values and synchrophasor measurements that require accuracy of the time synchronization in the range of 1 μs.

2.4.7 Command Messages

In many existing and smart grid-related applications, system operators or other utility personnel, such as maintenance crews, may need to use command messages to issue control commands from a local or remote HMI. If such functions require a higher degree of security, they may require some form of secure communication via authentication and encryption. Additionally, RBAC may be needed to ensure that only authorized users can send the command message.

The development of the smart grid will for sure identify new types of messages that will have to be added to the existing list above.

CHAPTER 3

Requirements for IEC 61850

3.1 General Requirements

Earlier in the book, we defined what we consider a smart grid and described its main functions and functional components, as well as the interfaces between them and the types of messages that are needed in order to perform the smart grid functionality. Based on the experience from the last few decades, not only are the computer and communications technologies changing continuously, but the electric power grid is also changing. That is why there were discussions in different industry working groups (International Council on Large Electric Systems (CIGRE) [1]) and task forces Utility Communications Architecture (UCA) 2.0) on the definition of several fundamental requirements that must be considered when defining any aspect of the IEC 61850 standard.

3.1.1 Interoperability

The first main requirement is the support of interoperability. This means the ability of devices from one or several manufacturers to exchange information and use this information to perform their own functions. Some people feel that interoperability is not sufficient and are asking for interchangeability instead. We need to understand that these are two completely different requirements and that the focus of the IEC 61850 standard should be to make sure that the devices can talk to each other and understand what is the meaning of the data that they receive. For example, if two devices are operating in a permissive overreaching transfer trip (POTT) scheme and one of them receives a message that the Zone 2 elements of the device at the other end has started, so it can accelerate its tripping to improve the performance for the required action, this means that interoperability is being enforced and is working. However, it should not be the goal of the IEC 61850 standard to describe what is the specific distance protection algorithm that is being implemented by the manufacturer and even what kind of characteristic is being used. If we would like to support the interchangeability, it means that we need to standardize on the protection algorithms that are used for each specific type of protection function and this is something that is considered very difficult to accomplish due to the specific preferences not only of the different manufacturers, but of the users themselves.

In order to support interoperability, the IEC 61850 standard has to define clearly what is being communicated and how it is being communicated. Defining what is being communicated means that all functions and their function elements need to contain data that uses data types and naming conventions with clearly specified semantical descriptions to avoid any potential misunderstanding when processing the data.

At the same time, it is necessary to specify what kind of communication services should be used for the specific types of interfaces described in the earlier chapters, as well as the rules for building the data sets to be used in the data exchange.

3.1.2 Free Configuration

The second main requirement for the development of the IEC 61850 standard is to support free configuration. This means that it should be possible to support any specific protection philosophy that is being implemented by a supplier and it should be also possible for any user to select their preferred methods of implementing PAC schemes. To a great extent, this is due to the fact that technology is allowing us today to have all PAC functions integrated within a single multifunctional IED that performs not only the digitization of the individual analog signals or status indicators from the substation primary equipment, but also all measurements, protection, control, recording, event analysis, and other digitalized functions. At the same time, the existing technology supports the implementation of exactly the same functionality using a centralized substation PAC system in which case the digitization may be done by process interface devices located in the substation yard that will be publishing the sample values over the substation communications network and then have some other device perform the different types of measurement functions that will provide an output to different protection elements located in another device. Today more people are talking about the use of cloud technology, where exactly the same functionality can be performed in a server located somewhere in a private or public cloud. All of this should be possible to do using the same description of the system function elements and the data flow between them, without the need for complete reengineering while transitioning from the use of one technology to a different one.

3.1.3 Long-Term Stability

The requirement for long-term stability is driven by the already-mentioned continuous evolution of computer and communications technology, as well as of the grid itself. In order to meet this requirement, we have to be able to seamlessly transition from a PAC system that is based on multiple IEDs exchanging information between themselves over the substation communications network to a centralized system with exactly the same functionality and signal flow. At the same time, it should be possible to interface an SIPS that normally will exchange signals with a thermal power plant to another implementation that is going to interface with a VPP. From the point of view of the SIPS, there should be no difference between the requirement to increase generation or to shed load; it is going to remain the same, regardless if it is implemented on a large synchronous generator or on a number of wind generators in a wind farm.

Part of the long-term stability requirements also has to do with the evolution of the smart grid and the fact that it is continuously expanding and covering more and more domains. At the beginning, the digitalization may start at the substation, but it should be possible to extend the same modeling and communications principles to any other domain such as thermal or hydro power plants, wind and solar farms, distribution automation, wide area PAC, and many other domains. This means that any specific type of communication messaging should be possible to implement within the substation, in a protected environment, but also over a wide area network when it might be exposed to cybersecurity threats.

3.1.4 Flexibility

A very challenging requirement is the requirement for flexibility because it contradicts the requirement for interoperability. In order to support flexibility, we need to try to envision all possible requirements in the modeling and communication capabilities of the IEC 61850 standard. However, as the standard should not define how a specific function or system should be implemented, this becomes dependent on the understanding and preferences of the developer and as the probability that the vision of every single PAC specialist will have something unique that is based on their personal experience, we may expect that there will be no two identical implementations of the same function or function element; even the data that might be available in the specific function element model may be different depending on the implemented functionality and vision. As a result, we may have the same function element that is not interoperable with a different function element that is not supporting the data that is produced by the publishing device. In this case, two devices that are compliant with the IEC 61850 standard may not be able to interoperate, but they meet the requirements for flexibility. Therefore, it should be possible to consider methods to reduce the flexibility in order to support interoperability and this means that development of specific profiles should be supported by the IEC 61850 standard.

3.1.5 Communication Support

In Chapter 2, we identified the different functions and components of the smart grid, together with their interfaces and the requirements for communications exchanges between them. This included communications within the substation, between substations, and between the substation and the system level.

The communication interfaces can be used to support PAC-related functionality. Because they play an important role in the smart grid environment, it is essential that the IEC 61850 standard should be capable to support them based on the specific communication services defined.

The same type of communications may be required to meet different performance characteristics depending on the function that it is used for, which is why in the IEC 61850 standard it will be necessary to identify specific performance classes that can be used for requirement specifications and to evaluate the performance of any specific system.

The different communication services should also be capable of transporting different types of data that that can be configurable by the supplier or the user depending on the implementation philosophy.

It is clear as well that, when the communications are within the substation, they are considered to be in a protected environment, but, once outside of the substation, they become much more vulnerable to cyberattacks and that is why the IEC 61850 standard or corresponding standards should identify specific methods for cybersecurity.

Commonly used communication types that are required to be supported by the IEC 61850 standard are:

- Event-driven peer-to-peer (P2P-E) communications within the substation;
- Streaming peer-to-peer (P2P-S) communications within the substation;
- C/S communications within the substation;
- Event-driven peer-to-peer (P2P-E) communications between substations;
- Streaming peer-to-peer (P2P-S) communications between substations;
- C/S communications between substations and control center;
- Reporting;
- Time synchronization;
- File transfers.

3.2 Modeling Requirements

From the discussions of the structure and the functionality of the smart grid, it is clear that the principal requirements for interoperability, free allocation, and long-term stability lead to a wide range of different kinds of specific requirements that have to be met by any standard that is designed to support the digitalization of the electric power system. There are several aspects that must be addressed, such as:

- Modeling of multifunctional devices;
- Modeling of their function elements;
- Modeling of the data that is contained within the function elements and exchanged between them to perform the different applications;
- The grouping of function elements in functions and their functional hierarchy;
- The definition of specific data types that are required to meet the specific function needs;
- The modeling of complete substation PAC systems including the description of the primary system;
- The association of the specific functions and function elements to the primary equipment and to the communications network in the substation that is used to perform all functions.

We also need to think about the impact of the digitization in digitalization on the operation of the protection and control systems and their maintenance. Considering that we are replacing the traditional hardwiring in the substation with digital communications, we realize that this is leading to a complete change in the way such a system will be tested and this is especially true when we talk about maintenance testing that is performed in a live energized substation.

We will have to identify ways to support the virtual isolation of the tested components of the system from the rest of the substation, as well as to be able to distinguish between the messages coming from actual substation protection and control devices from similar or identical messages coming from the test equipment.

Another critical requirement is to support time synchronization that is required to support the time-related requirements of PAC systems.

3.2.1 Modeling Requirements for Multifunctional IEDs

The modeling of a complex multifunctional protection IED such as a modern transformer protection device is possible only when there is good understanding of the problem domain. At the same time, we should keep in mind that the models apply only to the communication's visible aspects of the IED and that it should allow the implementation of different PAC modeling philosophies.

The functions in relatively simple IED, such as a low-end transformer protection relay on a two-winding transformer, are fairly easy to understand and group together in order to build the object model. That is not the case for the more complex devices such as an advanced transformer protection IED on a three-winding transformer.

Figure 3.1 shows a three-winding transformer with a breaker-and-a-half configuration on the high side that is represented by two virtual subdevices, one associated with each of the breakers (HV1 PAC and HV2 PAC) that will be interfacing with the instrument transformers and switch gear associated with them. Then we have another virtual subdevice associated with the medium-voltage winding (MV PAC) and another one associated with the low-voltage winding (LV PAC). Each one of these subdevices will contain its own protection, measurements, control,

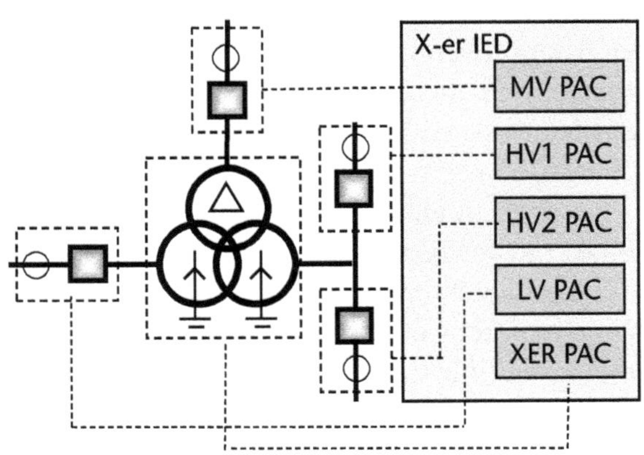

Figure 3.1 Three-winding transformer IED.

monitoring, recording, and other functions that will be grouped in a functional hierarchy.

In the model, we also need to include a virtual subdevice that will represent the functionality related to the transformer (XER) itself (XER PAC), such as its differential protection, sudden pressure protection, condition monitoring, and other functions.

However, this is just one of the possible ways of grouping the individual function elements when developing the model of the transformer PAC device. Another approach will be, instead of starting from the interfaces with the different voltage levels of the transformer, starting with the specific functions, so we have a protection function and then, within this protection function, we will have the transformer protection, the protection of the two high-side voltage breakers, the protection of the medium-voltage breaker, and the protection of the low-voltage breaker. The same will apply to the measurements and the automation, condition monitoring, and recording.

Both are valid modeling approaches, and they have to be supported by the definitions of the IEC 61850 standard.

Each of the above-described functions can be divided in subfunctions that represent groupings of related functional elements.

In the MV PAC virtual device of the example, we have a protection function that will include an overcurrent protection subfunction, with another layer of subfunctions, such as:

- Phase overcurrent protection;
- Ground overcurrent protection;
- Negative sequence overcurrent protection;
- Sensitive ground fault protection.

Each device subfunction then can be split into functional elements. Functional elements can be defined as the smallest functional unit that can exist by itself and also can exchange signals or information with other elements within a device or a system.

An example of a protection functional element is a nondirectional ground overcurrent element with extremely inverse characteristic. Instantaneous and definite time-delayed ground overcurrent elements are added to provide a more advanced overall overcurrent characteristic that will support faster fault-clearing time, with each function element used to represent the different steps in an overcurrent protection subfunction.

Figure 3.2 shows an example of part of the protection function hierarchy of a multifunctional transformer IED.

The above-described functional hierarchy needs to be appropriately supported by the modeling hierarchy of IEC 61850. This means that the model should allow a device to contain multiple functions, each of which may contain one to many layers of subfunctions. This is especially important from the point of view of the management of the mode and behavior of the functions, subfunctions, and function elements in an IED. This approach also will support the ability to share a function element with multiple other elements at a specific level of the function hierarchy,

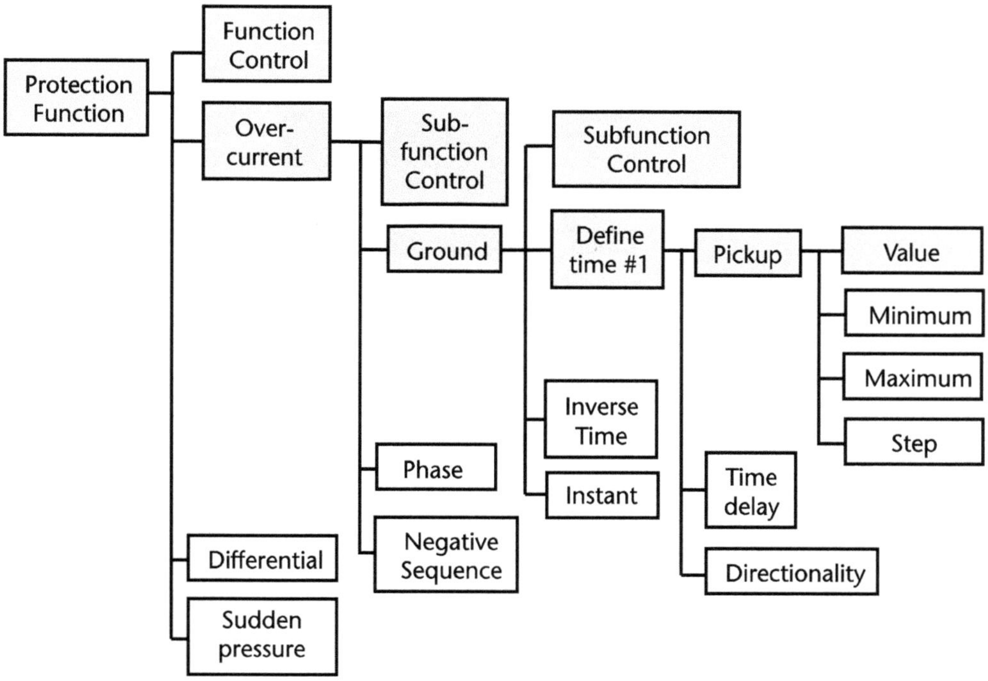

Figure 3.2 Protection functional hierarchy.

for example, that several overcurrent elements may share a single directional element for their directional supervision.

One of the main advantages of such a hierarchical model is that it allows the efficient control of the behavior of the components of a function with a multilayer functional hierarchy. This means that the IEC 61850 standard needs to define a method for mode control at each level of the functional hierarchy and some inheritance rules. If, for example, we set the mode of a specific level of the hierarchy to off, this should be inherited by all subfunctions and their function elements regardless of their original mode-setting values.

Once we define the requirements for the modeling of the functional hierarchy, we need to look at the details of the modeling of the function elements.

3.2.2 Function Elements Modeling

As we described earlier, the function elements are the building blocks of the model and each one of them is an object that contains data. The data in each specific type of function element is of different data types and depends on the role that it plays in the system.

Figure 3.3 shows the simplified model of a function element with its different interfaces to the PAC system. Each of these different interfaces is related to a specific data type that needs to be supported by the IEC 61850 model.

If we start with the input to a function element and consider a simple time-delayed overcurrent protection function element with an inverse time characteristic, its input is going to define what specific type of overcurrent protection it is going to perform. If it is a phase overcurrent element, the input will be pointing to the three

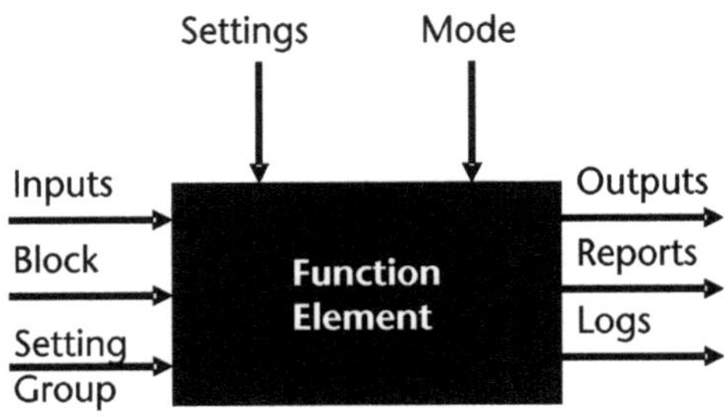

Figure 3.3 Function element simplified model.

phase A, phase B, and phase C currents coming from a measuring element that has calculated them based on the streaming sampled values from the instrument transformer interfaces. Given that the function element is to operate as a ground overcurrent element, the input will be a zero sequence current.

In some cases, the overcurrent protection element needs to be possible to block, for example, in the reverse blocking scheme for a distribution bus protection application. This is why there is a block input that will bring a signal from another overcurrent element to perform this blocking.

In order to improve the sensitivity of the protection system in some cases, depending on a change in the substation topology, it may be necessary to switch to a different setting group and there has to be an input to the function element that allows different criteria to be used to switch from one setting group to another depending on the setting selection analysis.

Another input to the function element is its settings, and if we look at Figure 3.2, we can see that the ground overcurrent element that is shown there has three different settings that are represented by data objects representing the pickup, the time delay, and the directionality of this overcurrent element. These are data objects representing the individual settings in the model. Like any complex model, the data objects contain multiple data attributes, so for the settings, some of the typical data attributes that are required are the set value that is going to lead to the start of the overcurrent element and then the value that is going to have a range defined by a minimum possible setting, a maximum possible setting, and a step size that specifies what values can be used when defining the setting.

Depending on the complexity of the function element, there might be many additional settings that will require specific data objects to be defined in the IEC 61850 standard to support them. To ensure the consistency of the implementation by different vendors, there will be a need to define specific common data classes that they are going to be used for the specific types of data objects in the model.

Another very important component of the function element model is the ability to control its mode. The user should be able to enable the function element to turn it off or to switch it to a test mode in order to be able to maintenance test the function element as part of a live system.

If a function element starts when the input signal has exceeded the pickup setting value, it may be necessary to use this change of state in some communications-based protection scheme. In other cases, this may be based on the operation of the function elements after the time delay has expired. This means that it is necessary for the standard to support these two states of a function element, started or operated, and send them in a communications message for other applications to use. It also may be required for the change of state events to be time-stamped for further analysis of the sequence of events or by the communications monitoring system.

The start or operation of a function element may also have to be reported or logged and it should be possible for the user to have the ability to define what specific data from the function element should be included in the data set used for reporting or logging.

The function element may also contain data objects representing the value of the input signal that resulted in the starting of the function element and the time when this value was reached. The function elements in a substation PAC system may play completely different roles and, because of that, may have different interfaces and data structure.

As we discussed earlier, protection function elements will require as an input, for example, phase or ground currents with their magnitude and phase angle. Phase currents are available from another function element that is performing measurements calculations based on the sampled values coming from a process interface function element. It is clear that the measurement function elements will look very differently from the data model of protection function element and a process interface function element will be completely different as well.

The IEC 61850 standard should support measurement elements that are performing different types of calculations and whose output is available not only for protection, but also for condition monitoring, visualization, and other functions. To improve the efficiency of the data acquisition process, the IEC 61850 standard should allow the representation of the system measurements as individual phase magnitudes and angles (Figure 3.4), sequence components, root mean square (RMS) values, and anything else that might be required by specific applications.

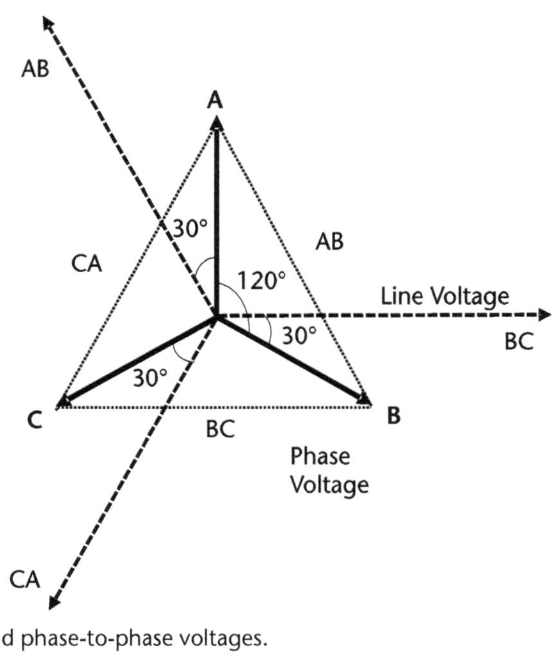

Figure 3.4 Phase and phase-to-phase voltages.

In most cases, the measured values will be available using streaming communication methods with a specific publishing rate depending on the requirements of the applications. However, in some cases, it might be more efficient instead of streaming the magnitude and angle of phase currents and voltages to send a new value using a report by exception mechanism as shown in Figure 3.5 where a dead band is configured in a way that, if the measured value is changing within the dead band, no new value is being published on the network. When the measured value crosses the dead band boundary on either side, the new value will be captured, time-stamped, and published for use by the interested applications. Both communication methods should be available in the IEC 61850 standard.

This example leads to another requirement related to the content of the data set that is used for streaming data. If the output is from a process interface function element, typically the data set will contain individual samples from a current or voltage waveform and the data will be published in intervals of maybe hundreds of microseconds. However, if the same streaming method is used for an output of the measuring function elements, then the data set may contain magnitudes and angles of currents and voltages and the data may be published once per cycle.

The digitization of the substation is performed by the process interface function elements that are connected to hardwired analog interfaces and perform the function of A/D. The more commonly used function elements for this purpose are connected to the secondary side of current and voltage instrument transformers. Because they are distributed in the substation yard and the data acquisition process is not controlled by any specific device, for the data to be usable by different applications, all these function elements must be accurately time-synchronized.

A requirement for the IEC 61850 standard related to the fact that the process interface function elements have different roles is to support different sensor sampling rates, different data sets, and different publishing rates.

3.3 Requirements for Engineering Support

The digitalization of the electric power grid is much more than the development of object models of PAC function elements and their integration in multifunctional devices. It also means that the complete engineering and maintenance process should be based on digital technology. That is why there are some requirements for the IEC 61850 standard that go beyond what is normally covered by a protocol and should support the complete engineering process.

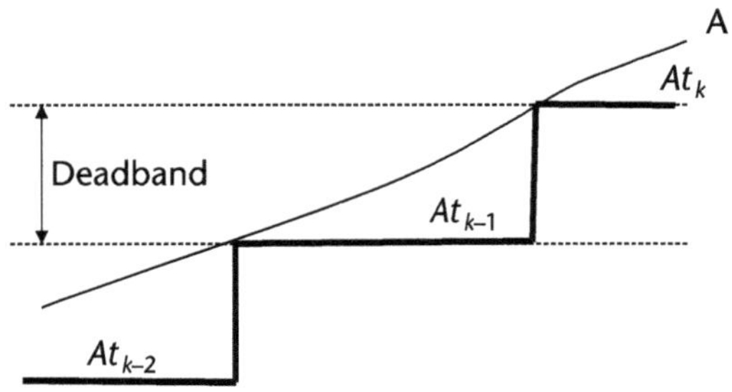

Figure 3.5 Report by exception.

Traditionally, the engineering of such systems has been done using different computer tools that are producing documents that have to be manually processed by human beings, but this is the way things have been done in the last few decades, while today, with the advancement of the artificial intelligence technology and the availability of machine-readable file formats, many engineering tasks can be completely automated.

Typically, at the start of a new project, the specialists put together a request for proposal that describes the functionality of the system that they would like to be supplied by the manufacturers. These are text documents accompanied by some drawings that, in many cases, define in generic terms what is expected to be delivered. These documents require significant efforts to be developed, but the main problem is that they may lack enough details that may result in unexpected surprises at different stages of the engineering process.

That is why one of the key requirements for the development of the IEC 61850 standard is to create an environment that will support the complete description of all components of a substation PAC system functionality. This needs to start with a detailed description of the connectivity of all the primary equipment, including busbars, disconnecting switches, circuit breakers, power transformers, instrument transformers, and any other specific components of the substation.

The next step is to define the PAC functionality from the building blocks of the substation, the different types of bays such as transmission lines, power transformers, busbars, and distribution feeders (Figure 3.6). This should include the functional hierarchy for each of the bays as well as the specific function elements that should be used to perform the required functions and their association to the primary equipment that they have to protect, monitor, or control.

It will be helpful if it is possible to describe as well how the user envisions the allocation of specific functions to physical devices.

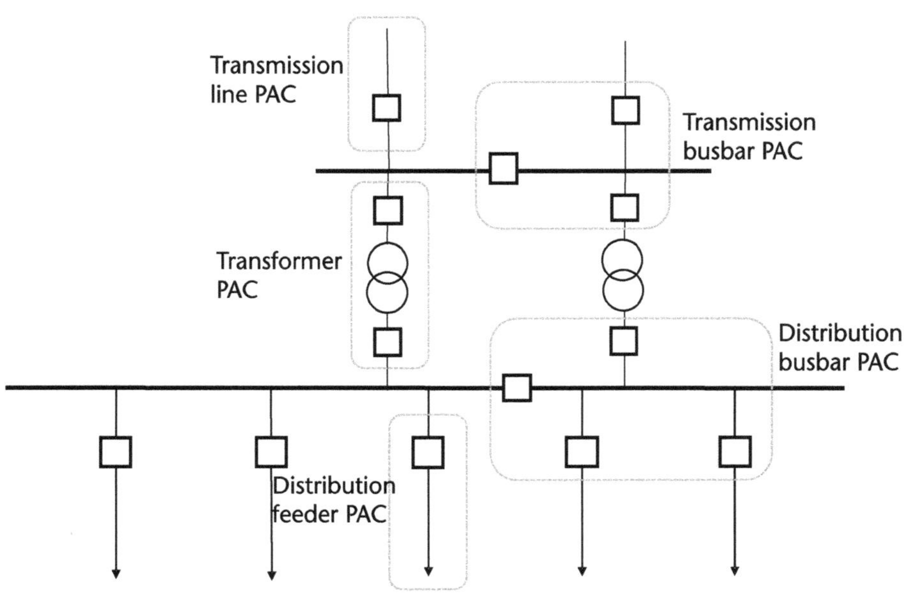

Figure 3.6 Substation one-line diagram with standard bays.

Another important part of the system description is the definition of the substation communications network and how the individual intelligent electronic devices are connected to it. The definition of the individual signals and the signal flow between the different components of the system should be possible to describe as well. All this information should be presented as an XML file that is human readable, but more importantly it is machine-processable. Such a file can be sent to the approved suppliers and then they can use their own tools to find out which specific devices that they produce can meet all the requirements presented in the project description file.

Such an approach can dramatically improve the efficiency of the engineering process and also reduce the possible errors that are almost inevitable in the traditional procurement process.

Once the manufacturer has found the devices that match the requirements of the system specification from the user, the manufacturer can replace the virtual devices representing the user specification with the complete capability description of their own devices and should be able to generate automatically the resulting system implementation file.

The IEC 61850 standard is required to define such a system configuration file that includes all the data models and the signal flow as well as the top-down engineering process to generate and maintain it. This will provide our industry with a very powerful tool for the life cycle of any project because it can be used later to support their condition monitoring, maintenance, testing, and cybersecurity of any substation PAC system.

3.4 Testing-Related Requirements

Some of the most critical requirements for the development of the IEC 61850 standard are related to testing. This is because traditional substation PAC systems are hardwired and have been tested for decades based on well-established testing practices and testing systems. The standard philosophy is that, for testing a protection and control device in a live substation, it must be taken out of service and physically isolated from the rest of the substation. Then it is connected to test equipment, the tests are performed, and it is returned to service after the successful completion of the test procedure. The isolation is performed by using dedicated test switches installed during the construction of the substation.

In a digital substation where all interfaces between the process and the different devices are communications-based, physical isolation is simply not possible, and this is a big concern for the protection and control community. That is why the IEC 61850 standard needs to take special care to ensure that any component of the PAC system can be virtually isolated and tested without any danger of undesired operation and tripping of substation equipment.

One of the most challenging issues related to testing is that typically it requires an outage that in today's utility environment is very difficult to get. That is why, because we are trying to improve the operation of the electric power grid based on digitalization, it is necessary to include in the IEC 61850 standard features that will support the performance of maintenance testing without taking out of service the complete bay or any protection and control device. This leads to a requirement for the IEC 61850 standard to support having in test mode not only the complete device, but also a single function, subfunction, or function element as required by the test.

Because we are testing a device or one of its components in a live substation, we should be able to distinguish between messages that are coming from a system component in a normal operating mode from one that is being tested. This means that we need to have in the model a data object indicating if its associated data in the data set is coming from a function element that is being tested.

Another challenge is when a function element being tested has a physical output to the process that may lead to a tripping of a breaker during the test. Therefore, the IEC 61850 standard should define a special mode for this kind of function element that will allow it to be tested without actually operating on the process.

In a live substation during the testing, we will have a mixture of actual messages coming from the substation PAC devices with messages coming from the test equipment. The IEC 61850 standard should include features that indicate if it is a real or simulated message and also to be able to tell the equipment being tested which messages it should process: the actual from the substation or the simulated from the test equipment.

Finally, the IEC 61850 standard should ensure that, for some cases, we should be able to distinguish that a test message is coming from dedicated test equipment instead of the simulation of equipment existing in the substation.

Reference

[1] WG 33.03 Communication Requirements in Terms of Data Flow Within Substations - Results of WG 33.03 and Standardization Within IEC, K-P Brand, Electra No. 173, August 1997.

CHAPTER 4

Development of IEC 61850

4.1 IEC International Standard Development Process

For a few decades, the world has been moving towards globalization, which is significantly changing the ways that everything is being done. Many users are shifting from applying products from a single supplier that they have been using for many years towards combining products from several manufacturers in their applications in order to meet the requirements for improvements in reliability and efficiency.

Because of that, the importance of standardization is continuously growing and this is where the international standardization organizations play a very important role. Each of these organizations has its own process that is followed in the development of an international standard and, because we are focused on the development of IEC 61850, we are going to examine what the IEC standard development process looks like [1].

To develop a new international standard, three main requirements need to be met:

- There needs to be a requirement for such development.
- There needs to be a clear understanding of what exactly needs to be standardized.
- There needs to be someone that is interested in doing the work.

The requirements for development of a global, international standard should be based on well-established technical rules defined by previous experience with proprietary solutions. They must ensure that all components of the systems that will be designed according to the new standard will be able to work together and meet the efficiency and performance requirements of the system.

The need for the development of a new international standard can be identified as a result of the work of different international industry organizations, such as the International Council on Large Electric Systems (CIGRE) and the International Conference on Electricity Distribution (CIRED) or national research organizations working in a specific domain.

Regardless of how the need for the development of a new standard is identified, it is well understood what will be standardized as an IEC International Standard is only developed when several IEC member countries are willing to be involved in its development. That is why IEC defines two types of member countries:

- A participating (P) member is an IEC member country that sends experts to participate actively in the technical work for the development of the standard.
- An observing (O) member is an IEC member country that is only acting as an observer.

What status a country has impacts the voting rights at the different stages of the standard development.

A requirement for the development of a new standard depends on the level of urgency by the industry for the specific standard. In some cases, it may be a long-term identified trend that indicates the need to start looking into the subject matter, but without being able to justify the need for immediate development. This is what is identified by IEC as a preliminary stage. It may be a way of collecting feedback from member countries that are interested in the topic, so they can put together some ideas about what the new item proposal may look like. A CIGRE working group can also be considered as a preliminary stage, because such working groups typically include experts from several countries interested in a specific topic and the publication of a CIGRE brochure identifying the needs and specific requirements can be used as the foundation for a future new work item proposal.

Once a stakeholder group from one or several member countries has identified the need for standard development and prepared a new work item proposal in the required special form, it has to be delivered by a National Committee to the IEC for consideration by the relevant IEC technical committee.

A proposal for new work can also be submitted by a technical committee or its secretariat, a liaison organization, the IEC Standardization Management Board (SMB), or the IEC General Secretary and chief executive officer (CEO).

A new work item proposal can be issued not only for a new standard, but also for an extension of an existing standard or for a technical specification.

A new work item proposal is then sent for approval to the member countries, which not only have to state if they approve or disapprove, but also must commit experts to work on the new document. For the proposal to be approved, it is necessary that two-thirds of the technical committee P-members approve the new work item and the required minimum number of experts from different countries is available. The new work item proposal is also sent to other technical committees that are requested to indicate their interest in this new proposal to the technical committee secretary.

Once the new work item proposal is approved, the technical committee establishes one or more working groups (depending on the complexity of the project), which starts the preparatory stage. This stage ends when a first committee draft (CD) is prepared by the working group and is considered ready for circulation to the members of the technical committee for comments and approval. The committee draft is registered by the office of the IEC CEO and, if approved, the technical committee may decide to publish it as a publicly available specification (PAS) if there is an urgent need by the industry.

This is the most important commenting stage in the development of a standard because it gives an opportunity to the interested national committees to submit all their detailed technical comments that they believe need to be part of the technical content of the standard. The comments are provided to the technical committee which then goes through the comments resolution in order to prepare the committee draft for vote (CDV). It is submitted to all national committees for a vote and a last chance to provide technical comments. The CDV of an international standard is approved if more than two-thirds of the P-members are in favor and the total number of negative votes cast by all national committees is less than 25% of the total votes.

What happens next depends on the results from the vote. If it has been approved without any recommended technical changes, then the CDV can be published directly. Otherwise, if technical changes have been requested, they are made by the technical committee and the revised version is sent to the IEC central office in Geneva for processing and to be published as a final draft international standard (FDIS). It is sent to all national committees for voting. If a country submits a negative vote, it must be accompanied by a technical explanation presenting the reasons for it. No comments are allowed with a positive vote. The approval requirements are the same as for the CDV.

After the FDIS is approved, the IEC International Standard is published by the IEC central office in Geneva.

At this time, typically a maintenance team is established with the responsibility of monitoring the development of the relevant technologies and evaluate if the standard can be reconfirmed as is, if it should be withdrawn, or if a new edition can be prepared.

4.2 The Development of the IEC 61850

Now, after we reviewed the IEC standard development process, we can review how the IEC 61850 standard itself has been developed.

The need for such a standard has become obvious based on the experience from the digitalization of the electric power grid that started with proprietary substation automation systems in the last few decades of the twentieth century. Many utilities were in the process of switching from electromechanical and solid-state protection devices to multifunctional IEDs integrating not only protection, but also different measurement, control, monitoring, and recording functions. Because not all utilities were willing to work with a single supplier, they were placing requirements for interoperability between devices from different manufacturers. The communication technologies were developing quickly as well and many industries were switching from manual manufacturing processes to automated systems based on programmable logic controllers and real-time communications between them.

These trends were not specific to any country or part of the world; they were global trends that obviously had to be addressed by the development of a global international standard.

When we were discussing the different stakeholders that participate in the preliminary stage of the standardization process, we discussed the different organizations that can be involved in defining the requirements for the development of a new standard; for example, a CIGRE working group 34.03 published in October 1990 a report on data communication in the HV substations: the protection and control interface. Unfortunately, this report is not available at the CIGRE central office, but a good summary of the work performed and the results are available in a CIGRE Electra article published in August 1997 [2]. The report identified the data flow and communications requirements for future substation PAC systems based on the concept of PICOM (PIece of information for COMmunication). The convener of the working group became a member of the working group developing the IEC 61850 standard and many of the ideas introduced in the report were integrated later in the standard itself.

At the same time in North America after the success of the Utility Communications Architecture (UCA) 1, it was decided that it was time to develop a standardized approach for the modeling and the communications exchange within substations, which lead to the start of the UCA 2.0 project.

The situation was not very different in Europe, and there was a proposal to initiate work on the definition of the communication requirements and methods to meet them for substation automation systems.

In the following sections, we will briefly describe the steps that were taken from the mid-1990s until today in the development of the IEC 61850 standard, followed by more details in the relevant chapters later in the book.

4.2.1 UCA 2.0

The UCA was an important project for the North American electric power industry. It started in 1986 with a workshop and continued with related work for about 5 years. The result was published as UCA 1.0, which specified Manufacturing Message Specification (MMS) (ISO 9506) in the application layer for utility communications. This work became the basis for IEC 60870-6-503 entitled "Telecontrol Equipment and Systems-Part 6-503: Telecontrol Protocols Compatible with ISO Standards and ITU-T Recommendations – TASE.2 Services and Protocol." It is also known as ICCP (Inter Control Centers Protocol) and was published in 1994.

That same year, John Burger from American Electric Power made a statement at the IEEE Power & Energy Society (PES) Power System Relaying Committee meeting that the industry cannot work efficiently in an environment where every manufacturer is using a different communications protocol and that, in an effort to solve this problem, his company has selected a token ring-based protocol, Profibus, to become the standard for their applications. This triggered an immediate reaction from many participants, as Profibus was a proprietary protocol. This is why the Electric Power Research Institute (EPRI) started the set of activities that were under the banner of UCA 2.0 with the goal to establish a standard communication protocol for the substation environment based on the MMS.

It was clear to all stakeholders that a decision needs to be made about the use of token-ring versus the Ethernet and which technology can meet the requirements for protection systems' performance in a realistic substation environment and, more specifically, the fault-clearing time during short-circuit faults.

The EPRI helped a lot by funding a major effort (as part of the UCA 2.0 work) to test the Ethernet and Profibus for a substation use case where more than 100 devices were all connected to the same network and had to be able to receive a fault signal generated by one device within 4 ms after a major fault that caused a network disruption. Several system configurations were staged at SISCO (at the time, SISCO was involved in both Profibus and Ethernet technologies):

- 10-Mbps shared media Ethernet;
- 10-Mbps switched Ethernet;
- 100-Mbps shared media Ethernet;
- 100-Mbps switched media Ethernet;
- Profibus.

The results from the testing discovered that a 10-Mbps switched Ethernet network and a 100-Mbps unswitched Ethernet network (shared media) could both meet the requirements. A 10-Mbps unswitched network (shared media) and Profibus could not meet the requirements. The 10-Mbps switched Ethernet network just barely met the requirement, while the 100-Mbps Ethernet networks surpassed the requirements with a significant margin. The EPRI also funded a statistical analysis of Ethernet that closely matched the observed results.

Once this testing was completed in 1996, it became clear that Ethernet was the right choice and all the work since then was about making Ethernet work for UCA 2.0 based on the Generic Object-Oriented Substation Event (GOOSE) for protection messaging and MMS-TCP/IP for supervisory control and data acquisition (SCADA).

Under UCA 2.0, we also started developing a document that was called GOMSFE, which stands for Generic Object Models for Substation and Feeder Equipment. A small group of protection experts was working together with some object modeling and communications experts in the development of the requirements for substation PAC systems and the better ways for achieving this. This is where many ideas came from, including the basic function element object "brick" and the P2P communications message GOOSE.

The latest version of GOMSFE that we had was 0.92. This work resulted in the publication of a UCA 2.0 document, which was published as an IEEE Technical Report 1550 in December 1998. Many of the concepts developed under UCA 2.0 became essential parts of IEC 61850. Their history is covered later in the book.

What was amazing was that, in parallel with the development of UCA 2.0, many manufacturers that were involved started implementing the technology in their devices to see if it worked, and we had multiple interoperability demonstrations to show that devices from different suppliers can exchange GOOSE messages over an Ethernet communications network. This demonstrated that the industry was seeing the tremendous potential of this emerging technology and were heavily investing their resources to make it work.

Since the beginning of the UCA 2.0 project, the UCA users group was responsible for coordinating some of the activities, such as the organization of interoperability testing of the GOOSE implementation between different manufacturers' devices.

4.2.2 IEC 61850

IEC TC 57 is one of the Technical Committees of the IEC, which develops and maintains international standards for power systems control equipment and systems including energy management systems (EMS), supervisory control and data acquisition (SCADA), distribution automation, teleprotection, and associated information exchange for real-time and nonreal-time information, used in the planning, operation, and maintenance of power systems.

We can consider that the journey started in 1993 at the plenary meeting of IEC TC 57 in Sydney, Australia, where Germany issued a green report asking to start investigating the standardization for substation automation. Based on that, an ad hoc working group (AHWG) "Substation Control and Protection Interfaces" was established. After four meetings at the beginning of 1995, it proposed four new work items:

- Functional architecture, communication structure, and general requirements;
- Communication within and between unit and substation levels;
- Communication within and between process and unit levels;
- Companion standard for the informative interface of protection equipment.

After the approval of these new work item proposals by the national committees, three new working groups (10, 11, and 12) were established within TC57 to work on the development of IEC 61850.

The fourth proposal lead to the development of IEC 60870-5-103 as a short-term solution.

In November 1995, Working Groups 10, 11, and 12 of IEC TC 57 met for the first time in Baden, Switzerland, to start work on the development of the IEC 61850 standard. Since then, over 25 years and more than 100 working group and editor meetings, the members of the working groups have produced a few thousands of pages of IEC 61850 standards and related reports.

The responsibility of Working Group 10 (as well as its title) was to define the functional architecture and general requirements (see Figure 4.1). Because the focus of IEC 61850 at this time was substation communications, the communications between IEDs located in different substations, as well as between the substation level and the control center, were considered out of scope.

The title of Working Group 11 and its responsibility were with regard to communication within and between unit and station levels (see Figure 4.2). It included both horizontal communications between multifunctional devices performing in distributed protection schemes, as well as vertical communications between the IEDs and the substation level. This later became commonly referred to as the station bus.

Working Group 12 was responsible for communication within and between process and unit levels (Figure 4.3). It included vertical communications between the process interface devices and the substation level. This was the level performing

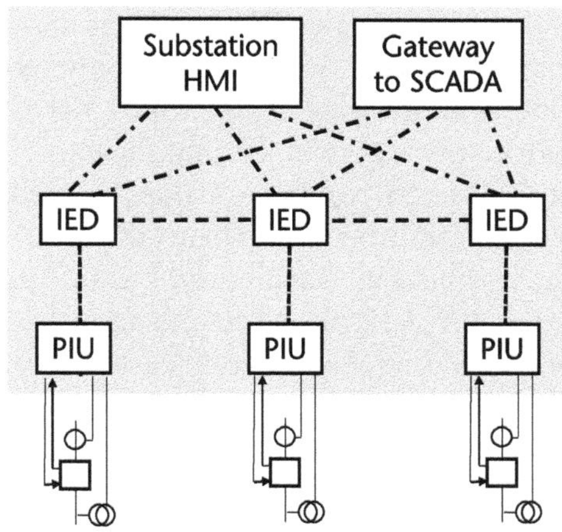

Figure 4.1 Working Group 10 responsibility.

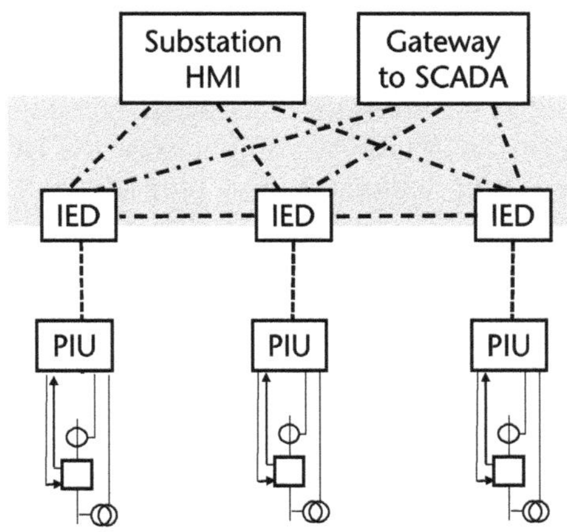

Figure 4.2 Working Group 11 responsibility.

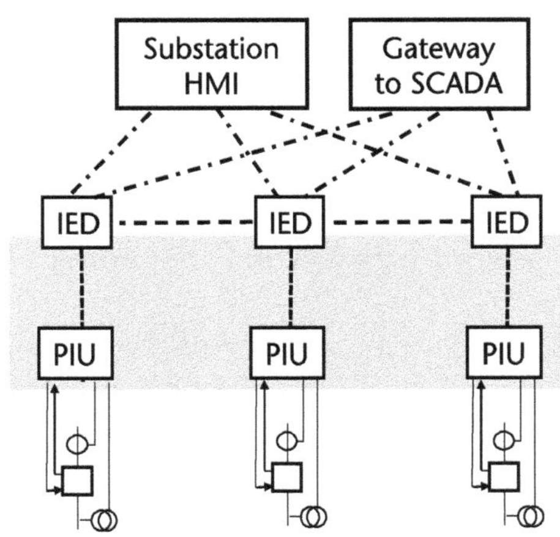

Figure 4.3 Working Group 12 responsibility.

the digitization of the analog signals and was initially concentrated on the current and voltage signals. This later became commonly referred to as the process bus.

At the beginning, all three working groups met separately. However, after more than 1 year of work, it was concluded that the work could be done more efficiently if members of the different working groups met together in joint task forces to ensure better coordination between their activities. As a result, after a coordination meeting that was held in Washington DC in April 1997 that defined the basic document layout of IEC 61850 and it was agreed to hold joint meetings and that some of the work would be done in joint task forces, each one including members from all three working groups.

In the development of IEC 61850, 1998 was an extremely important year. As mentioned earlier, at the same time, a group of experts was developing GOMSFE under the UCA 2.0 umbrella in the United States. It was becoming clear to the industry that it did not make sense for two groups of international experts to be developing at the same time two different standards for the same purpose. The efficient solution would be to stop the development of GOMSFE/UCA 2.0 and bring the North American experts working on it into the IEC 61850 working groups. After a harmonization meeting in Atlanta, Georgia, in January 1998, such an agreement was reached at a follow-up meeting in Tampa, Florida, in February 1998.

At this time, the UCA Users Group became the UCA International Users Group. The work on GOMSFE was published as IEEE TR 1550 in 1999 [3].

From there on, it took a lot of work and many meetings all over the world to publish CDs, CDVs, and FDIS until finally Parts 3 and 4 of IEC 61850 were the first published at the beginning of 2002. They were followed by different other parts. Between May and July 2003, the key IEC 61850 Part 7 was published as an international standard. The publication of Part 10 as a standard in May 2005 completed Edition 1, 10 years after it officially started.

With the major work on IEC 61850 being completed (see Figure 4.4), at the IEC TC 57 plenary meeting in Montreal, Canada, in October 2003, it was decided to merge the three working groups into Working Group 10. It was given the responsibility for the future maintenance of the standard and, considering the need to expand the standard outside of the substation to cover other smart grid domains, received a new title "Power System IED Communication and Associated Data Models."

At the same meeting, two new working groups were established:

- Working Group 17: distributed energy resources;
- Working Group 18: control of hydro power plants.

To celebrate the tenth anniversary of the development of the standard in August 2005 (see Figure 4.5), we had meetings in Baden, Switzerland, and in Klaus, Austria. On the way from Baden to Klaus, we visited the first two substations that used IEC 61850. Both were in Switzerland:

- A 16-kV distribution substation as greenfield project in Winznauschachen;
- A 380-kV transmission substation as retrofit in Laufenburg.

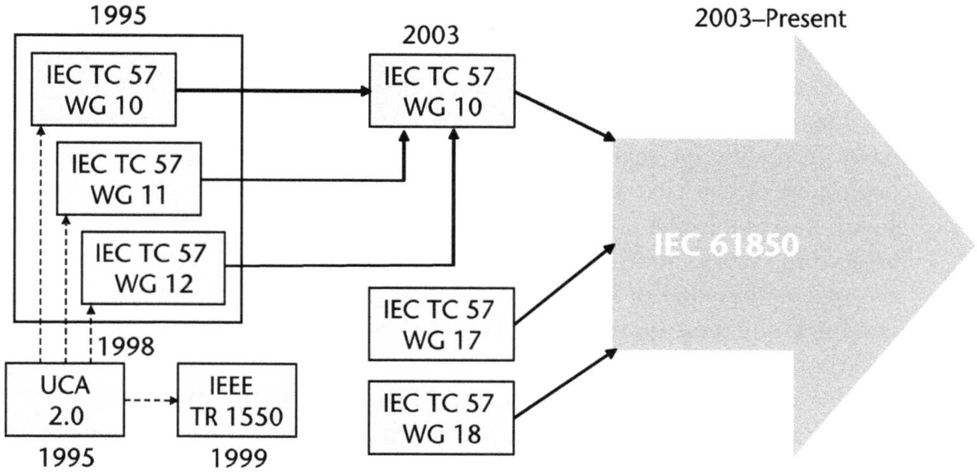

Figure 4.4 IEC 61850 contributors.

Figure 4.5 Working Groups 10, 11, and 12 members at the tenth anniversary celebration in Klaus, Austria.

This made all of us feel very proud that all the effort and time that we have invested in the development of the standard was not wasted and that the industry has started to see its benefits and then implement it, starting a trend that has resulted in the digitization of many thousands of substations all over the world.

4.3 The Insider's View

The previous sections of this chapter described the IEC standard development process and briefly summarized the key moments in the development of Edition 1 of the standard.

However, not many people, other than the members of the working groups, have any idea how the standard actually is developed, so I will try briefly to explain how it really works based on my many years of involvement in it.

I stated earlier that it took many working group meetings to develop the standard. In reality, we schedule three working group meetings every year: one at the beginning of the year (typically in February), one later in the spring (usually in June), and one in the fall (in October). The meetings are held in different countries around the world in order to make it more convenient for members of the working groups, so they do not have to travel long distances every time. In many cases, we will have one meeting in North America, one meeting in Europe, and one meeting in Asia or Latin America.

For each meeting, we have to find a host company or organization that will provide the facilities required for the meetings of the whole working group, as well as multiple smaller rooms for parallel meetings of the different task forces working on various parts of the standard or technical reports.

The meeting typically starts on Monday morning with the plenary session where all the participating members of the working group get together to receive an update on the latest developments. After that, we split into separate task forces that will usually have a half-day or sometimes a full-day meeting discussing the development of a CD or CDV and working on the comments resolution following a vote on a specific document.

At the end of the week on Friday, we typically have another plenary session where we all get together and the leaders of the different task forces present a short report on what they have accomplished during their session.

When we are in an early stage of the preparation of a CD, many times different members of the task force make presentations related to the topic to be covered, presenting ideas of what they believe should be included in the document. The actual work on writing the document itself is performed between the working group meetings by members who had volunteered to work on a section of the document according to its accepted outline. The contributions are submitted to the leader of the task force or the editors of the documents and distributed between the members before the meeting. During the meeting, they are reviewed and then edited when necessary.

We all spend a significant amount of time together and, considering how special the members of the working group are, it is easy to understand why we have developed very close relationships and have become really good friends. Reading this text, one may decide that this is a very smooth process. However, in many cases, it can be quite challenging, because many of the members of the working group are not only brilliant people, but also very opinionated. So even though we are good friends, during the heat of the discussion of a specific topic, we may have very serious arguments and disagreements. What is good is that none of this is personal, so after arguing with each other and reaching some acceptable resolution during the day, we may get together in the evening to have dinner and a glass of wine.

This gives you an idea of the environment of the different working group meetings. So, many times we will end up in a room that has a setup with multiple tables around (Figure 4.6). People sit across from each other and have very intense discussions.

Figure 4.6 Working group meeting in Nuremberg, Germany, in 2006.

We are going to talk a little bit about the Ethernet and how it impacts the development of this technology. At the time when we started UCA 2.0 and IEC 61850, the Ethernet technology was collision-based, so there are collisions in the system when multiple devices try to send a message at the same time. The same way we also had collisions in our meetings, when multiple people started talking at the same time because they were rushing to express their opinion about the subject matter of the discussion. Therefore, we used a certain tool that is similar to tools used in technology to solve these issues. The tool that we were using to solve our collision issue was a Beanie Baby frog, named Fred. Fred was to be the token (see Figure 4.7), so any time when we start these heated discussions and the convener or the chairperson who is leading the meeting decides that people cannot hear each other, he or she will switch to token-ring mode and only the person who is holding Fred will be the person who may talk. When somebody else wants to talk related to the discussed subject, he or she will raise a hand and the person holding Fred will throw it across the room. This required some skills, especially considering that there were bottles and glasses on the tables, so we were having some fun with that. Fred really played a very critical role in the development of the standard. If it were not for Fred, probably we would be still having all these discussions and nothing would have been published.

Considering that the standard has been extended to cover other domains of the smart grid to support its digitalization, today we still have the three meetings of Working Group 10, but we also have three meetings of Working Group 17, focused

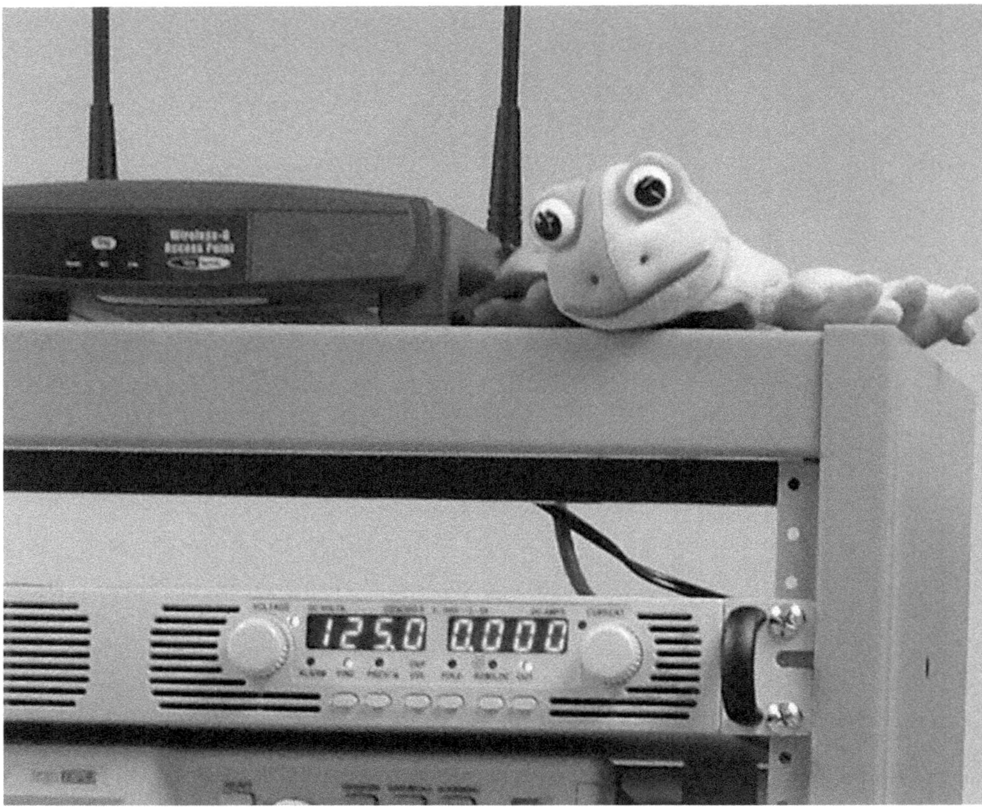

Figure 4.7 Fred, the token.

on the integration of distributed energy resources and other smart grid technologies, such as distribution automation and electric vehicles. This doubles the number of meetings that members of both working groups need to attend, which is why typically these meetings are held in consecutive weeks at the same location, making it easier to travel, but, at the same time, taking us away from our families and our everyday job responsibilities. Because we all have these everyday tasks that do not go away, a lot of the work for the development of the standard is done on our own time during the weekend, in the evening, or even when we are on vacation.

One thing that many people may not understand is that all travel expenses and the time for the development of the standard is paid for by the companies or organizations providing the experts who participate in the working groups. If some experts are self-employed, they pay all of it themselves. What I find really interesting is that, after all the time and money were spent to develop the international standard, once it is published, we and our employers who paid for all of it have to buy it ourselves, which seems unfair.

References

[1] https://www.iec.ch/standards-development/stages

[2] "WG 34.03 Communication Requirements in Terms of Data Flow Within Substations: Results of WG 34.03 and Standardization Within IEC, K-P Brand," *Electra*, No. 173, August 1997.

[3] IEEE--SA TR 1550—1999 IEEE-SA Technical Report on Utility Communications Architecture Version 2.0, 1999.

CHAPTER 5

The IEC 61850 Standard and Related Documents

5.1 Introduction

IEC 61850 is a continuously evolving standard due to the continuously changing industry and technology. The engineering, installation, testing, commissioning, and maintenance of thousands of substations that have been installed and are operating around the world also help to identify challenges with the interoperability between devices and tools, gaps in the standard or definitions that may be subject to misinterpretation. All of that requires continuous work on the improvement of the different parts of the standard.

At the same time, the experience of using IEC 61850 has demonstrated its significant benefits in comparison with other technologies, making it a fundamental technology for the smart grid. That is why members of different IEC TC 57 working groups are working on the extension of the standard in other domains.

The goal of this chapter is to provide an overview of the different parts of the standard and their current status as well as to inform about the different technical specifications and technical reports related to the implementation of the standard that have already been published or are under development by the different working groups. Details about the contents of each of the parts of the standard are not included in this chapter because they are discussed in detail in the following chapters in relation to specific modeling or implementation and application issues.

5.2 IEC Standard Document Types

In Chapter 4, we discussed in detail the development of international standards. Now we focus on the content of the IEC 61850 standard and different documents related to it (Figure 5.1). In order to understand what each of these documents represents, we need to briefly describe the different types of documents that are produced as part of the standard development process and what it takes to approve them.

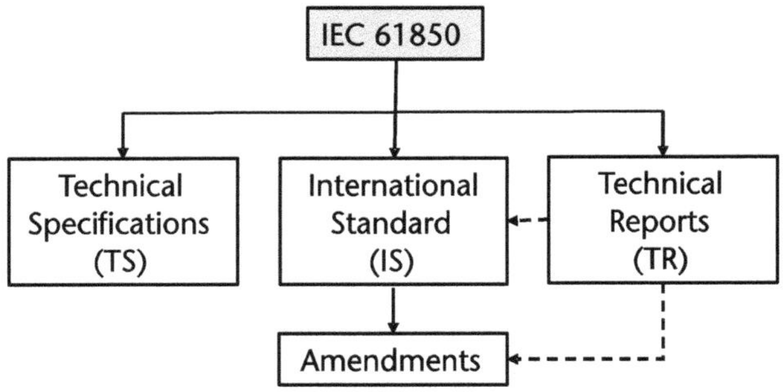

Figure 5.1 IEC 61850 document types.

Other than an international standard, there are several technical specifications (TS) that in many ways are similar to an international standard but have not passed through all approval stages because standardization is seen to be premature. A technical specification is a normative document that is approved by two-thirds of the P-members of an IEC technical committee. The final vote takes place at the draft technical specification (DTS) stage.

In this chapter, we have a list of many technical reports (TRs). They play a very important role in supporting the implementation and the application of the standard, because they focus on a specific issue that is not directly covered in the standard. Within IEC 61850, TRs are used to gather all of the requirements that may impact multiple standards so that they can be reviewed and balloted in their entirety instead of piecemeal.

The TRs typically consider multiple-use cases that help to identify gaps in the existing standard and propose extensions or modifications to the models and services. They are approved by a simple majority of P-members of an IEC technical committee; TRs are not normative documents, but many of their recommendations typically become part of an amendment to the standard or its next edition.

Speaking of amendments, they are normative documents that modify or extend the technical content of an existing international standard. They are developed following a similar process as international standards. The IEC Central Office may prepare, for user convenience, a consolidated version consisting of the base edition with the amendment(s) and designated as, for example, Edition 2.1 (i.e., the second edition of the standard consolidated with the first amendment).

5.3 The Core IEC 61850 Standard Parts

As described in previous chapters, the IEC 61850 standard was developed as a set of general parts defining the requirements for modeling and communications related to substation automation and later extended to cover a broader range of power utility automation applications. Edition 1 of the standard contained 10 main parts (Figure 5.2), some of which are generic, defining the general principles and a set of requirements, followed by the abstract modeling parts defining communication

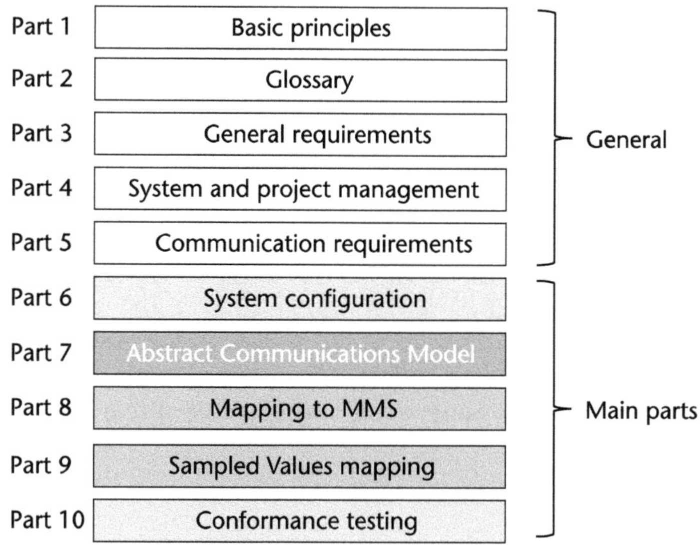

Figure 5.2 Parts of the core IEC 61850 standard.

services, data types and objects, function elements, and, finally, a set of parts defining mapping to specific protocols to be used in actual real-life implementations and their conformance testing. Some of the parts contain subparts, making the total of documents included as part of the publication of Edition 1 of the Standard 14.

Based on the feedback from the industry and the evolution of requirements and technology, the standard has evolved with an expanded scope to cover all smart grid-related domains and support its digitization and digitalization.

The following sections briefly describe the content of the different parts of the core IEC 61850 standard and their current status.

5.3.1 Part 1

IEC 61850 Communication Networks and Systems in Substations—Part 1: Introduction and Overview is a TR that was published on April 28, 2003, as an introduction and overview of the IEC 61850 standard series. It described the philosophy, the work approach, the contents of the other parts, and the documents of other bodies that have been reviewed.

IEC TR 61850-1:2013 Communication Networks and Systems for Power Utility Automation—Part 1: Introduction and Overview, Edition 2, was published on March 14, 2013, and is applicable to power utility automation systems and defines the communication between IEDs in such a system and the related system requirements. This part still gives an introduction and overview of the IEC 61850 standard series, but also includes significant technical changes compared to Edition 1 and resulting from the extended application scope of the IEC 61850 standard. In order to support smart grid-related applications, it covers distributed generation monitoring and automation, power quality, and other considerations. Substation-to-substation communications are also included.

5.3.2 Part 2

IEC 61850 Communication Networks and Systems in Substations—Part 2: Glossary is a TS and was published on August 7, 2003. It contained a glossary of specific terms and definitions used in the context of substation automation systems, which are standardized in the various parts of the IEC 61850 series.

Due to the expansion of the scope of the standard, it was revised and published on April 17, 2019, as *IEC TS 61850-2:2019 Communication Networks and Systems for Power Utility Automation—Part 2: Glossary* and is a technical revision that canceled Edition 1.

It includes new definitions of terms used in the new edition of the IEC 61850 standard series, as well as updates of existing definitions related to the expansion to cover power utility. Some definitions have been removed, and others have been corrected or clarified.

5.3.3 Part 3

IEC 61850 Communication Networks and Systems in Substations—Part 3: General Requirements was published on January 15, 2002. It applies to substation automation systems and defines requirements for the communication between IEDs in the substation including quality requirements (reliability and availability), environmental withstand requirements, and auxiliary services.

A revised document was published as *IEC 61850-3:2013 Communication Networks and Systems for Power Utility Automation—Part 3: General Requirements, Edition 2*, on December 12, 2013, and expands the requirements to cover IEDs applied for protection and control of electric power systems and power plants.

Some key additions related to product safety were added on the basis of IEC 60255-27, as well as EMC requirements in line with the IEC 60255 series and IEC 61000-6-5.

5.3.4 Part 4

IEC 61850 Communication Networks and Systems in Substations—Part 4: System and Management was published on January 17, 2002. This describes the requirements of the system and project management process and for special supporting tools for engineering and testing. The main topics covered in the document are the engineering process and the life cycle of the substation protection and control system, as well as the quality assurance process covering the complete life cycle.

IEC 61850-4:2011 Communication Networks and Systems for Power Utility Automation—Part 4: System and Project Management, Edition 2, was published on April 4, 2011, followed by Amendment 1 on November 3, 2020, published as a consolidated version. It is a technical revision of Edition 1 related to the change of focus of the standard from substation automation systems to power utility automation. Some of the changes are used to align the document more closely with the other parts of the IEC 61850 series.

5.3.5 Part 5

IEC 61850 Communication Networks and Systems in Substations—Part 5: Communication Requirements for Functions and Device Models was published on July 24, 2003, and applies to substation automation systems and the requirements for standardization of the communications between IEDs and other components of substation automation systems. It describes the different types of functions and the requirements for achieving interoperability based on the logical node modeling approach and pieces of information for communications (PICOM) description. Message and system performance requirements are also defined.

IEC 61850-5:2013 Communication Networks and Systems for Power Utility Automation—Part 5: Communication Requirements for Functions and Device Models, Edition 2, was published on January 30, 2013, and applies to power utility automation systems, but still maintains the understanding that substation automation systems are the foundation for utility automation.

The major technical changes compared to Edition 1 are the result of the expansion of the standard to cover other smart grid domains and the related communications outside of the substation. This includes interfaces for communications between substations and the substation and system-level applications.

5.3.6 Part 6

IEC 61850 Communication Networks and Systems in Substations—Part 6: Configuration Description Language for Communication in Electrical Substations Related to IEDs was published on March 23, 2004. It is one of the key parts of the standard that demonstrates that it is defining not only a communications protocol, but also a completely new engineering environment suitable for different smart grid applications. It introduces the concept of a substation configuration language based on XML schema and an object model that covers all the aspects of substation automation systems, including the connectivity of the primary equipment in the substation, a communication system model, and the modeling of the different IEDs. The model is related to IEC 61850-5 and the IEC 61850-7 series.

It specifies several types of files for describing communication-related IED configurations and IED parameters, communication system configurations, switchyard (function) structures, and the relations between them. The purpose is to exchange IED capability descriptions, and substation automation (SA) system descriptions between IED engineering tools and the system engineering tool(s) of different manufacturers in a compatible way.

IEC 61850-6:2009 Communication Networks and Systems for Power Utility Automation—Part 6: Configuration Description Language for Communication in Electrical Substations Related to IEDs, Edition 2, was published on December 17, 2009, followed by Amendment 1 on June 7, 2018, published as a consolidated version. It is a technical revision of Edition 1 related to the change in other parts of the standard, especially in different parts of IEC 61850-7.

Other than clarifications and corrections, the main changes compared to Edition 1 are the result of the new focus to power utility automation and functional extensions added on the basis of related changes, especially in IEC 61850-7-2 and IEC 61850-7-3.

Other extensions address requirements for configuration data exchange between system configuration tools related to substation-to-substation communications and additional functionality for the engineering process.

5.3.7 Part 7-1

IEC 61850 Communication Networks and Systems in Substations—Part 7-1: Basic Communication Structure for Substation and Feeder Equipment—Principles and Models was published on July 23, 2003. This provides an overview of the architecture for communication and interactions between substation devices such as protection devices, breakers, transformers, and substation hosts. It uses simple examples of functions to describe the concepts and methods applied in the IEC 61850 series. It also describes the relationships between other parts of the IEC 61850 series and defines how interoperability is achieved.

IEC 61850 Communication Networks and Systems for Power Utility Automation—Part 7-1: Basic Communication Structure—Principles and Models, Edition 2, was published on July 15, 2011, followed by Amendment 1 on August 31, 2020, published as a consolidated version. It is a technical revision of Edition 1 related to the change in other parts of the standard, especially in the other parts of IEC 61850-7.

Edition 2 and the amendment still have as their main goal introducing the modeling concepts expanded to cover power utility automation systems and their applications. Some smart grid-related additions that improve the efficiency of the system are the introduction of hierarchical logical devices and its impact on the testing features. The definition of logical node inputs and how they support the modeling of the data flow and testing, modeling for time synchronization and statistical data are also introduced. Namespaces and their definition and applications are discussed as well. Multiple new logical nodes related to the expansion of the scope of the standard are introduced together with clarifications for certain issues.

5.3.8 Part 7-2

IEC 61850 Communication Networks and Systems in Substations—Part 7-2: Basic Communication Structure for Substation and Feeder Equipment—Abstract Communication Service Interface (ACSI) was published on May 12, 2003. The ACSI defines the abstract interface describing communications between a client and a remote server, as well as the abstract interface for high-speed communications between functions in multiple devices and for the transmission of sampled measured values.

IEC 61850-7-2:2010 Communication Networks and Systems for Power Utility Automation—Part 7-2: Basic Information and Communication Structure—Abstract Communication Service Interface (ACSI), Edition 2, was published on August 24, 2010, followed by Amendment 1 on February 10, 2020, published as a consolidated version. It is a technical revision of Edition 1 related to the expansion of the standard to cover utility automation and smart grid-related functionality.

The major technical changes compared to Edition 1 include the removal of data types that were identified as not required, the additional service tracking for control blocks, and role-based access control (RBAC) viewing concept. Security issues are addressed by the IEC 62351 series.

5.3.9 Part 7-3

IEC 61850 Communication Networks and Systems in Substations—Part 7-3: Basic Communication Structure for Substation and Feeder Equipment—Common Data Classes was published on May 12, 2003. This part of the standard specifies common data classes and common attribute types related to substation applications. The common data classes apply to status information, measured information, controllable status information, controllable analog set point information, status and analog settings, and attribute types used in these common data classes. These common data classes and attribute types are used in the modeling of functions in substation and feeder equipment.

IEC 61850-7-3:2010 Communication Networks and Systems for Power Utility Automation—Part 7-3: Basic Communication Structure for Substation and Feeder Equipment—Common Data Classes, Edition 2, was published on December 16, 2010, followed by Amendment 1 on February 10, 2020, published as a consolidated version. It is a technical revision of Edition 1 related to the expansion of the standard to cover automation functionality in other smart grid-related domains.

This second edition defines new common data classes used for new standard-defining object models for other domains based on IEC 61850 and for the representation of statistical and historical data.

5.3.10 Part 7-4

IEC 61850 Communication Networks and Systems in Substations—Part 7-4: Basic Communication Structure for Substation and Feeder Equipment—Compatible Logical Node Classes and Data Classes was published on May 13, 2003. It specifies a range of logical nodes representing function elements in substation automation systems. It defines compatible logical node names and data names for communication between IEDs, including the relationship between logical nodes and data objects. It also defines a list of abbreviations that can be used in creating data object names. In order to support better interoperability, this part also provides a detailed semantical description of the data object names used.

IEC 61850-7-4:2010 Communication Networks and Systems for Power Utility Automation—Part 7-4: Basic Communication Structure—Compatible Logical Node Classes and Data Object Classes, Edition 2, was published on March 31, 2010, followed by Amendment 1 on February 12, 2020, published as a consolidated version. It is a technical revision of Edition 1 related to the expansion of the standard to cover automation functionality in other smart grid-related domains.

Compared to Edition 1, the revision includes corrections and clarifications addressing the technical issues raised by the users of the standard. Edition 2 defines new groups of logical nodes of general interest from IEC 61850-7-410 and IEC 61850-7-420, as well as the addition of new logical nodes in the existing groups to meet the needs of smart grid-related applications, such as monitoring, power quality, substation-to-substation communications, and extensions to the model for statistical and historical statistical data.

5.3.11 Part 8-1

IEC 61850 Communication Networks and Systems in Substations—Part 8-1: Specific Communication Service Mapping (SCSM)—Mappings to MMS (ISO/IEC 9506-1 and ISO/IEC 9506-2) and to ISO/IEC 8802-3 was published on May 25, 2004. Following the abstract definitions in Part 7, it specifies a method of exchanging data through substation local-area networks by mapping ACSI to MMS and ISO/IEC 8802-3 frames. MMS services and protocol are specified to operate over full 7-layer OSI stack and TCP-compliant communications profiles for client-server communications and support both centralized and distributed architectures. This standard includes the exchange of time-critical and nontime-critical real-time data, control, and reporting and provides mappings for the services and objects specified within IEC 61850-7-2, IEC 61850-7-3, and IEC 61850-7-4.

IEC 61850-8-1:2011 Communication Networks and Systems for power Utility Automation—Part 8-1: Specific Communication Service Mapping (SCSM)—Mappings to MMS (ISO 9506-1 and ISO 9506-2) and to ISO/IEC 8802-3, Edition 2, was published on June 17, 2011, followed by Amendment 1 on February 21, 2020, published as a consolidated version. It is a technical revision of Edition 1 related to the expansion of the standard to cover power utility automation functionality in other smart grid-related domains and support communications over wide area networks. It also inserts into the standard some of the findings of technical reports.

Because not all ACSI services defined in IEC 61850-7-2 can be mapped to MMS, this part defines additional protocols. It addresses many issues that have been raised since the initial publication of the standard and also reflects the evolution of communications technology, such as Gb Ethernet and parallel redundancy protocol (PRP) and high-availability seamless redundancy (HSR). It also supports the communication of GOOSE and sampled value messages over wide area networks for smart grid-related applications.

Other changes compared to previous editions are the extension of the length of the object reference and support for IPv6 and the new COMTRADE file format. The time synchronization is extended with an IEEE 1588 profile defined with IEC/IEEE 61850-9-3.

As can be seen, there are many extensions compared to early edition, but there is also the deprecation of the unicast sample value model and the Generic Substation State Event (GSSE) model.

5.3.12 Part 8-2

IEC 61850-8-2:2018 Communication Networks and Systems for Power Utility Automation—Part 8-2: Specific Communication Service Mapping (SCSM)—Mapping to Extensible Messaging Presence Protocol (XMPP) was published on December 14, 2018, and is intended to meet the needs of different smart grid-related applications over web services. It should support the exchange of data through a variety of networks, including public networks. It supports a limited number of the abstract services specified in IEC 61850-7-2 and, at this moment, only supports client-server communications and time synchronization services.

The mapping is to XML messages that are transported over XMPP. This part of the standard describes the use of the protocol and how specific features of XMPP are used for the mapping. It also defines the XML payloads corresponding to the ACSI services and how to meet the requirements for end-to-end security.

5.3.13 Part 9-1

IEC 61850 Communication Networks and Systems in Substations—Part 9-1: Specific Communication Service Mapping (SCSM)—Sampled Values over Serial Unidirectional Multidrop Point to Point Link was published May 12, 2003. It defines the specific communication service mappings for the communication between the process level and the bay level using a serial unidirectional multidrop point-to-point link in accordance with IEC 60044-8, which introduced the merging unit as an interface to electronic current and voltage transformers, as well as the data object as an output. This data is provided over the communication between the merging units and devices such as protection IEDs.

IEC 61850-9-1 specifies a serial communication interface between the merging unit and other equipment using a subset of the abstract communication services defined in IEC 61850-7-2 and are mapped on an ISO/IEC 8802-3 based communication link. IEC 61850-9-1 has been withdrawn.

5.3.14 Part 9-2

IEC 61850 Communication Networks and Systems in Substations—Part 9-2: Specific Communication Service Mapping (SCSM)—Sampled Values over ISO/IEC 8802-3 was published on April 20, 2004. This part defines the specific communication service mapping for the transmission of sampled values according to the abstract specification in IEC 61850-7-2. The mapping of the abstract model is based on direct access to an Ethernet link for the transmission of the sampled values in combination with IEC 61850-8-1. This is the foundation of what is commonly known as process bus. It specifies the externally visible functionality of implementations and conformance requirements but does not specify implementations or products.

IEC 61850 Communication Networks and Systems for Power Utility Automation—Part 9-2: Specific Communication Service Mapping (SCSM)—Sampled Values over ISO/IEC 8802-3, Edition 2, was published September 22, 2011, followed by Amendment 1 on February 12, 2020, published as a consolidated version. It is a technical revision of Edition 1 related to the expansion of the standard to cover power utility automation functionality in other smart grid-related domains and support any sensor logical node from group T that has a stream of sampled values as an output. It also inserts into the standard testing-related functionality that was missing in Edition 1.

Because the standard does not define specific communications architectures in order to support different applications, Edition 2 defines both unicast and multicast-based communications interfaces; however, the unicast was depreciated in Amendment 1. Considering the critical impact of the process bus interface on the functionality of the system, redundancy protocols such as PRP and HSR are supported. The time synchronization of the merging units adds references to IEC 61588:2009 and IEC/IEEE 61850-9-3 for SV synchronization.

5.3.15 Part 9-3

IEC/IEEE 61850-9-3:2016 Communication Networks and Systems for Power Utility Automation—Part 9-3: Precision Time Protocol Profile for Power Utility Automation was published on May 31, 2016, and defines a utility profile for time synchronization based on IEEE 1588 [1] that can achieve precision better than 100 ns and meet the highest time synchronization classes defined in IEC 61850-5 and IEC 61869-9.

5.3.16 Part 10

IEC 61850 Communication Networks and Systems in Substations—Part 10: Conformance Testing was published on May 30, 2005. This specifies standard techniques for testing of conformance of implementations, as well as specific measurement techniques to be applied when declaring performance parameters. The use of these techniques will enhance the ability of the system integrator to integrate IEDs easily, operate IEDs correctly, and support the applications as intended. This part is developed in coordination with the UCA International Users Group, which performs the IEC 61850 conformance testing.

IEC 61850-10:2012 Communication Networks and Systems for Power Utility Automation—Part 10: Conformance Testing, Edition 2, was published on December 14, 2012, as a technical revision to Edition 1. Compared to the previous edition, it updates the server device conformance test procedures and adds several new test procedures for client device conformance testing, sampled value device conformance testing, engineering tool-related conformance testing, and GOOSE performance testing.

5.4 The IEC 61850 Standard-Related Documents

The previous section presented the different parts of the core IEC 61850 standard that have been published so far and their status. This section includes a list of the currently published additional parts of the standard and related documents with their status as of the time of this writing.

When there is a year in the title of the document, it indicates that it has been published and the year of publication. If there is no year in the title, it means that this is a document that is still under development and has not been published yet.

If there is TS in the title, this means that it is a technical specification, and if there is TR in the title, it means that it is a technical report. If there is no TS or TR, it means that it is a standard document, and if there is AMD, it means that there is an amendment to the last edition with the year of its publication.

61850-6-2 Configuration Description Language Extensions for Human Machine Interfaces

61850-6-100 Guideline for Function Modeling in SCL for Substation Automation

IEC 61850-7-410:2012 Hydroelectric Power Plants—Communication for Monitoring and Control

IEC 61850-7-410:2012/AMD1:2015 Hydroelectric Power Plants—Communication for Monitoring and Control

IEC 61850-7-420:2009 Communications Systems for Distributed Energy Resources (DER)—Logical Nodes

IEC TR 61850-7-5:2021 IEC 61850 Modelling Concepts

IEC TR 61850-7-500:2017 Use of Logical Nodes to Model Functions of a Substation Automation System

IEC TR 61850-7-510:2012 Hydroelectric Plants—Modelling Concepts and Guidelines

61850-7-520 DER—Modelling Concepts and Guidelines

IEC TR 61850-7-6:2019 Guideline for Basic Application Profiles

IEC TS 61850-7-7:2018 Specification of Schema for Namespace Definition Files

IEC 61850-10-210 Interoperability Tests for Hydro Equipment Based on IEC 61850

IEC 61850-10-3 Functional Testing of IEC 61850 Based Systems

IEC TS 61850-80-1 Guideline to Exchange Information from a CDC Based Data Model Using IEC 60870-5-101/104

IEC TR 61850-80-3:2015 Mapping to Web Services—Requirement Analysis and Technology Assessment

IEC TS 61850-80-4:2016 Mapping Between the DLMS/COSEM (IEC 62056) Data Models and the IEC 61850 Data Models

IEC 61850-80-5 Mapping Between Modbus and IEC 61850

IEC 61850-80-6 Using IEC 61850 for the Communication Between Substations and Control Centres

IEC TR 61850-90-1:2010 Using IEC 61850 for the Communication Between Substations

IEC TR 61850-90-3:2016 Using IEC 61850 for Condition Monitoring

IEC TR 61850-90-4:2020 Network Engineering Guidelines for Substations

IEC TR 61850-90-5:2012 Using IEC 61850 to Transmit Synchrophasor Information According to IEEE C37.118

IEC TR 61850-90-6:2018 Using IEC 61850 for Distribution Automation

IEC TR 61850-90-7:2013 Object Models for Photovoltaic, Storage and Other Inverter Based Applications

IEC TR 61850-90-8:2016 Object Models for Electrical Vehicles

IEC TR 61850-90-9:2020 Object Models for Electrical Energy Storage

IEC TR 61850-90-10:2017 Object Models for Schedules

IEC TR 61850-90-11:2020 Methodologies for Modelling of Logics for IEC 61850 Based Applications

IEC TR 61850-90-12:2020 Wide Area Network Engineering Guidelines

IEC TR 61850-90-13:2021 Deterministic Network Topologies

IEC TR 61850-90-14 Using IEC 61850 for FACTS and Power Conversion Data Modelling

IEC TR 61850-90-15 IEC 61850 Based DER Grid Integration

IEC TR 61850-90-16:2021 Requirements for System Management

IEC TR 61850-90-17:2017 Use of IEC 61850 to Transmit Power Quality Data

IEC TR 61850-90-18 Modeling Alarmhandling for IEC 61850

IEC TR 61850-90-19 Applying Role Based Access to IEC 61850

IEC TR 61850-90-20 Guideline for Redundant IEDs with IEC 61850

IEC TR 61850-90-21 Use of IEC 61850 for Traveling Wave Fault Location System

IEC TR 61850-90-22 Network Autorouting

IEC TR 61850-90-23 Model Extensions to IEC 61850 to Support Microgrids

IEC TR 61850-90-24 Mapping of IEC 62351-7 on IEC 61850

IEC TR 61850-90-25 Model Update Based on Users Feedback

IEC TR 61850-90-26 IED Specification Description

IEC TR 61850-90-27 Model for (Distributed) Thermal Energy

61850-11-xxx Set of BAPs to Support DER Operational Functions

Reference

[1] IEEE 1588:2008 IEEE Standard for a Precision Clock Synchronization Protocol for Networked Measurement and Control Systems, 2008.

CHAPTER 6

Substation Communication Basics

6.1 Introduction

Electric power system protection has been around since the start of electricity use and, since this time, it has been based on single function devices performing simple dedicated tasks. With time, it became necessary to increase its sophistication and complexity using different PAC schemes that required exchanges of signals between individual devices. With the development of technology and increased requirements for the improvement in the stability, reliability, and security of the electric power grid, they became more reliant on communications.

However, many of the experts who were involved in the development of PAC systems did not study communications and did not have any experience with communications technologies. That is why the goal of this chapter is to discuss the communication basics in general and how these fundamental concepts are related to substation communications in order to support the substation PAC systems.

This also applies not only to substation communications, but also to inter-substation communications and wide area communications for system integrity protection schemes or wide area monitoring systems as well.

That is why we start first by discussing what do we mean by communications in general and what is required in order to successfully communicate.

6.2 Communication Requirements

Communication is simply the act of exchanging information between two or more objects. An object can be anything: people, groups of people, or devices.

Every communication involves at least two objects, one sender and one recipient, but both the numbers of senders and recipients can vary depending on the specific use case. This sounds very simple, but it is actually a very complex subject. In order to achieve successful communication, it is necessary to answer many different questions:

- Who is sending and who is receiving the information?
- What is the information that we need to exchange?

- What is the reason that we need to exchange information?
- How can we exchange the information?

I believe that when we are looking at any new subject it is very helpful to compare it with what we are familiar with in our everyday life, because it is based on experiences that we all share regardless of the fields in which we work.

6.2.1 Who Is Sending and Who Is Receiving the Information?

The simpler case is when one person is sending the information and one person is receiving, that is, one-to-one communication (Figure 6.1). In the substation environment, a simple example of this type of communication is the feeder protection relay sending a trip signal to its breaker to clear the fault.

Another typical case is when someone is presenting a paper at a conference. In this case, during the presentation, it is one-to-many communication. In the substation, an example of such communication is the underfrequency protection relay sending a trip signal to several distribution feeder breakers to shed load.

The third common case is when a radio host is talking to whoever is interested in listening to him; this is the case of broadcasting communication. There are no typical cases requiring broadcasting in the substation.

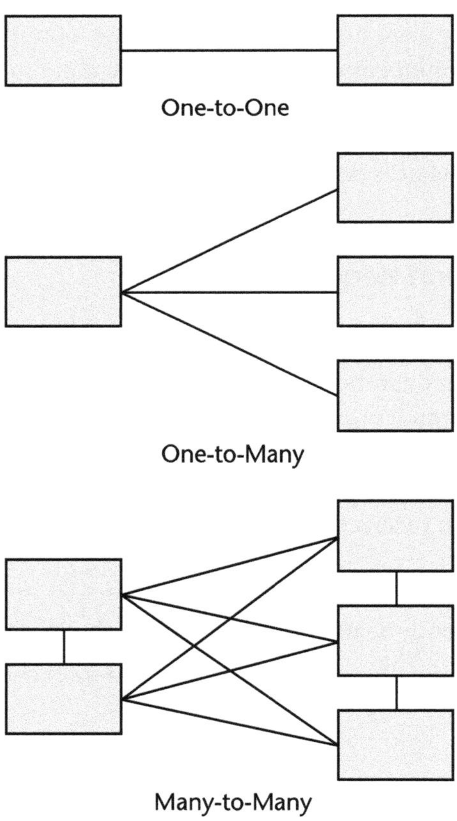

Figure 6.1 Types of communication.

Things get more complicated when people are in a meeting room discussing a specific topic. In this case, we have many-to-many communication. This is not a typical case in protection schemes because they typically are based on one-to-one or one-to-many communications.

6.2.2 What Is the Information That We Need to Exchange?

The information that is exchanged also has an impact on the communications. For example, it may indicate a change of state (a traffic light) or a different value of a certain parameter of interest (the speed limit). It can also be weather information or information about a new task that we are given. This limited number of examples shows the wide variety of information that might be communicated and the different requirements that it will impose on the communication methods and the timing of the required actions.

In the substation environment, we also can have a variety of different types of information, for example, the change of state of a circuit breaker, variations in system frequency, and the operation of protection relays.

6.2.3 What Is the Reason That We Need to Exchange Information?

The reason to exchange information is very important because it has an impact on the actions that need to be taken. In the traffic light example above, depending on the change of the color of the light, for example, from green to yellow and red, will require us to first slow down and then stop. This is critical information because it has an impact on the safety of many people that are driving. At the same time, a traffic sign indicating a speed limit is something that we better obey in order to avoid getting a speeding ticket, but we can maintain a speed not necessarily exactly at the number indicated by the sign.

The weather information gives us an indication of what do we need to wear in order to feel comfortable under the specific conditions. It is not that critical in some cases, for example, in the spring when it is not really hot or cold, but it may be critical in the summer or winter if the weather is very hot or cold and it becomes a safety issue.

It is similar in the substation environment when the reason to exchange information is to clear a short-circuit fault; the information from the relay to the circuit breaker has to be delivered immediately in order to reduce the fault-clearing time as much as possible. However, if the information is used to inform a frequency load-shedding scheme what is the loading of a specific feeder, it is not that critical from a timing point of view but may have an impact on the decision-making by the scheme. Weather information can have a significant impact on the control of renewable DERs because it may help to determine what kind of storage capacity will be required.

6.2.4 How Can We Exchange the Information?

This is the most complex question to answer regarding communications. The starting answer is "It depends!" and it depends on the answers to all the previous questions.

First, the simplest form of communication is between two objects, a sender and receiver. They can exchange information using a direct point-to-point interface. For example, in the human world two people can just sit across each other and talk. However, this requires that they speak the same language and their communication capabilities are similar. If one of the persons is expressing what he has to say by speaking and the person he is talking to has a hearing problem the communication is not going to be successful. At the same time, if the person is sharing information by showing an image on a computer screen to a blind person, this also is not going to be a successful communication. Even if the person who is speaking is talking to a person with good hearing and both are using the same language, there might be some words that mean different things to both people, which may lead to misunderstanding, that also is going to be a failure.

Things get even more complicated when there are more than two people that have to exchange information. First, they have to share some media to which all have access, like being in the same room. They have to agree on the language that they are going to use if they come from different countries and they also have to agree on the rules to follow when they need to speak to avoid people speaking at the same time, which will make it difficult to follow. These examples show how communications between people and communications between devices in practice have very similar requirements.

At the beginning of using communications in electric power system protection, it was a simple exchange of binary signals using point-to-point connections with copper wires between the two devices. With the development of communication technologies, this has changed significantly and now we are talking about communications between multiple devices over communication networks. That is why next communication networks, their components, and the different methods that can be used to exchange information are discussed.

6.3 Communication Nodes

A communication network is simply a collection of connected objects. In this case, we refer to the objects as nodes. The physical network nodes are electronic devices that are attached to a network, and depending on their functionality are capable of creating, sending, receiving, or transmitting information over communication links.

Fundamental properties of any network are the total number of nodes and the total number of links between nodes. Depending on the network topology, each node has a certain number of links that connect it to other nodes.

Depending on their functionality nodes can send or receive or both send and receive data. These nodes can be end nodes or intermediary nodes. End nodes are the ones that are the starting point in the communication or the endpoint in the communication. If two devices need to communicate with each other the device, which initiates that this communication is an end device and the device intended as the final recipient of the communication is an end device. These two end devices are going to communicate with each other over a communication network with the help of some intermediary network devices.

In our everyday lives, we typically deal with computer networks, systems in which multiple computers are connected to each other to share information and resources using communication links. The communication elements (nodes) can be computers, mobile devices, switches, routers, and the communication links can be coaxial cables, optical fiber cables, and wireless local area network (LAN). In a substation, the communication elements can be computers, different types of process interface devices, multifunctional IEDs, switches, routers, and the communication links can be shielded twisted pair cables or optical fiber cables, but, for some purposes, a wireless LAN can be used as well.

The computers and IEDs are end nodes, while the switches and routers are intermediate nodes. Each node needs a network interface controller (NIC) for communications in a network. The NIC is a hardware component that connects a node to a communication network. They are also known as a network interface card, network adapter, LAN adapter, or physical network interface.

When connected to a network, every node must have a media access control (MAC) address. The MAC address is a unique identifier assigned by device manufacturers to a NIC for use as a network address in communications within a network segment.

The degree of connectivity of a node is the measure of the number of connections a node has with other nodes.

6.4 Transmission Modes

6.4.1 Definition

The transmission or communication mode means the way that data is transferred between individual devices that are interconnected directly or over a network.

There are three types of transmission mode (Figure 6.2):

- Simplex;
- Half-duplex;
- Full-duplex.

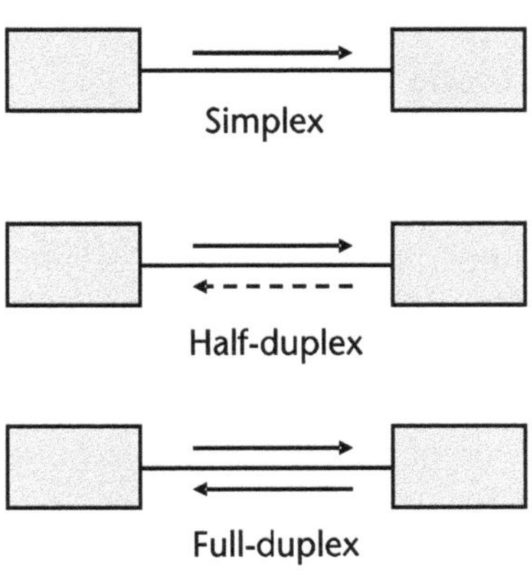

Figure 6.2 Transmission modes.

6.4.2 Simplex Mode

In the simplex mode, the communication is unidirectional, similar to the traffic on a one-way street. Only one of the two devices on a link can transmit; the other can only receive. The simplex mode can use the entire capacity of the link to send data in one direction.

Computers use the simplex mode with their interface devices: keyboard, microphone, monitor, and speakers. The keyboard and microphone can only provide input, while the monitor and speakers can only give an output.

In the substation, the simplex mode can be used with the sensor or the trip coil of a breaker. The sensor can only provide input to an IED, while the trip coil can only give an output.

6.4.3 Half-Duplex Mode

In the half-duplex mode, each node can both transmit and receive, but not at the same time. When one device is sending, the other can only receive, and vice versa. The half-duplex mode is used in cases where there is no need for communication in both direction at the same time. The entire capacity of the link can be used for each direction.

This is similar to the case of a bridge that is so narrow that only one car then go through at the time so while one carriage is crossing, a car coming in the opposite direction has to wait to go after the bridge is free of traffic.

In people's communications, an example is using a walkie-talkie. The two people talking cannot do it at the same time. They have to take turns talking and listening.

In the substation, this mode can be used for applications that are not time-critical but it is required to send information in both directions and it is allowed for the communicating devices to take turns.

It is clear that the half-duplex mode is cheaper because it requires less resources to implement; however, the performers over the communication link may not satisfy some performance requirements. The half-duplex mode is used in older networks with legacy bus topologies due to the constraints of the network medium (coax cable).

6.4.4 Full-Duplex Mode

In the full-duplex mode, both nodes can transmit and receive simultaneously. In this mode, signals going in one direction share the capacity of the link with signals going in the other direction. This mode is usually based on the link containing two physically separate transmission paths, one for sending and other for receiving. This is similar to a two-way street with one lane for traffic in each direction.

Full-duplex mode is used when communication in both direction is required all the time. In people's communications this is similar to talking on the phone, when both can talk and listen at the same time.

In the substation environment, the full duplex mode is used for protection-related applications when all protection devices need to be able to transmit and receive from other devices at the same time for the protection schemes to operate efficiently.

6.5 Communication Media

In data communication terminology, a transmission medium is a physical path between a sender and a receiver, the link through which data is sent from one node to another. Transmission media is broadly classified into two main types: wired media and wireless media.

6.5.1 Wired Media

It is also referred to as guided transmission media. Signals are transmitted over a narrow pathway by using physical links. Its advantages are high speed and security. The disadvantage is that it can be used for relatively shorter distances.

There are three major types of guided media:

1. *Twisted pair cable:* It consists of two separately insulated conductor wires wound about each other that can be used for half-duplex communications. Two such pairs are bundled together in a protective sheath to support full-duplex communications. Twisted pair cables are the most widely used transmission media. A twisted pair has two types:
   - *Unshielded twisted pair (UTP):* This type of cable can block some interference and does not depend on a physical shield for this purpose. It is used for home, office, and similar applications. It is the least expensive and easy to install, but is susceptible to external interference and is usable for short-distance transmission due to attenuation.
   - *Shielded twisted pair (STP):* This type of cable uses a special jacket to block external interference. It has better performance at a higher data rate in comparison to UTP, but at the same time is more expensive and more difficult to manufacture and install. It is used for short distances, for example, in some substation applications.
2. *Coaxial cable:* It has an outer plastic covering containing an outer conductor shield and an inner conductor, each having a separate insulated protection cover. It provides high bandwidth and good noise immunity, while being relatively inexpensive.
3. *Optical fiber cable:* Optical fiber cables transmit data using the reflection of light through a glass or plastic core surrounded by a less dense glass or plastic covering called the cladding. It is used for the transmission of large volumes of data. The cable can be unidirectional or bidirectional. It allows wavelength division multiplexing (WDM) and supports unidirectional and bidirectional modes. The optical fiber can be single-mode or multimode. In the case of single-mode, the fiber enables one type of light mode to be propagated at a time, while the multimode fiber can propagate multiple modes. The differences between single-mode and multimode fiber is due to the different fiber core diameter, wavelength, and light source, resulting also in differences in bandwidth, distance, and cost. Single-mode fiber is

mostly used for long-distance applications, while multimode optical fiber is applied for short distances.

The main advantages are that it is lightweight and immune to electromagnetic interference, while at the same time having increased capacity and bandwidth. Some disadvantages are higher cost and more difficulty in installation and maintenance.

6.5.2 Wireless Media

In our everyday lives, we are usually connecting wirelessly using mobile devices or laptop computers. It is more convenient, and the speed and quality have improved significantly over the years. However, wireless networks are more susceptible to interference and present cybersecurity challenges.

Wireless media are also referred to as unguided transmission media because no physical medium is required for electromagnetic signals transmission. Different wireless media use different frequency bands that have an impact on data transmission speed and range: the higher the frequency, the faster the data transmission and shorter the signal range.

There are three types of signals transmitted through wireless media:

1. *Infrared:* Infrared waves are used for very short-distance communications and also they cannot penetrate through obstacles, which prevents interference between systems. Their frequency range is 300 GHz to 400 THz. It is used for communications between devices in very close proximity.
2. *Microwaves:* Microwave communications require direct line-of-sight for transmission between the sending and receiving antennas that need to be properly aligned with each other. The distance covered by the signal is directly proportional to the height of the antenna. Their frequency range is 1 to 300 GHz. They are mainly used for mobile phone communication and television distribution, as well as in electric power utility communications.
3. *Radio waves:* They are easy to generate and can penetrate through buildings. They do not require alignment between sending and receiving antennas. Their frequency range is 3 kHz to 1 GHz. AM and FM radios and mobile phones and devices use radio waves for transmission.

6.6 Network Area

Communication networks can be broadly classified in the following categories according to scale or the area of a network:

- Personal area network (PAN);
- Local area network (LAN);
- Metropolitan area network (MAN);
- Wide area network (WAN).

6.6.1 PAN

Such network serves the needs of a person or a family and is used to connect their personal computers and mobile devices. The range of coverage can vary from few tens of square meters to several hundred square meters. In such a network, both wired and wireless communication media are used. The PAN is usually connected by a modem to an internet service provider and to a wired/wireless router that supports all personal devices.

6.6.2 LAN

LANs cover a small geographic area, such as a very large home, office building, or campus. They are widely used to allow computers in company offices to exchange information or share resources, for example, printers. LANs also use both wired and wireless communication media and today run at speeds of 100 Mbps up to 10 Gbps.

LANs are also used for industrial automation in factories or substation PAC. Depending on the applications and performance, reliability, and security requirements, they can be designed using different network topologies.

6.6.3 MAN

This type of network covers a larger geographical area than a LAN, from a few blocks in a city to an entire city. MANs also use wired or wireless communication channels supporting high data rates.

MANs might be owned and operated by the city as a public utility, but they also can be owned and operated by a single organization. They are used to interconnect LANs, and the area covered by a MAN can be tens of square kilometers.

6.6.4 WAN

As the name indicates, these are the largest size networks that cover a large geographical area, typically a country or maybe even a continent. They interconnect MANs and LANs and also use combinations of wired and wireless communication media.

WANs typically have a mesh network topology. WANs are used for wide area monitoring and control, as well as for system integrity protection schemes.

6.7 Network Topology

6.7.1 Network Topology Definition

As we have already seen, there are many different factors that we need to consider when designing the communication networks for a digital grid. It not only depends on the requirements of the applications, but also on the technology being used for the communication links, as well as the topology of the communications network selected.

For electric power system applications, wired networks are the preferred choice because they ensure the best connection, security, and reliability. However, just selecting a wired network is not quite enough. We need to carefully consider the different types of network topologies.

The network topology identifies how the nodes are connect to each other (Figure 6.3). Each of the different types of network topologies have their respective advantages and disadvantages. Before choosing a topology, we need to consider the size of the network, its cost, the applications that we will be running, the performance, reliability and security requirements, and the potential for future growth. Common types of topologies are discussed next.

6.7.2 Bus Topology

Bus topology is the one that was used at the early stages of development of computer networks. It represents the simplest design in which nodes are in a linear order. Each node in a bus topology setup connects to a single cable. It is important to note that, in a bus topology, data is not transmitted directly between two nodes. It is based on what is known as shared media access. When a device sends data over the bus, the electrical signals representing it reach all devices connected to the backbone cable. All devices in principle can be senders and receivers, but not at the same time, which means that it uses half-duplex communications.

Every network topology has advantages and disadvantages. The main advantages of a bus topology are lower cost and ease of installation and setup. Because a bus topology connects via one primary cable (the backbone cable), the cable costs will be lower than in other topologies. However, we should be aware of the disadvantages of bus topologies:

- Bus topology is not great for large networks because, as more nodes are connected, the performance of the network slows down because of data collisions.

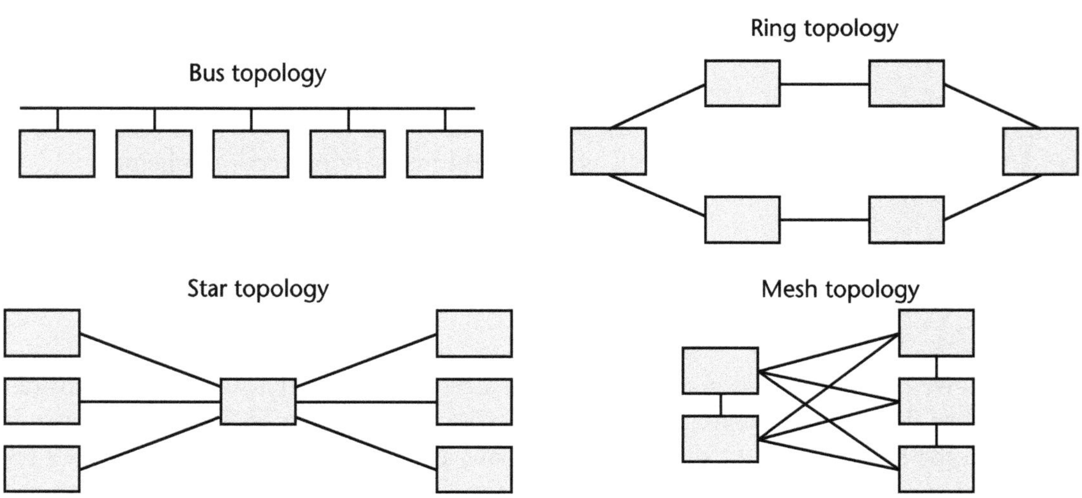

Figure 6.3 Network topologies.

- If the backbone cable fails or gets damaged, the whole network will fail and the identification of problems is difficult.
- Every node on the network sees the data on the network that may be a security risk and troubleshooting of individual device issues is hard. There is also a need for terminators at both ends of the backbone cable.

6.7.3 Ring Topology

Ring topology also has a very simple design. As the name indicates, the nodes are arranged in a ring, where each device has two adjacent nodes. From the sender, the data travels from one node to another until it reaches the desired destination.

Ring topologies can be unidirectional or bidirectional. If we need it to be bidirectional, we have two connections between the network nodes called dual ring topology.

Similar to the bus topology, ring topologies are also very easy to install, manage, and expand. Because the data flow is in one direction and only the node that has the token can send data, the chance of packet collisions is practically nonexistent.

The ring topology also has some disadvantages. If one node or one link fails, the entire network goes down. Troubleshooting is more difficult, especially with more nodes transmitting data in both directions. Due to the unidirectional ring, a data packet (token) must pass through all the nodes. The addition and removal of any node while the network is in operation are difficult and may cause issues in network activity.

6.7.4 Star Topology

The star network topology, as the name indicates, looks like a star. All end nodes are connected to a node (a hub) which is the center of the star. This gives this topology several advantages. First, it makes it more reliable than the bus and the ring, because the failure of any end node or its connection does not affect the rest of the network. Also, the fact that each node has its own connection to the central node makes troubleshooting much easier. The performance is faster as well because data does not have to go through each node before reaching its destination and there are no collisions. Adding or removing devices when the network is in service is not a problem. However, the number of devices in a star network depends on the number of ports that the central node has.

Like all other network topologies, the star is not perfect. There are two disadvantages to consider: the cost and the central hub. Because star topologies need more cables and a central hub, they are more expensive to set up and run. Also, its performance and reliability depend on the central hub. If it fails, the whole network will fail as well and, if it is slow, the communications between the end devices will be slow.

6.7.5 Mesh Topology

A mesh topology is one in which each node is connected directly to other devices with point-to-point links. In a full mesh network, all the nodes within the network are connected with every other node. In a partially connected mesh, all the nodes are not connected with all the other ones in the network.

Mesh networks are mostly used for their reliability. This is because there are few or no problems caused by data traffic or failure of a node or a link. Even if it happens, the rest of the network will work fine. It is also preferred when there are requirements for improved security and privacy, because all the connections are point-to-point.

The main disadvantages are the higher cost of the hardware, installation, and maintenance.

6.7.6 Tree Topology

Tree topology is also referred to as hierarchical topology, because it has a hierarchical structure that requires what are called root nodes, which then connect to subroot nodes and continue expanding to other nodes at the lower levels of the hierarchy. The main benefit of tree topology is that it combines the reliability of bus and star topologies. It is less expensive than mesh topology. The data is transmitted through branching cables and there are no loops, which makes troubleshooting very simple.

The main disadvantage of tree topology is that a failure of a root node has an impact on the affected part of the network. It is also more complex and expansive because of the root nodes.

6.7.7 Hybrid Topologies

Hybrid topologies combine two or more of the earlier described topologies. They can be beneficial, but they require significant expertise and experience in order to avoid potential problems in their installation, setup, operation, and maintenance.

6.8 Functional Relationships

6.8.1 Relationships

When we think about how nodes communicate over a network, one of the factors to consider is the relationship between them. In many ways, this is like the relationships in human interaction and, because of that, the names of the different relationships in computer networks reflect what we know and understand from people's communications. Some of these relationships have significant similarities and for many people they are all the same. In principle, during a conversation, one person initiates it and then we have requests and responses as part of the dialogue. This is when we can identify the differences, which becomes clear when we look at the three typical communication relationships (Figure 6.4) that are used in the electric power system-related applications discussed next.

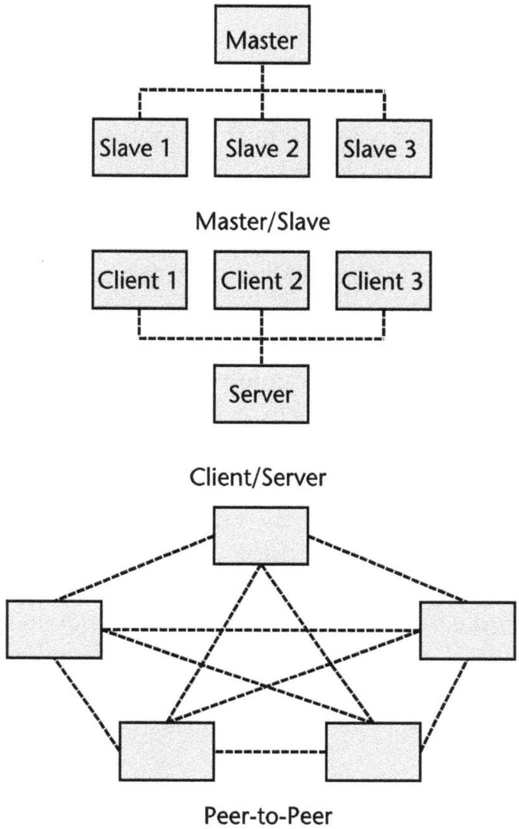

Figure 6.4 Functional relationships.

6.8.2 Master/Slave

Master/slave is a type of asymmetric communication or control where one "master" node controls one or more "slave" nodes and serves as their communication hub. It has been used for decades, but, in the twenty-first century, the terminology has been challenged due to its association with slavery and some organizations have even banned the use of the name. One company replaced them with "controller/agent."

The name is controversial and should be avoided as much as possible, but it adequately represents the relationship between the devices engaged in the communications. In the master/slave model, there is one node (the master) that controls the communications with one or more other nodes (the slaves). It is very important to remember that, as in the time of slavery, there was one owner to whom one or more slaves belonged. In the same way in this communication relationship, there is only one master and only it can initiate communication and send requests to the slaves. The slaves can only send replies to the master. They cannot initiate communication, either to the master or to other slaves. The master can address each slave individually or all slaves simultaneously.

In electric power systems, the master/slave communications have been used in the early days of SCADA for data polling or breaker operation controlled by the master, with measuring or switchgear control devices acting as slaves.

6.8.3 Client/Server

Many people believe that master/slave and client/server communications are the same and, to a certain extent, this is true, because in both cases one node sends a request and another responds to it. If we look at the master/slave communications, we see that there is a single master, so each of the slaves can respond to requests only from it. In the case of the client/server communications, it is the opposite; one server can respond to requests from multiple clients. This is like our everyday life case of going to a restaurant; we have one server responding to the requests of multiple clients on the same table or at different tables in the restaurant.

Typically, the servers wait to receive a request from a client. However, in some applications using client/server communications, we can also have unsolicited notifications. This capability makes it possible for clients to receive notification of application-specific events as they occur, without having to request notification explicitly in real time. For this to happen, a request for automatic notification of a specific event is registered with the system. Once this is done, the client is informed by the server whenever the specified event occurs.

Client/server communications are also used in electric power systems for data polling or breaker operation controlled by different clients, with measuring or switchgear control devices acting as servers. To make communications more efficient, unsolicited notifications can be used to report the change of state of a circuit breaker or the change of a measurement value outside of a predefined deadband.

6.8.4 P2P

P2P is a decentralized communications model in which, as the name indicates, each party has the same rights and can initiate a data exchange. It can be described as a relationship between two nodes on the same network that are able to exchange information without a third note being involved. This is the typical communication relationship that we have in our everyday life with our friends and colleagues.

In electric power systems, P2P communications are typically used for protection applications when multifunctional IEDs have to exchange information in a specific protection scheme that has high-speed performance requirements.

6.9 Communication Protocols

6.9.1 Protocol Definition

Any data communications must be governed by some set of rules and we call these rules a protocol. It determines:

- What is communicated in the network;
- How it is communicated in the network;
- When it is communicated.

There are five main elements of a protocol:

- Message encoding;

- Message formatting and encapsulation;
- Message timing;
- Message size;
- Message delivery options.

6.9.2 Message Encoding

Message encoding is related to the transmission medium. When we have a wired medium, which is also called a link, the communicating device (a computer or an IED) converts the data into electrical signals and sends the signals on the transmission medium, the cable. However, if we are not using a computer but a smartphone or tablet, it is connected to the network with the help of wireless medium. In this case, the communicating device converts the data into waves, because the transmission medium is wireless. In both cases, it converts the data into signals or waves by appropriately identifying the medium to which it is connected.

6.9.3 Message Formatting

Message formatting and encapsulation are related to the requirement that both the sender and the receiver must mutually agree upon a common format so that the communication becomes understandable. At the same time, some encapsulation is also done by adding information to the data such as the source information and the destination information to support the transmission of the information from the sender to the receiver.

6.9.4 Message Timing

Message timing is something else that is handled by a protocol. If the sender is very fast and the receiver is slow, it obviously cannot handle the flow, so there is a chance for the data to get lost. To avoid such loss due to high speed, sending flow control must be ensured by the protocol.

For example, after sending the data, the sender will wait for a certain period of time in order to receive an acknowledgment that the data is received by the destination. If that acknowledgment is not received on time, the sender understands that there are some laws and it retransmits that data again so that there is no loss by the receiver.

6.9.5 Message Size

The next element of the protocol is the message size. For example, if there is a big file but the link capacity is small, the sending device breaks this big file into smaller segments and each segment is numbered sequentially according to the rules of the protocol. These smaller segments can be transported over the network and segment's number will help the receiver to identify if there are missing packets after receiving all the smaller packets. The receiver will reassemble all the packets with the help of the numbering and, if there are no missing packets, it will have exactly the file from the sender.

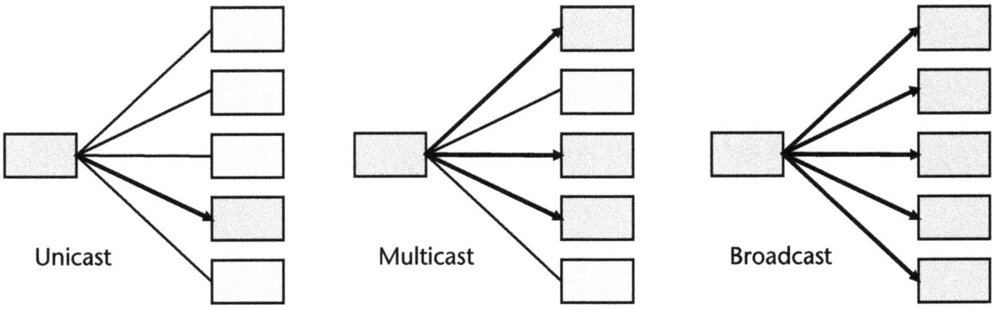

Figure 6.5 Message delivery options.

6.9.6 Message Delivery Option

Finally, the message delivery option deals with the issue of who is to receive the information from the sender (Figure 6.5):

- If it wants to send exactly to one destination, this is a unicast communication.
- If the communication is one sender to several receivers, this is multicasting.
- If the information from the sender is intended for all receivers, this is the case of broadcasting.

CHAPTER 7

Technology Fundamentals

7.1 Introduction

As discussed in previous chapters, digitization and digitalization did not start in the electric power industry but have been a significant trend in all areas of our private and professional lives. That is why the transition of our industry and the development of the IEC 61850 standard did not occur in an empty space but took advantage of many different technologies that have been developed and used over the last few decades.

Considering that many PAC specialists do not have a computer or communications background, it is important to introduce the fundamentals of some of the technologies that have a direct impact on the digitization and digitalization of the grid.

In this chapter, the following sections provide a brief overview of different components of some of these technologies as follows:

- Open systems interconnection (OSI) model;
- Manufacturing message specification (MMS);
- Abstract Syntax Notation One (ASN.1);
- Transmission Control Protocol/Internet Protocol (TCP/IP) and User Datagram Protocol/Internet Protocol (UDP/IP);
- Ethernet;
- Parallel Redundancy Protocol (PRP) and High-Availability Seamless Redundancy (HSR) Protocol;
- Unified Modeling Language (UML);
- Extensible Markup Language (XML).

The information contained in these sections is intended just to highlight the key characteristics of each of them and show how they impact the digitization of the grid.

7.2 The OSI Model

The OSI model [1] was published in 1984 and is a framework to support the development of interoperable communication products. Understanding the model helps to understand the operation of communications-based systems.

The conceptual OSI model (Figure 7.1) has 7 layers that are typically described starting from the application layer that we use in our everyday life, from the top layer down. The functionality of the different layers are provided by a mixture of hardware and software components that, at the end, ensure the successful transmission of data between devices connected to a communications network.

The application layer or Layer 7 provides network services to the end users that work with the data that is transmitted or received using interactive applications. These services allow the application layer to supply data to and receive data from the presentation layer.

Layer 6 is the presentation layer, which performs syntax processing and converts data from the application layer's format to the network format and vice versa. After the data is made available in the required format by the presentation layer, it is passed to the session layer or the application layer depending on whether it is transmitting or receiving.

The session layer or Layer 5 can handle multiple types of connections and is responsible for authentication and reconnection after the session is established. It passes the data to or from the session or transport layer depending on the direction of the data flow.

Layer 4 is the transport layer, which takes care of the transmission of data across the network connections and coordinates how the data is sent and where it goes. For internet applications, the two most used protocols are the TCP and UDP. When it completes its function, the transport layer passes the data to or from the network layer.

The network layer or Layer 3 is responsible for the routing of the data. It receives incoming transmissions and examines the data to conclude if the data has

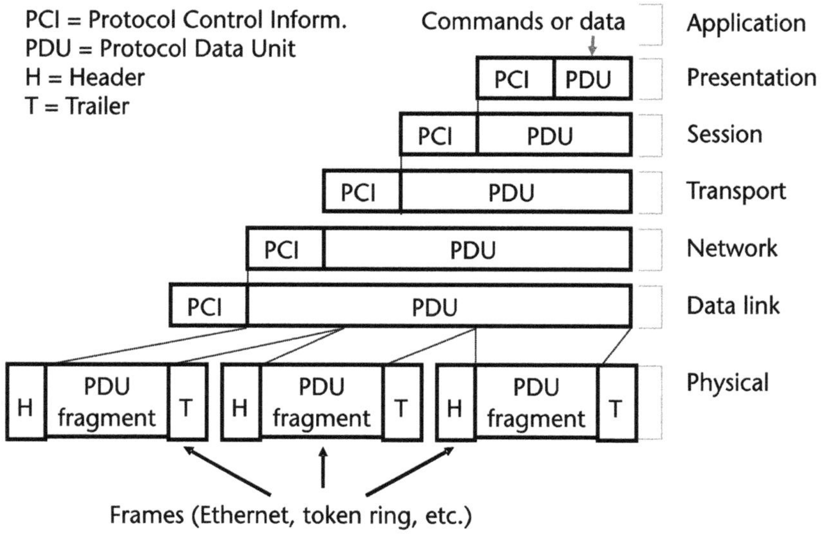

Figure 7.1 OSI stack.

reached its destination. It also sends data to the correct destination on outgoing transmissions. The IP (IPv4/IPv6) is one of the main protocols used at this layer and is the commonly known network layer for the internet. This layer also manages the mapping between logical addresses and physical addresses before the data is passed to the data link layer.

Layer 2 is the data link layer and it handles the links across the physical local area network (LAN). At this layer, the data passes to or from the physical layer. The most used data link protocol for local area networks is the Ethernet.

The physical layer or Layer 1 is at the bottom of the OSI model and is the electrical or physical layer of the model: the network cables, power plugs, wireless radio frequencies, pulses of light, and electric voltages.

Each layer in the OSI model uses a protocol data unit (PDU) to exchange information, which can only be read by the peer layer on the receiving device and is then passed to next upper layer after stripping. A service data unit (SDU) is a piece of information passed by a higher layer to the one below it for transmission using the service of that layer. To transport the SDU, the current layer encapsulates the SDU by adding a protocol control inform (PCI). The combined PCI and SDU is known as a PDU belonging to that layer. This forms the SDU of the layer below. This process is known as encapsulation and is shown in Figure 7.1.

In Layers 5 to 7, PDU is referred to as data. In Layer 4, it is a segment; in Layer 3, it is a packet; in Layer 2, it is a frame; and in Layer 1, PDU is a bit.

7.3 MMS

As we discussed in earlier chapters, most of the IEC 61850 standard defines abstract models and services that are required to represent the PAC functionality in electric power systems. However, to implement this in the real-life environment, we need to map the abstract model to actual protocols and other technologies.

Starting from the application layer of the OSI model, for client/server applications, the mapping is done to MMS [2–4], which is an international standard originally designed for the automotive industry for the automation of the manufacturing process using the remote control and monitoring of devices such as programmable logic controllers (PLCs), robot controllers (RCs), and remote terminal units (RTUs). It defines common functions for distributed automation systems in any industry.

The first version of MMS was published in 1990 by ISO TC 184 (Industrial Automation) and was an output from a General Motors' initiative, Manufacturing Automation Protocol (MAP). The goal was to move in the direction of standardized and interoperable communications between devices from different manufacturers, eliminating the challenges of integrating devices with proprietary communication interfaces.

The current version was published in 2003 and was confirmed in 2020. It includes two parts:

- Part 1: ISO 9506-1 Service Definition (2003);
- Part 2: ISO 9506-2 Protocol Specification (2003).

Part 1 describes abstract services that are provided to support remote client/server information exchange and data flow between devices used for different automation applications and control or supervision systems. It can be used over any network supporting full-duplex communications. The services are generic in nature and can be used in different industries, including electric power systems. The implementation in any specific device will use a limited subset of the services defined in the standard depending on its functionality.

Part 2 defines the MMS protocol in terms of messages described with the ASN.1 notation, which defines the syntax. The services allow the remote manipulation of variables, programs, semaphores, events, and journals.

In the core of MMS is the virtual manufacturing device (VMD) model (Figure 7.2), which specifies the communication-visible behavior of MMS devices from an external MMS's client application point of view. Any device or application can act as both a client and a server.

The VMD model contains different objects and services supporting the client and server functionality and defining their behavior. The information exchange includes named and unnamed variables or named variables lists and allows the type description of the variables. The supported services include read and write of simple or complex variables such as a floating point or an array. Both variables and types can be managed by clients using services such as define and delete.

At the same time, servers can provide unsolicited reports for changes in the values of variables.

The MMS model defines an object representing a resource in a VMD as a domain and a group of domains can be contained within a program invocation whose execution can be monitored and controlled through different services such as start and stop.

Event objects defined in the MMS model represent the state of an event, the action that should be taken by the VMD in case of a change in the state of the event condition and which MMS clients should be notified when it happens.

The MMS model also includes a journal-named object representing a log of time-stamped event data.

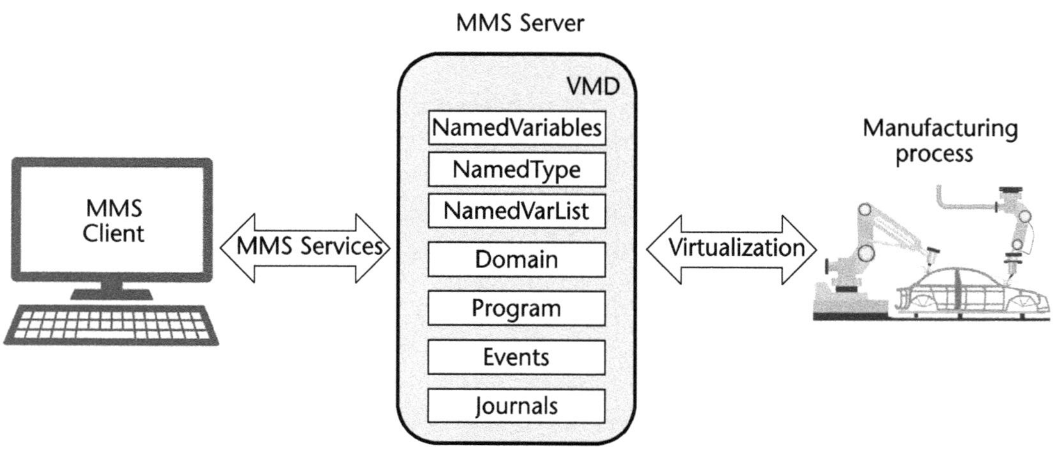

Figure 7.2 MMS virtualization.

Because of its versatility, MMS has been used in the electric power industry initially in IEC 60870-6-TASE.2 (Inter Control Center Communication) and later in the Utility Communication Architecture (IEEE TR 1550) and later in IEC 61850 and IEC 61400-25 (Communications for Monitoring and Control of Wind Power Plants).

7.4 ASN.1

As mentioned in the previous section, ASN.1 [5] is used in MMS at the presentation layer of the OSI model to specify the syntax of the messages, ASN is an acronym for Abstract Syntax Notation and has a long history. It is a platform-independent standard language used in different kinds of communication systems to define data structures for various applications. One of its characteristics is that it is human and machine-readable. ASN.1 is a joint standard of the International Telecommunication Union-Telecommunication Standardization Sector (ITU-T) and International Organization for Standardization (ISO)/IEC and is used to define a large number of protocols.

It was first published by ITU in 1984 as Recommendation ITU-T X.680 and, following many revisions and amendments, ISO/IEC 8824-1 was prepared by a Joint Technical Committee ISO/IEC JTC 1, Information Technology, Subcommittee SC 6, Telecommunications and Information Exchange Between Systems, in collaboration with ITU-T. The latest revision was published in 2021.

ISO/IEC 8824-1:2021 Information Technology—Abstract Syntax Notation One (ASN.1): Specification of Basic Notation—Part 1 defines a number of simple types, such as Booleans, integers, and character strings with their tags. It specifies a notation for referencing them and for specifying their values. It also defines mechanisms for constructing new types from basic types, for example, structures and lists, and specifies a notation for defining them, assigning them tags and specifying their values.

One of the main benefits of ASN.1 is that it is independent of a particular computer or programming language. An important factor for its success is the definition of encoding rules in ISO/IEC 8825-1:2021 Information Technology—ASN.1 Encoding Rules: Specification of Basic Encoding Rules (BER), Canonical Encoding Rules (CER), and Distinguished Encoding Rules (DER)—Part 1. It is a formal notation that removes ambiguities by describing data structures for representing, encoding, transmitting, and decoding data and provides a set of formal rules for describing the structure of objects independent of machine-specific encoding techniques. It uses a human-readable notation and supports a compact, encoded representation of the same information used in communications protocols.

The encoding rules defined in this standard may be applied to values of types defined using the ASN.1 notation. The application of these encoding rules produces a transfer syntax for such values. It is as is clear from the title of the ISO/IEC 8825-1 standard that it defines three sets of encoding rules:

- Basic encoding rules (BERs);
- Canonical encoding rules (CERs);

- Distinguished encoding rules (DERs).

The BERs give various choices for data values encoding. In contrast, the CERs and DERs select just one encoding from the options allowed by the BERs, so they can be considered as profiles of the BERs.

CERs and DERs differ from each other in the set of restrictions that they place on BERs. The DER is more suitable for cases when the encoded value is small enough to fit into the available memory and does not include nested values. The CER is preferred when there is a need to encode large values that cannot readily fit into the available memory or it is necessary to encode and transmit a part of a value before the entire value is available.

ASN.1 is closely related to more specific encoding rules that specify how to represent a data structure as a series of bytes. An example is the standardized XML Encoding Rules (XERs) that allow ASN.1 modules to be used as ASN.1 schemas against which XML documents can be validated.

BERs and XERs are used in IEC 61850 protocols, including the new IEC 61850-8-2, while DERs are used for X.509 certificates for security.

7.5 TCP/IP and UDP/IP

In our everyday lives, we spend a lot of time on the internet, and it is amazing that there has now been a half-century of its development. In the background of the World Wide Web, file transfers, email, and many other applications is TCP/IP. Its origins are in the Specification of Internet Transmission Control Program published as a Request for Comments (RFC) 675 in December 1974 to describe the functions to be performed by the internetwork Transmission Control Program and its interface to programs that require its services. It was later divided into a modular architecture consisting of TCP and IP and known as TCP/IP. The Internet Engineering Task Force (IETF) defines TCP in the RFC standards document number 793.

According to the reference 7-layer OSI model, TCP is at the transmission layer 4, with IP at the network layer 3. However, the TCP/IP model (Figure 7.3) combines the application, presentation, and session layers of the OSI model into an ap-

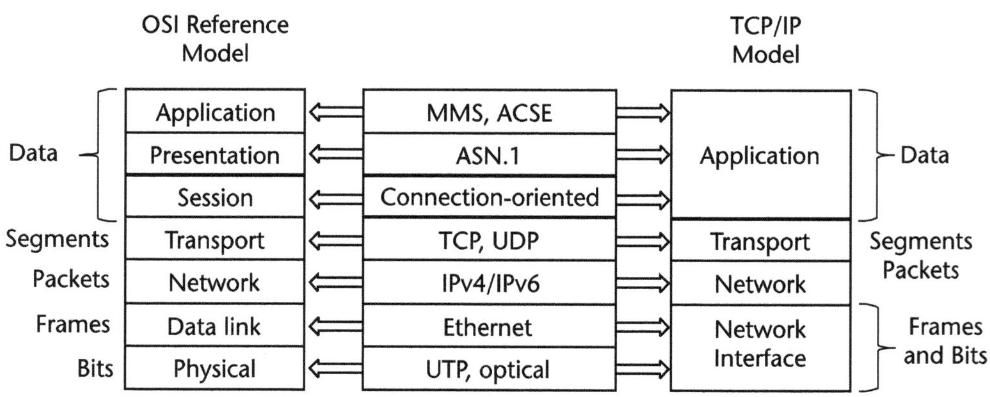

Figure 7.3 OSI versus the TCP/IP model.

plication layer interfacing with TCP. The data from the application layer is sorted in the transport layer into TCP segments.

TCP is a connection-oriented protocol that requires establishing a connection before transmitting the data and closing the connection once the transmission is complete. It determines how to break application data into packets that it sends to the network layer, supports the sequencing of data, and can guarantee its delivery to the destination router. Extensive error checking on the packets that it accepts from the network layer and acknowledgment of data are used in the communications and retransmission is performed in case of lost packets. The price of guaranteed delivery is slower speed, which limits its application together with the fact that it does not support broadcasting.

An alternative to TCP is the UDP published as RFC 769 in August 1980 and is used together with IP as the underlying protocol. The goal of UDP is to support the transmission of messages by application programs with a minimum of protocol mechanism. Like the TCP in the reference 7-layer OSI model, UDP is at the transmission layer 4, interfacing with the IP at the network layer 3.

The UDP is a connectionless protocol with no requirements for opening, maintaining, or terminating a connection. It uses messages defined as datagrams and does not support data sequencing. The UDP checked for integrity on the arrival of messages based only on a basic error-checking mechanism using checksums. Because there is no retransmission of lost packets, it cannot guarantee delivery of data to the destination. The lack of acknowledgment means that the sender has no information that its message was delivered, but it supports a continuous packet streaming, while the nonexistence of retransmission delays makes it much faster than the TCP and thus suitable for real-time applications. Because it avoids the unnecessary overheads of the TCP transport mechanism, it is very efficient in terms of bandwidth requirements. The UDP provides support for multicasting and broadcasting.

As mentioned earlier, both the TCP and the UDP use the IP as the protocol at the network layer 3. The commonly used protocol for several decades is IPv4, which was first mentioned in IEN 41 Internetwork Protocol Specification Version 4 that was published in June 1978 as a draft specification of the IP. The IP specification was then published in September 1981 as RFC 791. It defined the IP designed for use in interconnected systems of packet-switched communication networks and support transmitting blocks of data (datagrams) from sources to destinations, identified by fixed length addresses. The IP specifies a set of rules, for addressing and routing packets of data so that they can be transmitted across networks and delivered to the correct destination based on IP information that is attached to each packet, which allows routers to send packets to the right place.

An IP address is a unique address that identifies a device on the internet or a local network. In IPv4 is a 32-bit string of numbers separated by periods. The IP addresses are expressed as a set of four numbers; an example address might be 192.167.0.1 (the default router IP address for certain routers). Each number in the set can range from 0 to 255. So the full IP addressing range goes from 0.0.0.0 to 255.255.255.255 providing a total of about 4.3 billion possible IPv4 addresses. At the time when it was established, it was assumed that it would be more than enough in the 1990s, but, since then, the internet has exploded and today we live in a digital world with billions of devices being connected. This required the transition to the next IP addressing, IPv6. IPv6 is the most recent version of IP and is based on IPv4. The concept of IPv6 was first introduced in 1998 in RFC 2460 and finally was published by IETF as RFC 8200 in July 2017 providing a single specification.

While IPv4 is a numeric addressing method using a dot (.) for separation, IPv6 is an alphanumeric addressing method that uses a colon (:). The transition from a 32-bit address to a 128-bit IP address allows the number of possible combinations to go up to 340,282,366,920,938,463,463,374,607,431,768,211,456, which should be enough for now. The actual implementation of IPv6 is starting to happen, but it will take some time because it requires significant investments by all stakeholders.

7.6 The Ethernet

The Ethernet [6] is the most widely used technology at the OSI data link layer 2. It has a half-century of history starting with the first experimental Ethernet system developed in the early 1970s by Bob Metcalfe and David Boggs of the Xerox Palo Alto Research Center (PARC) to interconnect their computers and laser printers at a data transmission rate of 2.94 Mbps.

In 1979, the Digital Equipment Corporation (DEC), Intel, and Xerox started work on standardizing an Ethernet system that anyone could use and, as a result, they released Version 1.0 of the first Ethernet specification. It defined a 10-Mbps Carrier Sense Multiple Access with Collision Detection (CSMA/CD) protocol known as thick Ethernet because of the thick coaxial cable used to connect the nodes on the network.

The first IEEE standard for Ethernet technology developed by the 802.3 Working Group IEEE 802.3 Carrier Sense Multiple Access with Collision Detection (CSMA/CD) Access Method and Physical Layer Specifications was published in 1983. After the release of versions supporting thin coaxial cable and later unshielded twisted pair and fiber optic cables, in 1995, the performance of Ethernet technology was improved by a factor of 10 with the release of the 100-Mbps 802.3u 100Base-T standard commonly known as Fast Ethernet over twisted pair cable or two multimode fibers.

In 1997, the IEEE 802.3x standard defined a full-duplex Ethernet operation and bypassing the CSMA/CD protocol to allow two nodes to communicate over a point-to-point link through a network switch.

Gigabit Ethernet supporting a speed of 1 Gbps (1,000 Mbps) was released in 1998 as the 1-Gbps 802.3z 1000Base-X standard, followed by 10-Gb Ethernet in the IEEE 802.3ae-2002 standard, which defines full-duplex point-to-point links generally connected by switches. The Ethernet in this case is not the hub-based collision detection protocol from many years ago. It is the advanced high-speed protocol of today. The use of switched Ethernet is probably the most important advancement in Ethernet networks. The switch picks up every transmission before it reaches another node and then forwards the frame over the appropriate segment, and because any segment contains only a single node, the frame reaches only the intended recipient, thus allowing many interactions to occur simultaneously on a switched network. The format of an Ethernet frame was defined in the original IEEE 802.3 standard as shown in Table 7.1.

Table 7.1 Ethernet Frame

Pre	*SFD*	*DA*	*SA*	*Length/Ether Type*	*MAC Data + Pad*	*FCS*
7	1	6	6	2	46 to 1,500	4

In Table 7.1:

- *Pre:* The Preamble is an alternating pattern (7 bytes) of 1 and 0 that tells receiving stations that a frame is coming.
- *SFD:* The start-of-frame delimiter (1 byte: 10101011) indicates that the next bit is the leftmost bit in the leftmost byte of the destination address.
- *Destination address (DA):* The DA (6 bytes) identifies which station(s) should receive the frame.
- *Source address (SA):* The SA (6 bytes) identifies the sending station.
- *Length type:* This is the number of media access control (MAC)-client data bytes that are contained in the data field of the frame.
- *MAC client data:* This is a sequence of n bytes ($46 \leq n \leq 1,500$) of any value. The pad contains (if necessary) extra data bytes in order to bring the frame length up to its minimum size. A minimum Ethernet frame size is 64 bytes from the destination MAC address field through the frame check sequence (FCS).
- *FCS:* The FCS is a 32-bit cyclic redundancy check (CRC) value.

The MAC address is the physical address of any device, such as the network interface controller (NIC) in a computer or on the network and has two parts, each 3 bytes long. The first 3 bytes identify the company that made the NIC, and the second 3 bytes are the serial number of the NIC itself.

A DA may specify an individual address (unicast) or a multicast address destined for a group of nodes. A DA of all 1 bits goes to all stations on the LAN and is a broadcast address.

The performance of different functions over the Ethernet is further improved through the availability of priority tagging defined, while improvements in security are the result of using a virtual LAN (VLAN). The VLAN is a group of devices on one or more LANs that are configured in such a way that they can communicate as if they were attached to the same wire, when they are actually located on a number of different LAN segments. The IEEE 802.1Q specification establishes a standard method for tagging Ethernet frames with VLAN membership information. The key for the IEEE 802.1Q to perform the above functions is in its tags. The 802.1Q-compliant switch ports can be configured to transmit tagged or untagged frames. A tag field containing VLAN (and/or 802.1Q priority) information can be inserted into an Ethernet frame as shown in Table 7.2.

The format of the IEEE 802.1Q header is shown in Table 7.3.

In Table 7.3:

- *Tag Protocol Identifier (TPID):* This is a 16-bit field set to a value of 0x8100 that identifies the frame as an IEEE 802.1Q-tagged frame.

Table 7.2 Extended Ethernet Frame

Pre	*SFD*	*DA*	*SA*	*802.1Q Header*	*Length/Ether Type*	*MAC Data + Pad*	*FCS*
7	1	6	6	4	2	46 to 1,500	4

Table 7.3 IEEE 802.1Q Header

TPID	*PCP*	*CFI*	*VID*
16 bits	3 bits	1 bit	12 bits

- *Priority Code Point (PCP):* This is a 3-bit field that refers to the IEEE 802.1p priority. It indicates the frame priority level from 0 (lowest) to 7 (highest) and can be used to prioritize different classes of traffic such as GOOSE or sampled value messages.
- *Canonical Format Indicator (CFI):* This is a 1-bit field. If the value is 0, the MAC address is in canonical format, the setting for Ethernet switches.
- *VLAN Identifier (VID):* This is a 12-bit field specifying the VLAN to which the frame belongs. A value of 0 means that the frame does not belong to any VLAN; in this case, the 802.1Q tag specifies only a priority and is referred to as a priority tag. A value of hex FFF is reserved for implementation use. All other values may be used as VLAN identifiers, allowing up to 4,094 VLANs.

If a port has an 802.1Q-compliant device attached (such as another switch), these tagged frames can carry VLAN membership information between switches, thus letting a VLAN span multiple switches.

7.7 PRP and HSR

Today, digitization has become a part of our everyday life and many industries are using robots and industrial automation as part of the manufacturing process. Considering how critical, expensive, and dangerous any failure in some key industries can be and that all these systems heavily rely on communications over LANs, it is clear that the reliability of these systems is extremely important. The traditional Ethernet does not provide inherent redundancy, which is why standardization bodies have been looking for solutions to provide redundancy of the communications over Ethernet networks.

There has been some interest in the Rapid Spanning Tree Protocol (RSTP), which supports the reconfiguration of the communications network in case of failure, thus providing some redundancy; however, the reconfiguration time is a factor that needs to be considered and is not applicable to systems that require high-speed performance.

To address this issue, IEC TC 65/SC 65C—Industrial Networks developed IEC 62439-3 Industrial Communication Networks—High Availability Automation Networks—Part 3: Parallel Redundancy Protocol (PRP) and High-Availability Seamless Redundancy (HSR) [7] and Edition 1 was published in 2010 to support the needs of high-availability automation networks based on the Ethernet technology. As the title indicates, this part of IEC 62439 specifies two redundancy protocols designed to provide seamless recovery in case of a single failure in the network based on the parallel transmission of duplicated information. This second edition of the standard was published in 2012 and the third edition was published in 2016.

Every PRP node has two ports and is attached to two separated networks of similar topology making it a doubly attached node (DAN). PRP can be implemented entirely in software and integrated in the network driver. The cost of PRP is higher because it requires duplication of all network elements; however, it provides significant improvements in performance, especially for time-critical applications. Nodes with a single port (singly attached node) are either attached to one network only, which allows them to communicate only with other nodes attached to the same network, or they are attached through a RedBox if they need to communicate with both networks. The RedBox (or Redundancy Box) is a device that behaves like a doubly attached node.

In PRP, each source node (DAN) simultaneously sends two copies of a frame, one over each port (Figure 7.4). The two frames travel through their respective LANs until they reach a destination node (DAN) with a certain time skew. The destination node accepts the first frame of a pair and discards the second (if it arrives). Therefore, as long as one LAN is operational, the destination application always receives one frame. Because there is no recovery time in PRP, it allows the system to continuously monitor the redundancy and detect potential problems or failures.

The HSR protocol uses the same principle of parallel transmission of duplicated information, but it is implemented in a very different way (Figure 7.5). Every HSR node is a switching node, meaning that it can receive a frame on one port and forward it to another port.

A source node sends the same frame over its ports to the neighbor nodes. In a fault-free state, a destination node should receive two identical frames within a certain time skew, pass the first frame to the application that needs it, and discard the second frame if it comes.

A node forwards a frame unless it detects a frame that it sent itself or it is a frame for which it is the sole destination (in case of unicast). This does not apply when traffic supervision is needed.

Because PRP and HSR use the same duplicate identification mechanism, PRP and HSR networks can be connected without a single point of failure and the same nodes can be built to be used in both PRP and HSR networks.

The cost of HSR is that nodes require hardware support to perform the switching function or discard frames within microseconds, but this cost is compensated because there are no Ethernet switches in the typically used ring topology.

7.8 UML

Object modeling is one of the foundations of IEC 61850. The models in the standard represent the abstracts of the essential and communications visible parameters of the complicated real electric power systems world. This process of virtualization requires the use of modeling tools that can present the complex functionality of a

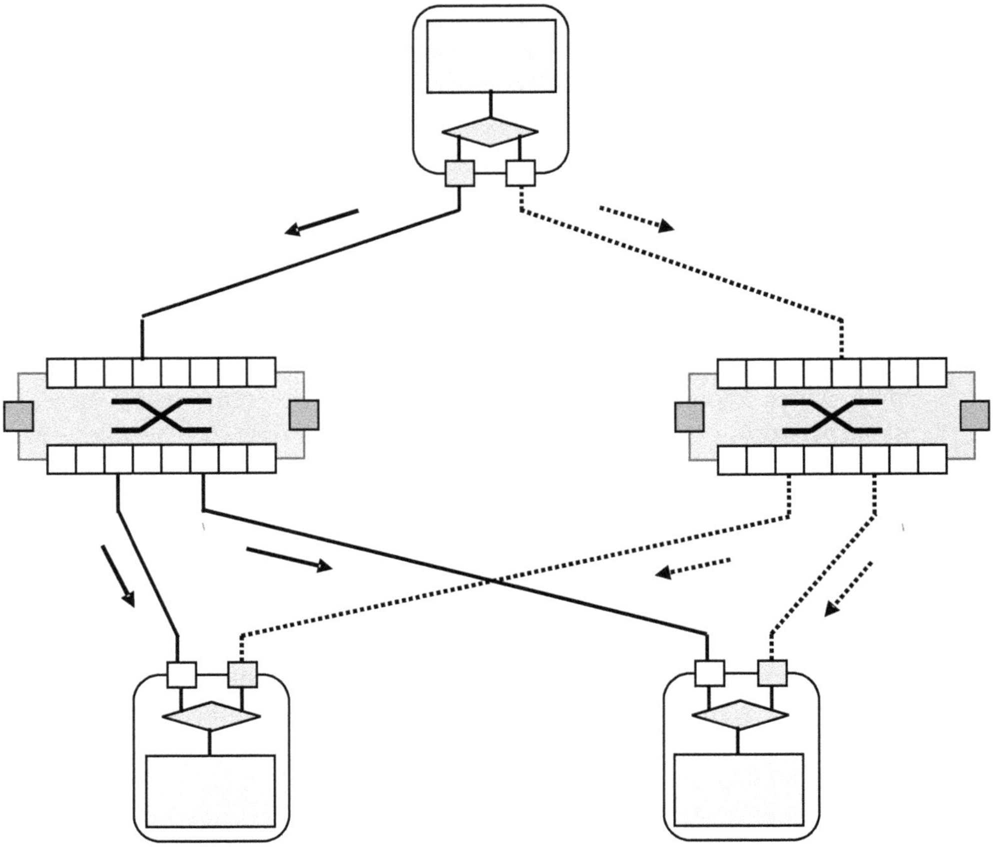

Figure 7.4 PRP principle.

substation and its protection and automation systems in a standardized way that is also easy to represent and understand.

In recent years, the UML [8] has become the common tool used in the modeling of any process, system, or device. It is a standard language for specifying, visualizing, constructing, and documenting different simple or complex systems.

Attempts to create an object-oriented modeling language began between the mid-1970s and the late 1980s. Different competing methods had advantages and disadvantages. The experience from their use led to the development of UML, which started in late 1994 when Grady Booch and Jim Rumbaugh of Rational Software Corporation began their work on unifying the Booch and object modeling technique (OMT) methods. UML 1.0, a modeling language that was well defined, expressive, powerful, and generally applicable, was submitted to the Object Management Group (OMG) in January 1997 as an initial RFP response. Since 1989, the OMG has been a nonprofit, international, open membership, computer industry consortium and UML is one of its standards. The current available version of UML is 2.5.1 and it was published in December 2017.

Several modeling tools are covered under the heading of the UML. It uses mostly graphical notations to express the design of software and other projects, systems, or structures. Different types of diagrams can be used to present data structures,

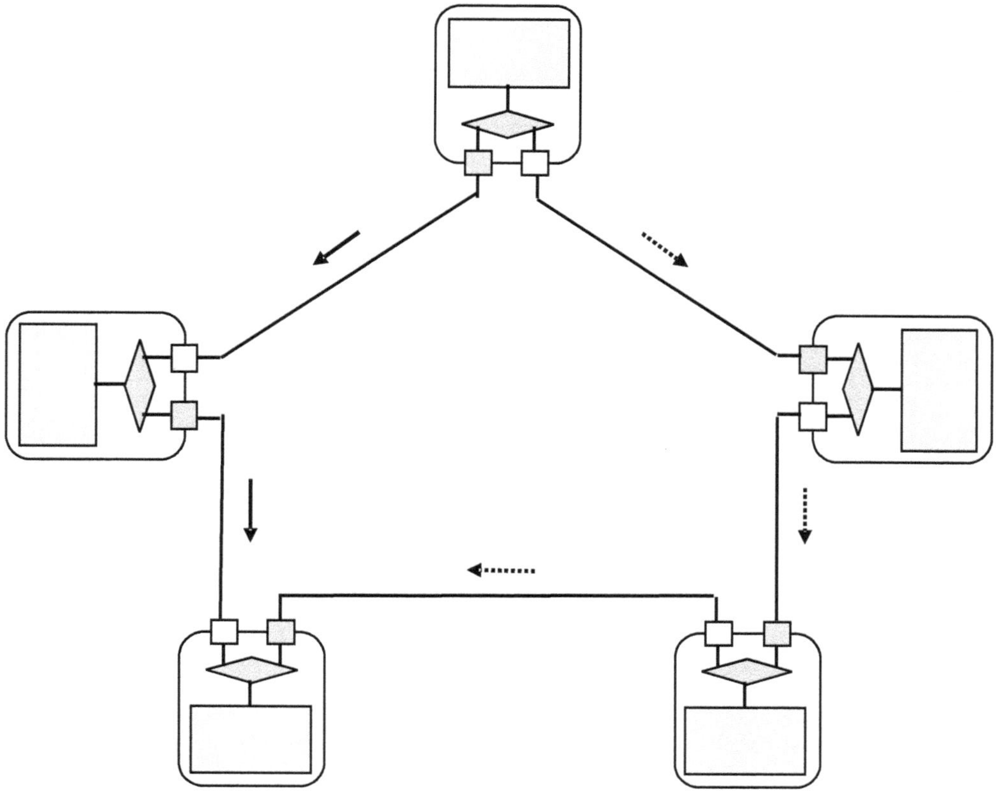

Figure 7.5 HSR principle.

device and operator interactions, or any other substation automation or protection-related process. Using the UML helps project teams to communicate, explore potential designs, and validate the architectural design of the system.

UML 2.0 defines 14 types of diagrams (Figure 7.6), divided into three categories:

- Structure diagrams include six diagram types that represent static application structure: profile diagram, class diagram, object diagram, component diagram, composite structure diagram, package diagram, and deployment diagram.
- Behavior diagrams include three diagram types that represent general types of behavior: use case diagram, activity diagram, and state machine diagram.
- Interaction diagrams are a subset of behavior diagrams and include four diagram types that represent different aspects of interactions and include the sequence diagram, communication diagram, timing diagram, and interaction overview diagram.

As the UML is applicable to object-oriented problem-solving, it is used in different parts of the IEC 61850 standard to represent in a standardized graphical way the complex models of multifunctional substation IEDs and their interface with the

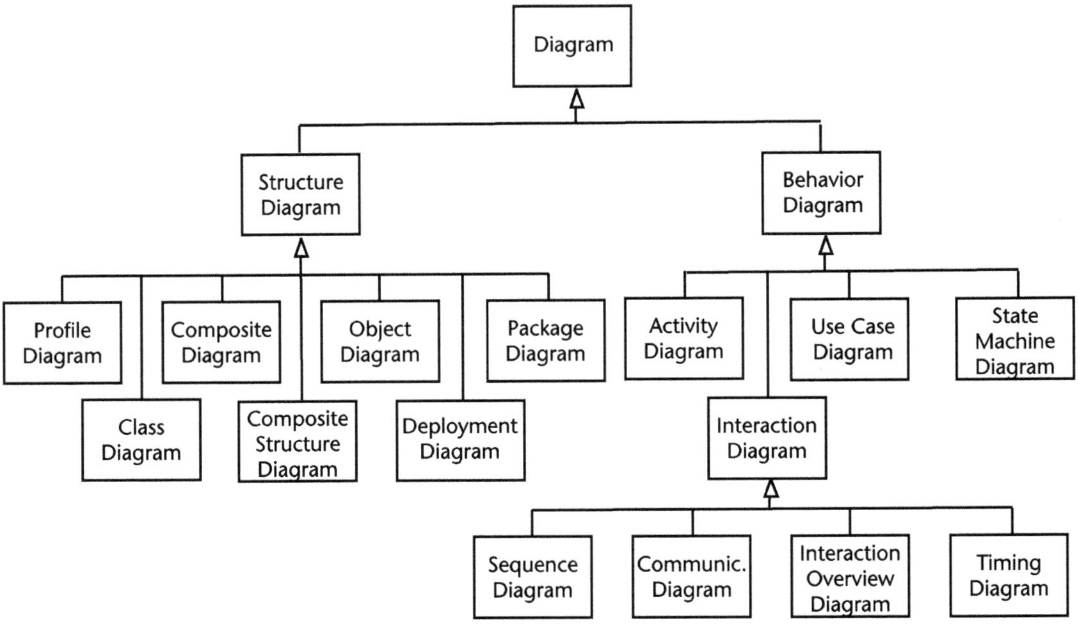

Figure 7.6 UML diagrams.

primary substation equipment and the communications network. The following is a list of the more commonly used diagrams in the standard:

- Use case diagrams;
- Class diagrams;
- Object diagrams;
- Sequence diagrams;
- State machine diagrams.

For example, the object model hierarchy is represented by class diagrams that show not only the relationship between the components of the model, but also their interfaces (Figure 7.7). These are modeled using:

- *Association:* This is a relationship between instances of the two classes. There is an association between two classes if an instance of one class must know about the other in order to perform its work. In a diagram, an association is a link connecting two classes.
- *Aggregation:* This is an association in which one class belongs to a collection. An aggregation has a diamond end pointing to the part containing the whole.
- *Generalization:* This is an inheritance link indicating one class is a superclass of the other. A generalization has a triangle pointing to the superclass.

Multiplicity of an association end is the number of possible instances of the class associated with a single instance of the other end. Multiplicities are single numbers or ranges of numbers.

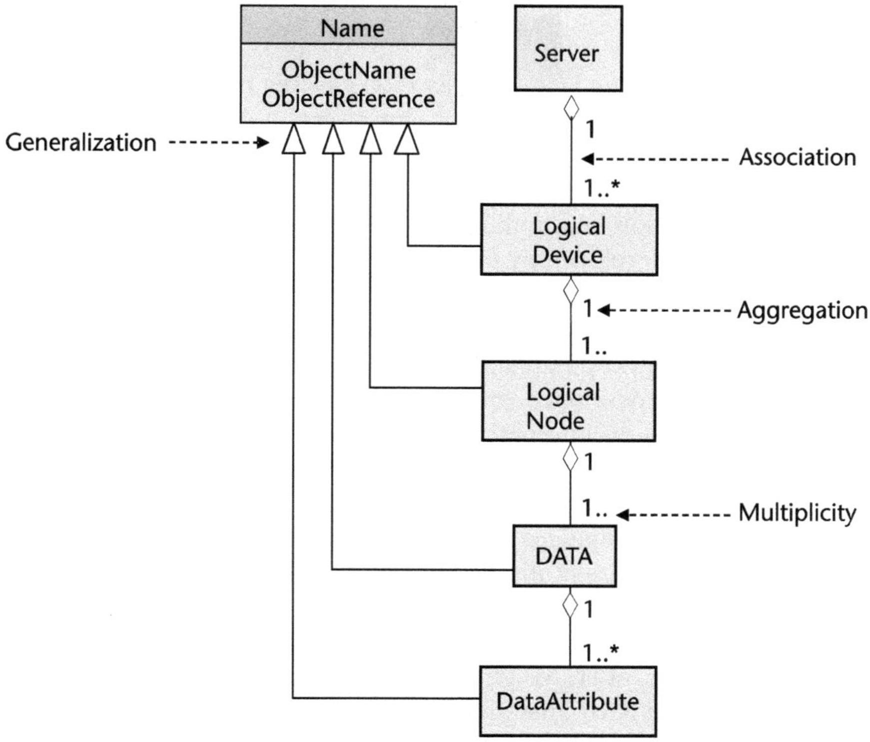

Figure 7.7 UML class diagram.

Object diagrams show instances instead of classes.

While class and object diagrams are static, to represent the dynamic behavior of the model, we use diagrams such as sequence diagrams (Figure 7.8) that are organized according to time and can help clarify a use case in order for it to be realized in software or state machine diagrams that show the possible states of the object and the transitions that cause a change in state. They are used in the design process to help with the transition from the analysis of the system to its implementation.

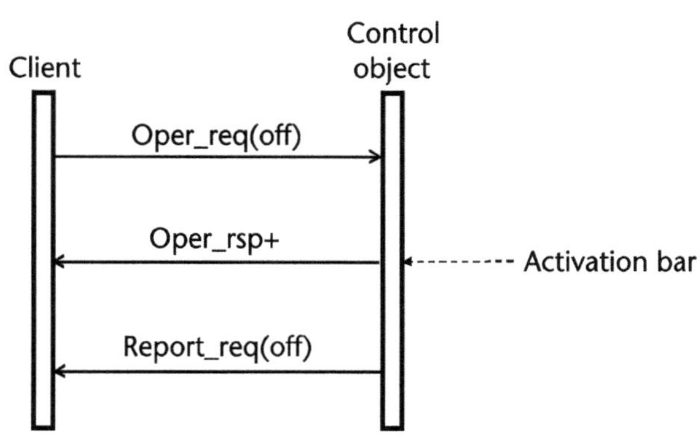

Figure 7.8 UML sequence diagram.

7.9 XML

XML [9] is a markup language based on existing markup languages that have been used for different applications for many years hidden behind the scenes. Even the Word files or PowerPoint presentations that we use every day have XML in the background. It is both human-readable and machine-readable. The latest version 1.1 (second edition) was published in September 2006.

XML was developed by members of the World Wide Web Consortium (W3C) and was released as a recommendation by the W3C in February 1997. It is a simplified version of the Structured Generalized Markup Language (SGML) and a cousin of Hypertext Markup Language (HTML).

SGML was developed to standardize the production process for large document sets and is an international standard (ISO 8879) that has been in use as a markup language primarily for technical documentation and government applications since the early 1980s.

The growing popularity of XML is the result of its flexibility and strength. It is extensible, because it allows extending the user's ability to describe the domain specifics of the document.

In appearance, XML is quite similar to HTML because they both use tags. The difference is that, in HTML, there is a specified set of tags that defines the format of the data, and, in XML, the user can create the tags required by the application domain. That is why XML is extensible; it extends the ability to describe a document, letting meaningful tags for applications be defined. For example, because any IED typically provides current measurements, for the phase A current measurement that is available as a floating point, we can create a tag called <PhsAf>. In a similar way, we can create as few or as many tags as our document needs. It is obvious that we are extending the tags to identify elements by what they are, not by how they look.

XML is a markup language because its purpose is to identify elements within the document. Without markup, the computer sees any document as one long string of text, with each character having equal importance to every other character. By marking up a document, we identify the bits and pieces in a way that gives them value and context.

The big advantage of XML is that it allows extensible markup, that is, we can mark up the document in ways that match our substation PAC system needs (Figure 7.9).

However, markup is nothing but a way of identifying information. It does not program the data to act in a certain way, to display in a certain way, or to do anything other than carry an identifying mark.

```
<DOI name="OpDlTmms" desc="Operate delay">
   <DAI name="setVal">
      <Val>300</Val>
   </DAI>
   <DAI name="maxVal">
      <Val>60000</Val>
   </DAI>
</DOI>
```

Figure 7.9 Example of settings in XML.

XML is a language because it follows a firm set of rules. It allows us to create an extensible set of markup tags, but its structure and syntax remain firm and clearly defined. This does not mean that it is a programming language; it is not used to program a set of actions, but for a well-structured markup definition.

XML applies structure to documents and data. The document structure defines the elements that make up a document, the information that we want to collect about those elements, and the relationship that those elements have to each other. The document structure is called the document tree where the main trunk of the tree is the parent and all the branches and leaves are children. Document trees are usually visually represented as a hierarchical chart.

A well-formatted document is not sufficient. It also has to meet some constraints in order to make sense in the problem domain. The constraints enforce rules that determine the presence of elements and their attributes, as well as the order of these elements. These rules are part of what is defined as a data type definition (DTD) or XML schema. This is the oldest schema inherited from SGML. An important property of XML schemas is that they are also extensible (i.e., if necessary the schema can be extended to meet new requirements).

XML schemas describe the structure of XML documents. The XML schema language is also referred to as the XML schema definition (XSD). XML schemas are much more powerful than DTDs.

Any XML schema is a template for the markup of the document. That indicates the presence, order, and placement of elements and their attributes in an XML document.

As XML defines the data structure, it will not display a page by itself. We must use a formatting technology, such as the Extensible Stylesheet Language (XSL), an XML-based language for expressing stylesheets. With XSL, the user can make context-sensitive display decisions in a Web browser.

References

[1] ISO/IEC 7498-1:1994 Information Technology—Open Systems Interconnection—Basic Reference Model: The Basic Model—Part 1, 1994.

[2] ISO/IEC 9506-1 (2003): Industrial Automation Systems—Manufacturing Message Specification—Part 1: Service Definition, 2003.

[3] ISO/IEC 9506-2 (2003): Industrial Automation Systems—Manufacturing Message Specification—Part 2: Protocol Specification, 2003.

[4] Overview and Introduction to the Manufacturing Message Specification (MMS), Revision 2, SISCO, November 1995.

[5] ISO/IEC 8824-1:2021 Information Technology—Abstract Syntax Notation One (ASN.1): Specification of Basic Notation—Part 1, 2021.

[6] IEEE 802.3-2018 IEEE Standard for Ethernet, 2018.

[7] IEC 62439-3 Industrial Communication Networks—High Availability Automation Networks—Part 3: Parallel Redundancy Protocol (PRP) and High-Availability Seamless Redundancy (HSR), 2021.

[8] https://www.uml.org.

[9] https://www.w3schools.com/xml/.

CHAPTER 8

Object Modeling and Virtualization

8.1 Introduction

According to the *Webster New World Dictionary of the American Language*, an object is ".. a thing that can be seen and touched; material thing that occupies space ..." Based on this definition, it is obvious that everything around us is an object, including any primary or secondary electric power system equipment. If we have a more abstract look at this definition, we will realize that objects can also be virtual, meaning that they occupy space in the memory of a computer system and they can be seen and touched using the different interface tools available to us.

As it is clear that the virtualization of the world around us is an important step in the digitization and digitalization of the electric power system and its protection and control, we have to take a look at the fundamentals of object-oriented technologies that are used in the process, which will help us better understand how these principles are used in the definition and application of IEC 61850.

That is why in this chapter we will have a look at the fundamental concepts of object-oriented analysis and design and how they are used to apply the principle of functional decomposition in the modeling of multifunctional protection and control devices. Once we understand these principles, we look at the functional hierarchy of complex IEDs.

When we complete the structural model of a multifunctional IED, we need to look at how the different functional elements that are part of it interact with each other to implement specific protection functions or schemes, as well as the data exchange with functions in other IEDs based on data sets containing data objects or attributes.

The semantical meaning of the communicated data is very important to be clearly defined, so we need to understand how the data naming can be defined and their semantical meaning can be established.

To help the reader completely understand how these principles of object-oriented design are applied, we look at the specific generic example of an abstract model of a virtual multifunctional protection IED.

8.2 Object-Oriented Design Principles

We should not forget that multifunctional protection IEDs are specialized industrial computers running software modules integrated based on object-oriented design principles and that is why they can be considered as being developed within an object-oriented programming (OOP) paradigm based on objects and classes. Objects represent information and behavior. Thus, they have properties (or components, attributes) and services (or methods and events). Properties are data that describe an object. Methods are things we can tell the object to do, while events are things that the object does.

One of the characteristics of OOP is encapsulation, meaning that the data is not directly available to the outside world but only to the methods that are within the class definition. Another characteristic of OOP is abstraction, hiding the implementation details while presenting the attributes to the external applications. One of the key features of OOP is inheritance, which significantly improves the efficiency of the software development process by enabling a derived class to inherit properties and behavior from a parent class, thus enabling code reusability as well as adding new features to the existing code.

Object-oriented analysis (OOA) is another important step that is focused on the assessment of the system requirements based on identifying the classes and objects and the relationship between them within the application domain from the point of view of what the system will do rather than how it does it.

The last stage in the process is the object-oriented design (OOD) that involves designing the objects, classes, and relationships between classes based on the results from the OOA.

As we already discussed, everything around us represents some kind of an object; we are not even talking about the devices that we use in our everyday lives. We are experiencing interactions with classes, objects, and instances when we use software tools, for example, to make a PowerPoint presentation. This is something that can help us understand how this works. If we are to prepare a new presentation, first we need to select what kind of slide we are going to use, which opens the window in PowerPoint showing us the predefined classes of slide objects (Figure 8.1). Once we select a specific class, for example, for a slide with the title and content (Figure 8.1), we can enter in it the specific text of the title of the slide and the bulleted text in it, thus creating an instance of this object as is shown in the right of Figure 8.1.

In our domain of electric power systems PAC, a protective relay can be considered as an object as it takes space and can be seen and touched. However, when we talk about digitalization, we are discussing a different type of objects, which in many cases can be seen and touched only through the communications interface of the IEDs with the help of a dedicated user interface. However, an analogy with the real relay objects can be very helpful in understanding the object models of multifunctional protective relays. This is our problem domain (i.e., the application or process that is being modeled by object-oriented representation (classes and objects)): power system protection and control.

Because the class defines the information and behavior of the instance, while the current state of the instance is defined by the operations performed on it, each instance has a unique identity. For example, a distance protection zone could be

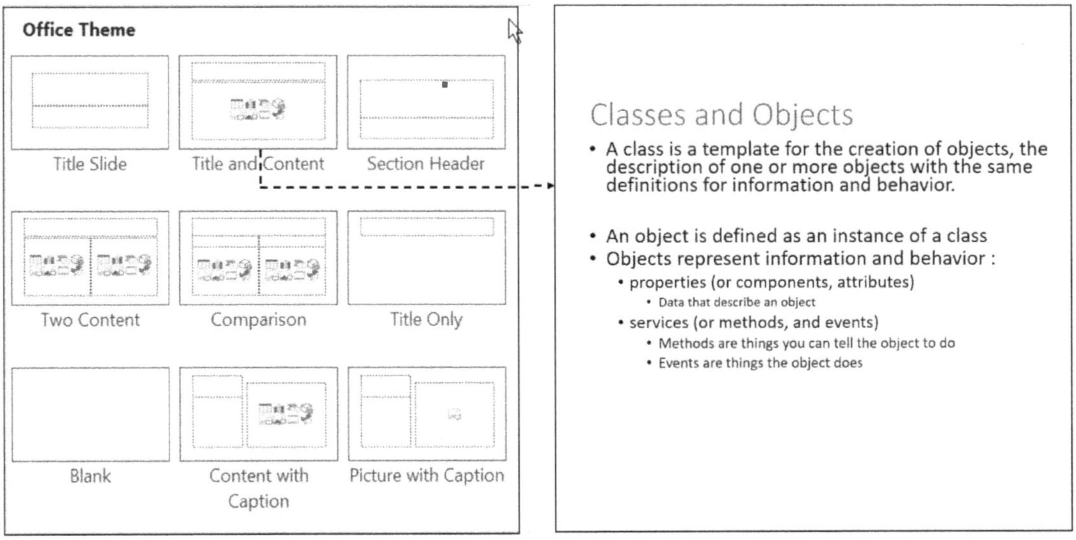

Predefined slide classes

Instance of a Title and Content class slide

Figure 8.1 PowerPoint object classes and instance.

defined as a class with phase and ground time-delayed steps as instances (object instantiations) of the class.

The standard class diagram contains three sections:

- Class name;
- Class attributes used to describe the qualities of the class;
- Class services used to describe how the class interacts with data.

In an object hierarchy, every child object inherits from its parent the attributes and behavior and adds new ones specific to the functionality that it represents.

8.3 Functional Decomposition

PAC systems are very complex in nature and contain a variety of components performing specific tasks and interacting with each other in order to achieve their functionality. Such systems can be implemented in many different ways, which depends on the implementation philosophy of the developer. To be able to define the requirements for the system, as well as to describe how it is implemented, it is necessary to use a method that can represent such complexity in the most efficient way. This method is functional decomposition, which is used to understand the logical relationship between the components of a distributed function, and is presented in terms of a function hierarchy that describes the functions, subfunctions, and functional elements with their interfaces.

Functional decomposition refers to dividing a system into components or modules to reduce its complexity. It can be viewed as a way of virtualization by matching encapsulated abstractions into basic components and takes place after the partitioning of a system into a hierarchy of building blocks.

The benefit of identifying reusable building blocks is that it significantly improves the efficiency of the development of the PAC system. Such building blocks can be with different levels of complexity and are defined in an object-oriented approach using a top-down method of analysis.

Figure 8.2 shows an example of the main building blocks of an abstract virtual distance protection. The interface with the transmission line being protected by the distance function is performed by the analog inputs module which digitizes the current and voltage signals from the secondary side of the current and voltage transformers. It produces a stream of sampled current and voltage values that are used by the waveform recording module to capture the transient behavior of the system during the fault condition and also by a measurement module that calculates from the sampled values three phase current and voltage phasors, as well as negative and zero sequence components. These calculated values become an input into the distance function module, which determines if the fault is within the zone of protection and makes a decision to operate or not. The distance protection scheme module is communicating with the remote end of the line in order to implement accelerated protection schemes such as permissive overreaching transfer trip (POTT) or directional comparison scheme.

Once the main modules of the distance protection are identified, the process of functional decomposition continues by looking into the functions that are performed within each of these modules. We can briefly look into the distance function module that will help us to understand how complex such functions can be.

The distance function is one of the more complex protection functions and includes a large number of building blocks, as can be seen from Figure 8.3, which shows a simplified block diagram of the different components and factors that have an impact on the definition, description, and configuration of the distance function model.

The configuration of the distance function is based on the impedance of the protected transmission line and setting criteria. The line impedance can be in the phase or sequence domain, depending on the implementation of the distance func-

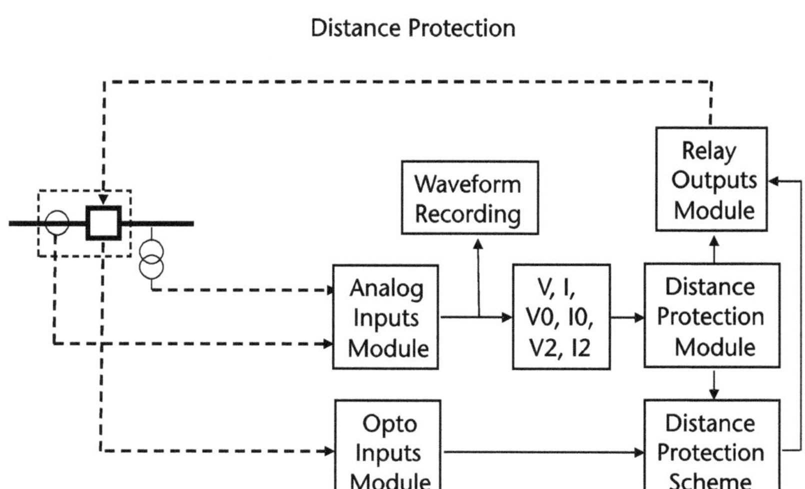

Figure 8.2 Functional decomposition of distance protection.

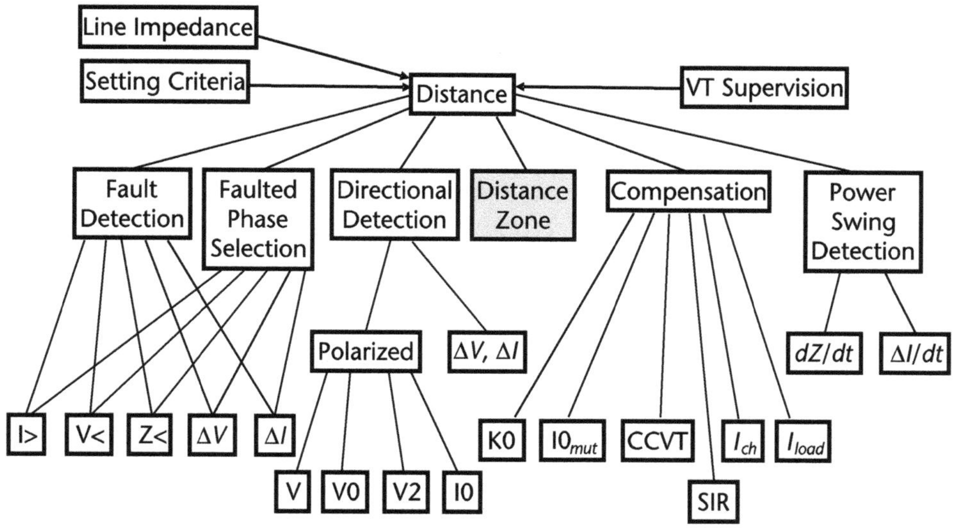

Figure 8.3 Distance function components.

tion. Voltage circuit supervision can be used to block the operation of the distance function in case of failure of any of the elements of the voltage circuit.

Advanced multifunctional distance relays can use event-driven logic (i.e., they initiate the distance protection function when they detect that a fault occurred in the system). The fault detection can be based on different operating principles: overcurrent, undervoltage, underimpedance, superimposed components of current, and voltage. This is modeled by the corresponding function elements.

The faulted phase selection is used to determine the type of fault and which of the loops is going to be used to identify if the fault is within the characteristic of the distance zone. Methods like the ones listed under fault detection are typically used.

The overall distance characteristic can be achieved using a single element, for example, an offset mho characteristic, or can be based on directional supervision of an element. The directional detection can be achieved using polarized quantities or superimposed component-based methods.

Power swings may result in the apparent impedance seen by the distance element entering the characteristic and resulting in undesired operation. Power swing detection can be used to block the operation of the distance elements. Different methods for power swing detection are used by relay manufacturers. Usually, they are based on the time that it takes for the apparent impedance to go through different characteristics used for this function. Some advanced relays use the changes in the superimposed currents to ensure the early detection of such a condition.

Considering the impact of different system configurations, instrument transformers, or system parameters on the behavior of the distance function, many vendors use different forms of compensation to improve the performance of the distance function. Some compensation examples are zero sequence, mutual coupling, capacitive voltage transformer (CVT) transients, and charging current.

Each of these function elements is considered as a simple object with a single instance within the distance function model. This is not the case with the distance characteristic. It is one of the key elements that require a lot of attention in the modeling process. It can be simple or complex and is typically based on the use of circle, blinder (resistive or reactive), or directional line elements. As the distance protection typically includes multiple phase and ground zones, each one of them will be included in the model as a separate instance of the distance characteristic object class. If we have a distance protection device that has three zones, we can expect to see six instances of the distance characteristic in the model: three for the phase distance subfunction, and three for the ground distance subfunction. It must be noted that such an object model defines access to attributes that characterize a function element but does not prescribe a specific algorithm to be used in its implementation.

8.4 Functional Hierarchy: Functions, Subfunctions, and Function Elements

The functional hierarchy shows the order in which objects in a system are put together and explains the relationship between the different parts of a system. Different functional components form a class of the hierarchy. In principle, a hierarchy class is composed of a base class (parent class) and derived classes (subclass) where a derived class inherits the properties of a parent class. Through such a hierarchy, a class can be composed of interrelated subclasses that can have their subclasses until the smallest level of components is reached. The best way of explaining this functional hierarchy principle is by looking at an example of something with which we are familiar: a transmission line protection device.

The functions in relatively simple IED, such as a low-end distribution feeder protection relay, are easy to understand and are grouped together in order to build the object model. That is not the case for the more complex devices such as an advanced distance protection for extra high-voltage level applications.

As we already discussed, the distance protection function has different components that need to be taken into consideration in the model. It is complex to represent advanced transmission line protection schemes, which typically exist in distance relays, and distributed functions based on high-speed P2P communications between multiple IEDs.

A multifunctional transmission line protection IED has a complex functional hierarchy that can be modeled as two main groups of functions: protection and nonprotection.

The protection functions can be further divided into main protection functions, backup protection functions, and protection-related functions. The main protection function in a distance relay is obviously the distance protection.

A local backup protection function example is a directional overcurrent protection, and breaker failure protection is a protection-related function.

Nonprotection functions are of several categories:

- Measurements;
- Control;
- Condition monitoring and diagnosis;
- Recording;
- Analysis.

Figure 8.4 shows an example of part of the functional hierarchy of a multifunctional transmission line protection IED.

Each of the above-described functions can be divided into subfunctions that represent groupings of related functional elements. In the protection functions, an example will be the distance and overcurrent protection subfunctions, each containing another layer of subfunctions such as phase and ground protection.

The bottom layer subfunction then can be split into functional elements. Functional elements can be defined as the smallest functional units that can exist by themselves and can exchange information with other elements within a device or a system. An example of a protection functional element will be a phase distance zone element used to represent the different zones in a distance protection, one per each zone.

Each function element is an object that contains multiple attributes describing its operating mode, configuration, status, and others, depending on its functionality.

8.5 Data Objects and Attributes

As we discussed at the beginning of this chapter, every object is described by specific information identifying its characteristics, status, and configuration. This information is represented by data objects, which are instances of specific data object classes. Each data object class may have a simple or complex data structure depending on what specific property of an object it represents.

A data object class may have hierarchical structure and contain multiple data attributes, each of which can be a simple one of a specific data type or with a hierarchical structure itself. To support interoperability and avoid any misinterpreta-

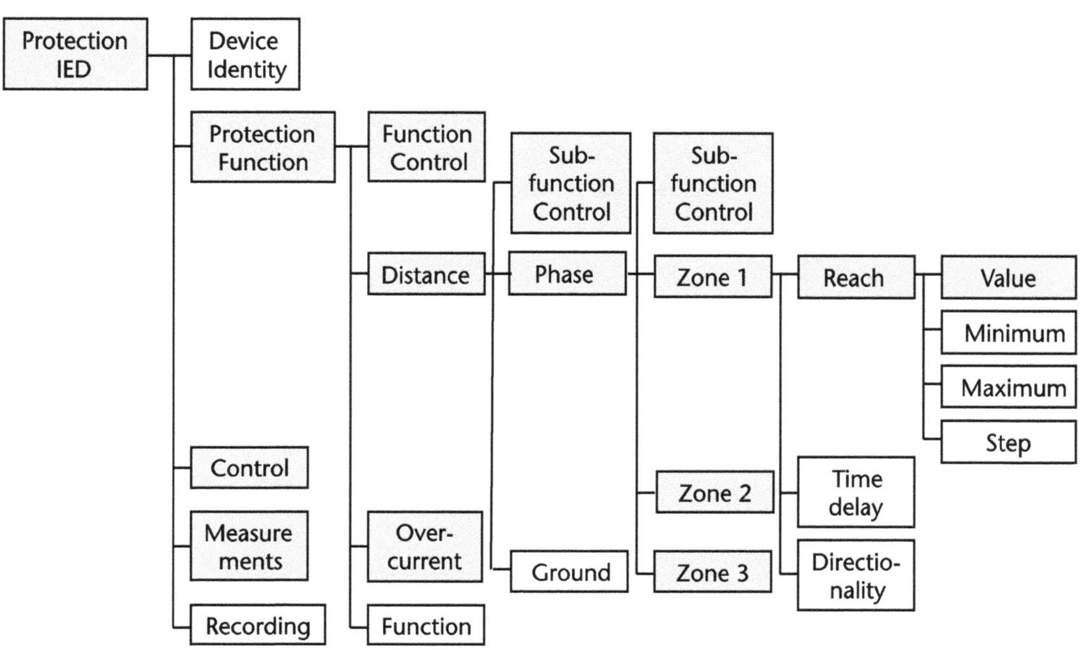

Figure 8.4 Transmission line protection IED hierarchy.

tion while processing the data, every data object must have a unique identification with a well-defined semantical meaning and structure.

For the function elements to interoperate over the substation LAN, it is necessary to model the data objects that are included in each of them. This may require multiple levels of modeling of data for the implementation of different substation automation functions. This includes the definition of data classes and their specific structure that includes one or more data attributes.

The function elements contain the information required by a specific function, such as measurements being calculated by an IED and provided to protection functions or a data acquisition system.

The data represents domain-specific information that is available in the devices integrated in a substation automation system. Any data object should include a data name and one-to-many data attributes. The data name is the instance name of the data object. Depending on the complexity of the model, it may be necessary to specify for both the data objects and data attributes if their presence is mandatory or optional in a specific instance.

The data objects in a function element can be instances of a simple data class or composite data classes and may contain data attributes that can be simple or nested, depending on the specifics of the model. When the data attribute has a hierarchical structure, each layer of the hierarchy contains data components. At the bottom of the data model hierarchy is the data type, which also can be primitive or composite.

The different data attributes can be grouped based on their specific use. For example, some may represent measurements, others specify the settings or indicate the status of the function element, and others are used for configuration or control.

We can consider as an example the data modeling of a measuring function that is using the output of the A/D converter function element and, from the stream of sampled values, calculating different measurements that might be used by protection and substation automation functions.

Figure 8.5 shows the measurement function that contains three measurement objects: the first calculating the three phase currents and voltages, the second calculating the phase-to-phase currents and voltages, and the third calculating the sequence components of the currents and voltages. Even though these three measurement objects are doing phasor calculations, they will contain different data objects

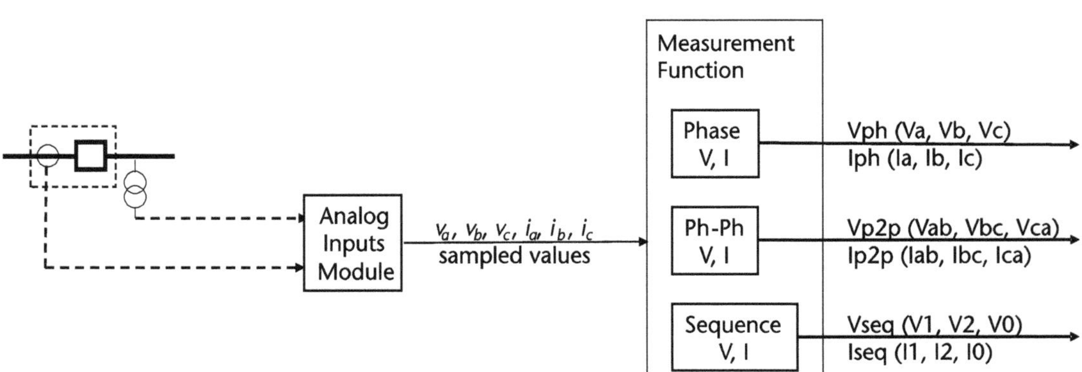

Figure 8.5 Data objects in a measurement function.

that can be accessed by different applications individually or based on some data object structure integrating the individual measurements.

If we look at the three-phase measurement element, it can be modeled as an object containing six measurands: Va, Vb, Vc, Ia, Ib, and Ic.

Because they are instances of a phasor data class, they are complex data objects that have two attributes: magnitude and angle. So when we read Va, we are actually reading Va(mag) and Va(ang). If we are not interested in the angle, we need to go down the data model hierarchy and read only Va(mag). The data type of Va(mag) and Va(ang) today probably will be a floating point, but in some device implementation it may be an integer. Each of the phasor attributes may contain multiple components, for example, description (of data type visible string), which can be used by a human-machine interface (HMI) application to display the utility standard name for the measured value.

If the measuring function element is used to calculate synchrophasors, it will contain an additional data object representing the time stamp for the phasor-measured values.

If we want to further improve the efficiency of interfacing with the measurement function element for the phase voltages or currents, we can define a complex data class Vph (Va, Vb, Vc) that will contain a collection of the three individual single-phase voltage measurements Va, Vb, Vc, so instead of reading each phase voltage by a separate request, we will have a single request for Vph that will bring back the magnitude and angle for the three phase voltage measurements.

8.6 Data Sets

After analyzing the object-oriented modeling of functions and function elements, as well as the data objects and their attributes that are used to represent the functionality and information available, we need to examine how the data is exchanged between the components of the PAC system in the digitized electric power utility environment. This is where we must look at data sets.

A data set is a grouping of data objects and data attributes that are defined as its members and are arranged in a specific order. Who the members of the data set are and what their specific order is needs to be known by all entities that are exchanging the information using that method. This can help to significantly improve the communications efficiency by limiting the payload in the message to an identifier of the data set and the values of the data contained in it.

Data sets can be related to the different levels of the functional hierarchy (see Figure 8.6). For example, we may have a data set containing a subset of the data available in a function element that is used for reporting any changes of their values during its operation. However, some of the same data can be used together with data from other function elements to form a data set representing a function or subfunction in the model's hierarchy.

If a member of a data set is just a data attribute, only its value is transmitted, but if the member of the data set is a data object, the values of all data attributes and their components that are defined in the data model will be transmitted.

The knowledge of the membership and its order in the data set not only is important for the interoperability between the communicating devices, it but also has

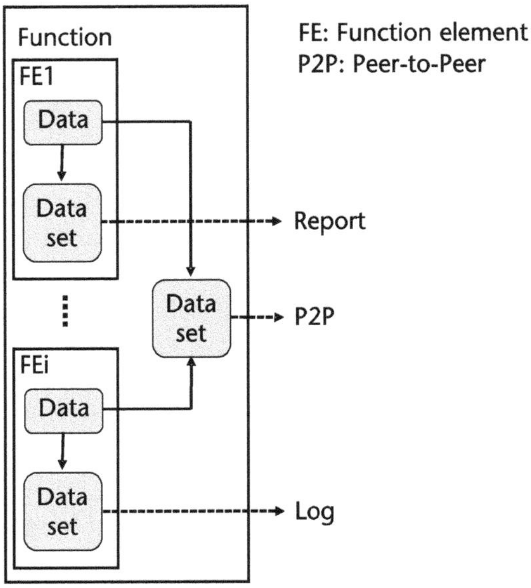

Figure 8.6 Data set use.

an impact on cybersecurity. If a hacker is able to get access to the communications link and capture the values from the message, it will not be very useful without the knowledge of what these values represent. That is why it is critical to protect the information about data set definitions.

Data sets not only are used in client/server applications such as reporting and logging, but they are also used to improve the efficiency of communications between protection and control devices using P2P communication services. In such cases, the data sets will contain information about the change of state of the different function elements used by a protection function such as its start or operation.

It is important to use predefined data sets as part of the engineering of a PAC system, so it can be properly tested as part of the commissioning and maintenance of the system. At the same time, it should be possible to dynamically create data sets by clients depending on the operational requirements for the system. This means that a client should have the ability to create or delete a data set and retrieve all values of its members.

The local reconfiguration of members of a data set, especially if it is related to protection and control operation, is very dangerous, because it may cause critical misoperations. To prevent unnecessary changes in the configuration of a data set, it is important first to try to avoid them as much as possible. If it is required for maintenance purposes, the changes should be performed by qualified specialists and subject to role-based access control, followed by testing to ensure that they meet the functional and performance requirements.

CHAPTER 9

IEC 61850 Model Details

9.1 Introduction

The process of virtualization requires the availability of tools that allow us to accurately represent all components and their behavior in a PAC system under different conditions. The IEC 61850 standard has been developed to support that in the most efficient way based on the object-oriented design principles. The model details at the beginning were based on the focus of Edition 1 on substation automation systems and the typical functionality of microprocessor-based protection devices and communications technology at the end of the last century.

With the evolution of protection and control technology, as well as the need to meet the requirements of the electric power grid with a dramatically increasing penetration of renewable DERs, the focus of the standard shifted from the substation towards power system automation. That resulted in changes in the IEC 61850 model not only covering different power system domains, but also being defined to improve the virtualization accuracy, as well as the reliability and efficiency of PAC systems.

This chapter examines the evolution of the IEC 61850 model intended to represent multifunctional IEDs working together over communication networks to implement various distributed functions. It starts with the object model of an IED as a server and the hierarchy of its components, followed by the data modeling based on data objects and data attributes. The semantically meaningful naming principles in IEC 61850 used in the object and data modeling are described as well.

The requirements and concepts for the IEC 61850 model are formulated in Part 5 of the standard. Based on this, the actual model is defined at three levels, each within a dedicated subpart of Part 7.

The first level is the Abstract Communication Service Interface (ACSI). It specifies the models and services for access to the elements of the specific object model, such as reading and writing object values or controlling primary substation equipment.

The second level defines common data classes (CDCs) and common data attribute types. A CDC specifies a structure that includes one or more data attributes.

The third level defines compatible logical node classes and data classes that are specializations of the CDCs based on their application.

Part 7-2 [1] specifies the first level of modeling: ACSI. Part 7-3 [2] covers the CDC, and Part 7-4 [3] defines the compatible logical nodes and data classes.

It is impossible in a single chapter to cover the definitions in the standard described in hundreds of pages. So what we will try to do is to describe the modeling principles and give some examples of how they are implemented in the standard. For all details of the actual model, the reader should refer to the standard itself.

While IEC 61850 has done a very good job in standardizing the logical nodes representing different function elements, as well as the data objects and data attributes based on common data classes, that is not the case with the logical devices. It is clear what they are and what their role is in the modeling of multifunctional devices; however, their use is not standardized but left to the developers of the devices. Because of that, there is a wide range of possible implementations of the model.

To clarify the challenges and demonstrate how they can be addressed, the second part of this chapter gives a detailed example of a possible model for a complex transformer protection multifunctional IED.

9.2 The IEC 61850 Logical Node Object Model

As we already discussed, the modeling of the electric power system and its PAC components is based on the principle of functional decomposition. In the foundation of this approach are the logical nodes representing the function elements used in the different intelligent electronic devices to implement any specific function. They provide the granularity of the model and allow us to describe the exchange of information between different functions and devices.

The concept of what became the logical nodes in IEC 61850 can be traced to the protection basic relay objects (building block PBRO) defined during the development of UCA 2 at the end of the last century.

In IEC 61850, the concept of logical nodes is first introduced in Part 5 where they are described as objects representing the smallest functional elements in a PAC system that can exchange information. Logical nodes are used to allow the exchange of information and make the functional element's behavior visible through communications. However, they do not standardize the algorithm or logic used to implement the function that they represent.

Any logical node is located in a physical device and can be used in a local or distributed function. As it is the building block and the smallest atomic part in the model that exchanges information, a logical node cannot be distributed between two devices. In the case of a local function, the data exchange between logical nodes is within the device over its internal digital data bus. When it participates in a distributed function, the data exchange is over a communications network.

The logical nodes belong to specific groups depending on their application, for example, protection or measurements. Part 5 of the standard defines the conceptual logical node groups that can be identified under two categories:

- Logical notes representing substation PAC functions;
- Logical notes representing primary substation or power system equipment.

These conceptual logical nodes from Part 5 are later defined as logical node classes in Part 7 of the standard.

While the groups of logical nodes defined in Edition 1 of the standards were concentrated on substation-related applications, with the shift of the focus of Edition 2 of the standard towards power system automation, the number of groups of logical nodes was extended to cover other smart grid-related domains.

Table 9.1 shows the groups of logical nodes defined in Part 7-4 of Edition 1 and Edition 2 of the standard.

The group indicators from Table 9.1 are used as the first of four letters in a logical node class name.

In defining the conceptual logical nodes, Part 5 of the standard provides a link between them and the IEEE function numbers, as well as the IEC function symbols. It is important to note that the conceptual logical nodes defined in this part are not always directly matched by the logical node classes defined in Part 7-4.

While most of them do match, for example, protection time overcurrent (PTOC) being used to represent protection time overcurrent function (51) in both these parts of the standard, the directional overcurrent protection function (67) in

Table 9.1 Groups of Logical Nodes

Group Indicator	*Logical Node Groups IEC 61850 Edition 1*	*Logical Node Groups IEC 61850 Edition 2*
A	Automatic control	Automatic control
B		Reserved
C	Supervisory control	Supervisory control
D		DERs
E		Reserved
F		Functional blocks
G	Generic function references	Generic function references
H		Hydro power
I	Interfacing and archiving	Interfacing and archiving
J		Reserved
K		Mechanical and nonelectrical primary equipment
L	System logical nodes	System logical nodes
M	Metering and measurement	Metering and measurement
N		Reserved
O		Reserved
P	Protection functions	Protection functions
Q		Power quality events, detection-related
R	Protection related functions	Protection-related functions
S	Sensors, monitoring	Supervision and monitoring
T	Instrument transformer	Instrument transformer and sensors
U		Reserved
V		Reserved
W		Wind power
X	Switchgear	Switchgear
Y	Power transformer and related functions	Power transformer and related functions
Z	Further (power system) equipment	Further (power system) equipment

Part 5 is modeled by PTOC and directional supervision is modeled by protection related directional (RDIR) in Part 7-4.

Another example is the directional Earth fault protection (67N), which in Part 5 is represented by logical note protection directional earth fault (PDEF). However, in Part 7-4, it is represented again by PTOC with directional supervision. The difference between the PTOC representing the phase overcurrent function element and the one representing the Earth fault is the input to the logical node. In the first case, the input is a three-phase current, while in the second case, it is the ground current. This way, we can use the same logical node class to also represent a negative sequence overcurrent element by just using the negative sequence current as the input to an instance of the same logical note PTOC class.

As the logical nodes represent function elements in the model, they provide a higher level of granularity compared to the IEEE function numbers, which allows better standardization. For example, the distance function in the IEEE model is represented by number 21, which does not indicate the number of zones available in the device. In the IEC 61850 model, the number of distance zones in the device is represented by the corresponding number of instances of LN protection distance (PDIS).

On top of all logical nodes representing PAC functions, Part 5 introduces two special logical nodes within the system logical nodes group. The first one is logical node physical device (LPHD), which represents the physical device. The detailed model in Part 7-4 includes three mandatory data objects: the physical device name plate information, the health status of the physical device in which it resides, and the proxy. The third mandatory data object in this logical node is used to indicate if it represents an external physical device, meaning that it acts as its proxy. This capability is very helpful when integrating devices supporting legacy protocols in an IEC 61850-based system.

The second special logical node identified in Part 5 is LLN0. It contains data related to the logical device in which it resides and can be used to control the behavior of all the logical nodes contained in this logical device. It also plays an important role in the modeling of the hierarchical structure of complex multifunctional devices.

The structure of all logical nodes defined in the standard is based on the definition of a general logical node class in Part 7-2 and includes attributes and services. This is shown in Figure 9.1.

Part 7-4 defines the detailed models of all logical nodes used in the standard. With the exception of LPHD, all other specific logical nodes inherit from the common logical note at least the mandatory data objects contained in it. This is the place to state that the inclusion of data objects in a specific logical note depends on the requirement for their presence, which can be mandatory, optional, or conditional.

When a data object is defined in the common logical node as mandatory, it means that this data object will be available in any logical node defined in the standard. An example of such data object is Beh (Behavior). The presence of Mod (Mode) is conditional, being mandatory in an LLN0 of the root LD of a hierarchy and optional in all other logical nodes (LN). The InRef (general input reference) is optional.

The data objects for nameplate data, health, mode, and behavior in a specific LLN0 contain information related to the logical device to which it belongs.

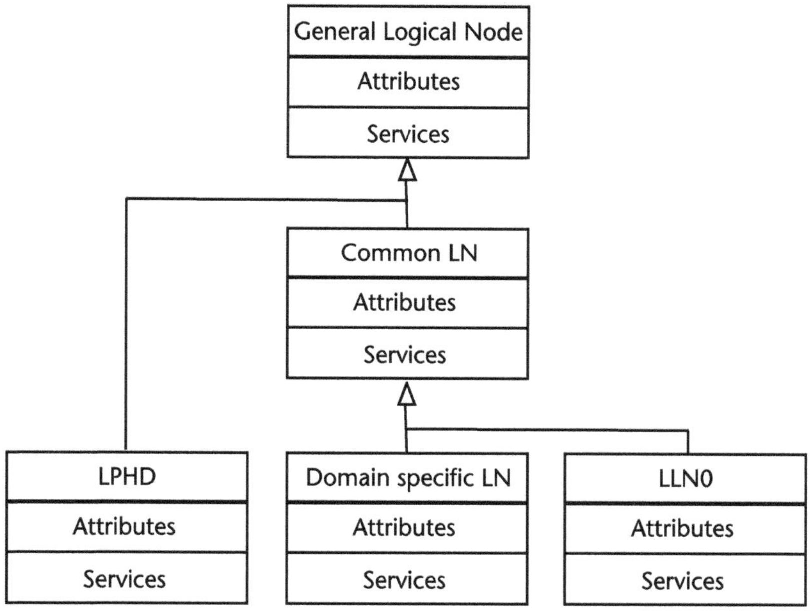

Figure 9.1 Logical node inheritance.

Another important data object in LLN0 is GrRef, which represents a reference to a higher-level logical device and was introduced in Edition 2 of the standard to support hierarchical structures of logical devices in complex multifunctional IEDs.

Each domain logical node class defined in Part 7-4 has a name that depends on the group to which it belongs and also represents an acronym related to its functionality. The class names are always four uppercase letters with the first letter being the group indicator. For example, the logical node class PTOC is representing a time overcurrent protection function element and belongs to group P (Protection functions).

The rules for composing logical node instance names are defined in Part 7-2 of the standard. The class name is followed by a numerical LN-Instance-ID. In many cases, there is also a LN-Prefix that starts with a letter. The total number of characters in the prefix and the instance identifier is limited to 12 characters. For example, we can name an instance of a logical node representing an inverse time ground overcurrent protection gndPTOC2

where:

- gnd is the prefix.
- PTOC is the logical node class name.
- 2 is the instance identifier.

Prefixes and instance identifiers are not allowed for LPHD and LLN0.

The logical nodes contain different data objects depending on their functionality. These data objects can be grouped in several categories such as measurements, status, settings, and control. The actual content of a specific logical node class is based on the inheritance hierarchy defined in Part 7-4.

Figure 9.2 shows an example of different data objects contained in the logical node PTOC, with each of the data objects in the example inherited from a different level of the hierarchy.

- Mod belongs to the Controls category and is inherited from the DomainLN.
- ClcMth belongs to the Settings category and is inherited from the StatisticsLN and specifies how the data attributes that represent analog or counter values have been calculated.
- Blk belongs to the Status information category and is inherited from the FunctionLN and indicates if the logical node has been dynamically blocked by another function.
- FltA belongs to the Measured and metered values category and is inherited from the ProtectionLN Fault current. Depending on how it is configured its value may represents the three phase current at the start or at operate of the logical node during the fault.
- Op belongs to the Status information category and is inherited from the CurrentProtectionLN and indicates if the function represented by the logical node has operated.

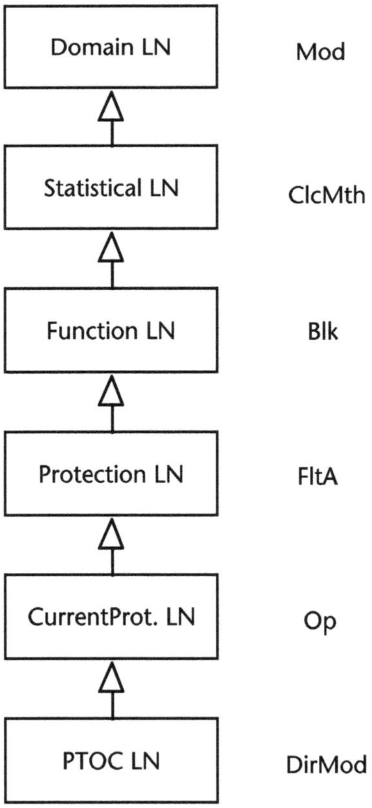

Figure 9.2 PTOC data object inheritance.

- DirMod belongs to the Settings category and is specific to PTOC. It represents the Directional mode of the overcurrent protection element and is used to enable operation when the direction of the fault matches the setting.

The implementation of any function in an IEC 61850-based protection and control system applies the principle of functional decomposition by distributing different tasks between logical nodes of different types. We use a simple example of an overcurrent protection function to illustrate this principle.

Figure 9.3(a) shows the traditional implementation of this function in an electromechanical relay. It is hardwired to the current instrument transformer if it is nondirectional and into the current and voltage transformer if it is directional (as shown in the figure). When the relay operates, it will close its output contact to trip the breaker and clear the fault. Figure 9.3(b) shows how this same function is implemented based on the functional decomposition principle.

The secondaries of the current and voltage transformers are connected to logical notes TCTR1 and TVTR1 that digitize the analog signals and publish current and voltage sampled values AmpSv and VolSv over the process bus.

The 3-phase measurement function represented by logical note MMXU1 subscribes to these sampled values and calculates the current phasors that become an input to the protection time overcurrent element represented by logical node PTOC1. When it operates, it sends a signal Op.general=true to the trip conditioning logical node PTRC1, which may send a trip signal through hardwiring to the trip coil of the breaker to clear the fault or a GOOSE message to a switchgear control unit to trip the breaker.

It should be noted that if a function element represented by a logical node in a specific vendor's device contains data objects or attributes that do not exist in the

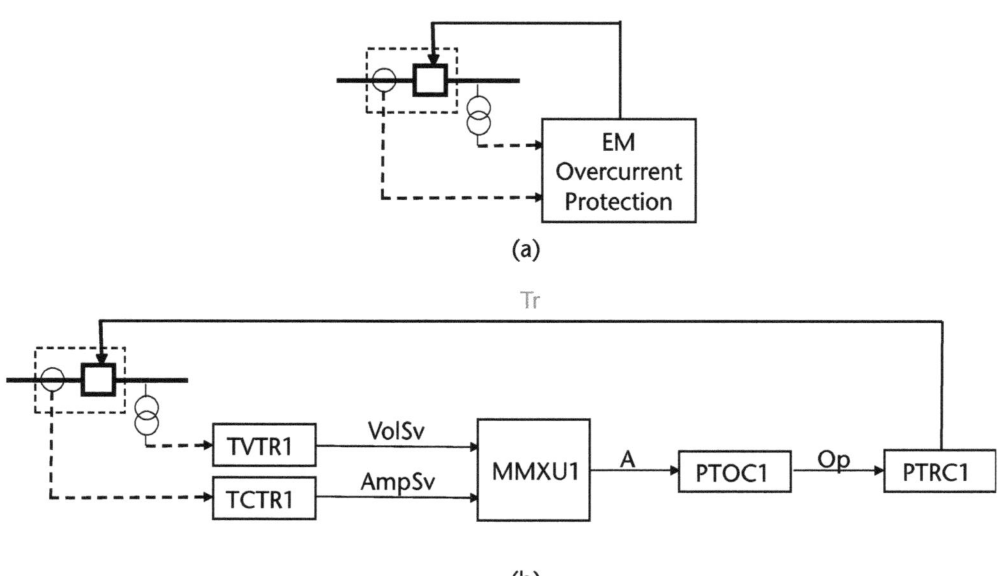

Figure 9.3 Protection function decomposition.

standard model, it is possible to extend the LN definition with proprietary data objects or attributes.

9.3 The IEC 61850 Logical Device Model

Part 7-2 defines a logical device as an entity that represents a set of typical automation, protection, or other functions in order to support the grouping of related functional elements. A simple example is a measuring logical device that will contain function elements calculating current and voltage phasors or current and voltage RMS values, as well as their sequence components.

If we apply the principle of virtualization, the logical device can be considered as a virtual representation of a panel in the conventional substation where we have protection panels, control panels, and measurement panels. For example, a distribution protection panel in the conventional substation will include multiple single phase or ground electromechanical overcurrent relays. In the IEC 61850 model, this will be represented by a virtual panel containing multiple instances of PIOC or PTOC logical nodes.

As can be seen from Figure 9.4, the Edition 1 model is relatively flat with the server containing logical devices that, in turn, contain logical nodes. However, as we have seen in Chapter 8, the functional hierarchy of advanced IEDs can have multiple layers that cannot be properly represented by such flat model. That is why, in Edition 2 of the standard, a more complex hierarchy (Figure 9.5) was implemented to support a more accurate representation of the functionality of advanced protection and control devices.

As we have seen, multifunctional devices are modeled using multiple types of logical nodes depending on the specific application of the IED. The logical nodes contain the information required by a specific function, such as function settings or measurements being calculated by an IED. Logical devices are introduced in the model to improve its structure and the efficiency of control of its behavior under different conditions. Another reason to group logical nodes in a logical device is to share some functionality, for example, directional supervision for multiple overcurrent or distance elements.

It must be noted that, although functions can be implemented by interactions between logical nodes located in different physical devices, even that typically a

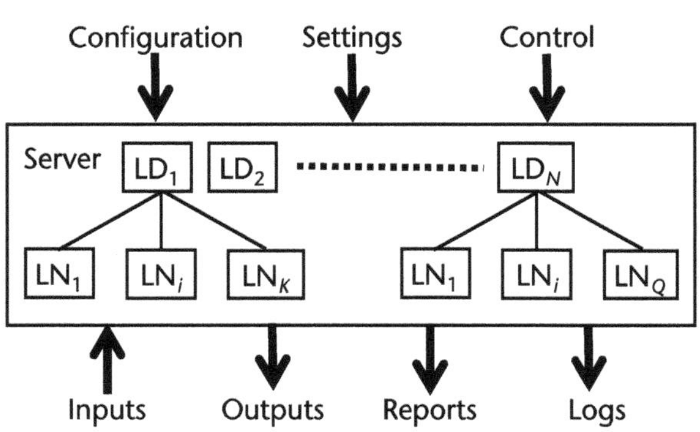

Figure 9.4 Server object model according to IEC 61850 Edition 1.

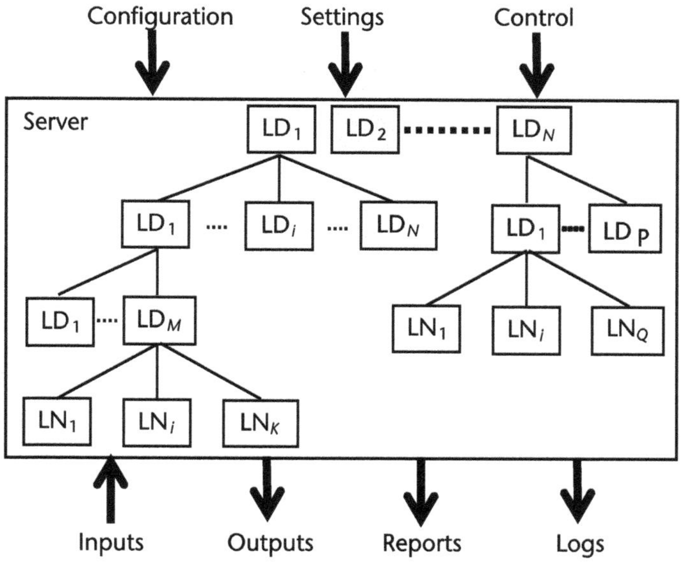

Figure 9.5 Server object model according to IEC 61850 Edition 2.

logical device contains functionally related logical nodes, a logical device is always implemented in one IED and contains only logical nodes from this IED.

Although the standard defines the logical device and how it can be used to group logical nodes, it does not standardize its implementation. This is left to the developers of the multifunctional protection and control devices and that is why we have seen a wide variety of ways that it is done. For example, in the early stages of the implementations of the standard, there were some devices that had a single logical device containing logical nodes representing many different unrelated functions, making it very difficult to understand its functionality.

Based on the experience in the last couple of decades, we have seen a trend between the different manufacturers in grouping together logical nodes in logical devices representing PAC, monitoring and recording functionality.

The development of the logical nodes, data objects, and data attributes has put a significant effort on standardizing their names to support semantical meaning. However, this is not the case with the logical devices. In order to further improve the interoperability between devices from different manufacturers, there is a need to make another step by standardizing the logical devices' configuration and naming for the most commonly used functions.

The general logical device class model in Part 7-2 includes as attributes the logical device name, reference, and all the logical nodes that it contains. The service in the model is GetLogicalDeviceDirectory and can be used by a client to retrieve the list of the object references of all logical nodes that it contains. The list will include one LLN0, most of the time one LPHD, plus all logical nodes representing function elements.

Some of the communication services in the standard are implemented at the logical device level, for example, GOOSE and sampled value exchange.

One of the roles of a logical device is to provide communications-visible information about the physical device to which it belongs. Instances of logical devices can also deliver information about external devices with which they interface. If the information provided by the logical device represents the physical device it is located in or an external physical device is determined by the value of the proxy data object in LPHD. When the value is false, it means that the information is for the host physical device, while, if the value of proxy is true, it means that it represents an external device. This feature of the standard allows the development of proxy servers or gateways and is a very powerful tool as part of a migration strategy from legacy to IEC 61850-based protection and control systems.

As we have already discussed, advanced multifunctional IEDs have a complex, multilayered functional hierarchy that could not be properly represented by the single-layer logical device model. That is why, in the modeling of such devices based on Edition 1 of the standards, we had to find a way to model this using the prefix of the logical node instance names to indicate the grouping of some of them. Because of that, in Edition 2 of the standard, we introduced the concept of nesting of logical devices to be able to model such multilayer hierarchy containing functions and nested subfunctions. This modeling is based on an LLN0 attribute GrRef. It is a setting data object of the common data class object reference setting (ORG) indicating that the logical device function refers to an upper-level functional group represented by the logical device to which it is referring. Because the modes of all functionally related logical nodes in a hierarchy might be related, these nodes may be switched on and off together, or their test mode is controlled as a group. This is explained further in the transformer protection IED example later in this chapter.

9.4 The IEC 61850 Server Object Model

The IEC 61850 object hierarchy defined in Edition 1 of the standard can be represented in a simplified way as shown in Figure 9.4. A server typically is any physical device that is being modeled as part of a substation PAC system.

The server represents the communications-visible behavior of an IED and can be a very simple or a very complex device and everything in between. An example of a very simple server is a merging unit that is publishing a stream of sampled values for currents and voltages over the substation process bus. At the other end of the spectrum is a very complex server such as a multifunctional transformer protection IED. Somewhere in between is a distribution feeder protection device.

Like any object model, the server model contains a set of attributes that describe what it is and services showing what can be done with it.

Data attributes included in the server model defined in Part 7-2 of the standard include:

- *Server access point:* An abstraction of an address that can be used to identify the server. There has to be at least one identified in the model, but it is possible to have more than one as well.
- *Logical device:* Identifies at least one logical device contained in the server, but it is possible to have more than one depending on the complexity of the functionality of the server.
- *File system:* Identifies a file system, if such exists in the server.

- *Two-party application associations:* It is used to identify a client with which the server communicates for bidirectional connection-oriented information exchange.
- *Multicast application associations:* It is used to identify a subscriber with which the server communicates for a unidirectional information exchange. The server acts as a publisher and may have one or many subscribers.

Using the GetServerDirectory service, we can implement the self-description principle allowing us to understand what functions are implemented in the server without having to look at the instruction manual of the device. In response to the GetServerDirectory request, the server returns a response containing the logical device names or the file system names from the server.

As mentioned earlier, the server model in Edition 2 of the standard evolved support to nested logical devices as shown in Figure 9.5. In the example in this figure, we can think of LD1 representing protection functions containing a distance protection subfunction (LD1), overcurrent protection subfunction (LD_i), frequency protection subfunction (LD_N). Then the distance protection subfunction contains phase and ground subfunctions, which contain multiple logical nodes representing the individual zones.

9.5 The IEC 61850 Data Model

9.5.1 Data Modeling Principles

In the previous sections, we discussed logical nodes and their applications. The examples included attributes of some logical nodes, so now it is time to look into more details of what these attributes are: data objects. Depending on the date object class definition, they can have a simple or complex structure themselves. It depends on the specific type of data that they represent.

It is impossible to describe all specific common data classes, date objects, and data attributes defined in the standard, so, in this section, we are just going to give some examples to describe the principles that are used in the definition of the data modeling. As most of the time, specialists from the protection and control community deal with measurements and protection functions, we are going to consider examples from these two groups.

The logical nodes contain the information required by a specific function, such as measurements being calculated by an IED and provided to protection functions or a data acquisition system. Logical nodes typically include not only data, but also data sets, different control blocks, logs, and others as defined by the standard.

The data represents domain-specific information that is available in the devices integrated in a substation automation system. The data can be simple or complex and can be grouped in data sets as required by the application.

Any data should comply with the structure defined in the standard and should include DataName, DataRef, Presence, and multiple DataAttributes. The DataName is the instance name of the data object. The DataRef is the object reference that defines the path name of the data object instance. The Presence is a Boolean type attribute that states if the data object is mandatory, optional, or conditional.

Each instance of a data class object must contain at least one DataAttribute. Instead of a DataAttribute, it is possible to have a Simple CDC or Composite CDC (both are specializations of the Data class). DataAttributes can be simple or nested. If they are nested, at each nesting level other than the first, the DataAttributeName is called DAComponentName. The DataAttributes are of a certain data type that can be primitive (BasicType) or composite (DAType).

The different DataAttributes can be grouped based on their specific use. For example, some indicate the status of the logical node, while others are used for configuration or measurements. The property of DataAttribute that shows its use is a functional constraint (FC). The standard defines many different functional constraints. Some more commonly used are:

- ST: Status information;
- CO: Control;
- SP: Set point that does not belong to a setting group;
- CF: Configuration;
- DC: Description (information intended for humans);
- SG: Setting group (for settings that belong to a setting group);
- MX: Measurements (analog values).

9.5.2 Data Modeling of Measurements

As an example, we are discussing the modeling of measuring and metering functions, so the last functional constraint from the above list is of specific interest to us. The MX functional constraint is used to indicate that the data attribute represents measurand information. The value of this data object can be read, substituted, logged, or reported. Probably the main characteristic of the data objects that fall in this category is that their value should never be written. The values of these DataAttributes are normally based on processed data from the IED.

The trigger option attribute specifies the condition that may cause the reporting or logging of the data. These may be due to change of the value or the quality of the attribute.

Part 7-3 of IEC 61850 defines the different CDCs used in the model. It should be noted that users of the standard are not allowed to extend existing CDCs or define new ones.

Here is a list of CDCs used in the modeling of measurand information:

- Measured value (MV);
- Complex measured value (CMV);
- Sampled value (SAV);
- WYE;
- Delta (DEL);
- Sequence (SEQ);
- Harmonic value (HMV);

- Harmonic value for WYE (HWYE);
- Harmonic value for delta (HDEL).

It should be noted that CDC can include data objects that are representatives of another CDC. For example, WYE is a collection of simultaneous 3-phase and neutral values of an electric system parameter: current and voltage. Each of these values is of the type complex measured value (CMV).

The CMV has a data attribute cVal (the value of this complex measured value) that is of type Vector with functional constraint MX.

Vector has two data attributes: mag (the magnitude of the complex value) and ang (the angle of the complex value). They are both of type AnalogValue.

AnalogValue may be represented as a basic data type INTEGER (attribute i) or as FLOATING POINT (attribute f). At least one of the attributes shall be used, but some devices might support both. IEC 61850 not only specifies the type, but also the precision (e.g., there are UNSIGNED and SIGNED INTEGERS of different precisions/ranges 8.16,32).

The most common logical node in all IEDs typically available in a substation automation system is the one used to model the multiple measured quantities by a device in a 3-phase system. It is mainly used to provide measurements from an IED to a substation HMI, any remote system operator, or other corporate client. It also can be used to model the function elements in a protection IED used to calculate the phase and sequence currents and voltages used by different protection function elements. The name of this logical node for 3-phase quantities is MMXU and it inherits all mandatory data from the relevant common logical node class in the hierarchy defined by the standard.

Table 9.2 shows the attribute names and types of different optional measured values included in MMXU.

As can be seen from the table, most of the attributes in this logical node are of WYE type, that is, they model the 3-phase and neutral values of measured voltages (PhV), currents (A), and impedance (Z).

Table 9.2 Measured Values Attributes in MMXU

PPV	DEL	Phase-to-phase voltages
PhV	WYE	Phase-to-ground voltages
A	WYE	Phase currents
W	WYE	Phase active power (P)
VAr	WYE	Phase reactive power (Q)
VA	WYE	Phase apparent power (S)
TotW	MV	Total active power (Total P)
TotVAr	MV	Total reactive power (Total Q)
TotVA	MV	Total apparent power (Total S)
TotPF	MV	Average power factor (Total PF)
Hz	MV	Frequency
PF	WYE	Phase power factor
Z	WYE	Phase impedance

The phase-to-phase voltages are modeled using PPV, which is of the DEL type.

The total values of active, reactive, and apparent power, as well as the total power factor, are of type MV.

Protective relays and power quality monitoring devices measure different system parameters that are used to determine unbalanced system conditions. Such measurements are modeled using logical node MSQI. Most of the measured values included in the logical node are of type MV. Some are WYE and DEL. The sequence components of the currents and voltages are modeled as attribute type SEQ.

Line differential relays calculate the differential current based on their local measurements, as well as using measurements received from the remote end or ends (if it is a multiterminal line). These measurements are available in logical node MDIF. The 3-phase measurements are represented as sampled values (type SAV).

Although most of the commonly used measurements are 3-phase, in some cases, we may need to represent measurements that are not phase-related. In this case, we need to use logical node MMXN (nonphase-related measurements). Most of the values in that logical node are of the MV type.

Advanced metering and power quality monitoring devices, as well as specialized energy metering devices, calculate the energy that is then used for billing or other purposes. Different energy values are available in the Metering logical node MMTR.

Power quality monitoring devices calculate hundreds of different system parameters, such as harmonics or interharmonics. Their modeling is based on a logical node dedicated to these measurements, MHAI.

The attribute types of the different measured values included in MHAI are HWYE, HDEL, WYE, DEL, or MV. This logical node can be instantiated for either harmonics or interharmonics depending on the value of the basic settings.

In the cases when the harmonics or interharmonics are calculated in a single-phase system with no phase relations, the MHAN logical node should be used in the device model. In this case, the attribute types for the measured values are of type MV or HMV.

After describing some of the typical data objects used to model measured values, we are finally at a point when we can give an example of a data path for a single-phase measurement of the current in phase B represented as a floating point:

MMXU1.MX.A.phsB.cVal.mag.f

where:

MMXU1 is an instance of the LN class MMXU defined in Part 7-4. This logical node represents function elements used for the calculation of currents, voltages, powers, and impedances in a 3-phase system.

MX is functional constraint for measurements.

A is an instantiation of the composite DATA class WYE (defined in Part 7-3) used to represent the three phase currents and the neutral current.

phsB is the value of the current in phase B as a Simple Common DATA class of type CMV (defined in Part 7-3).

cVal is the complex value of the current in phase B (of the Common DataAttribute type Vector).

mag indicates that this object represents the magnitude of the complex value (of type AnalogValue, defined in Part 7-3).

f is of the basic type FLOATING POINT (defined in Part 7-2), which is currently restricted to a Single Precision floating point range.

All measurements in multifunctional IEDs are modeled in a similar way and grouped into the special logical nodes described earlier.

9.5.3 Data Modeling of Protection-Related Data Objects and Attributes

Because protection functions are very different from measuring functions, we can expect that the attributes of protection logical nodes will be data objects of different common data classes.

The following is an example of one of the attributes of the protection time overcurrent logical node PTOC representing its status. This data attribute is the same in all protection logical nodes and plays a very important role in the modeling of different functions because it can be used to initiate tripping of circuit breakers, initiate breaker failure protection or autoreclosing, and trigger disturbance recording.

PTOC1.ST.Op.general

The status information (FC ST). data object Op is mandatory and indicates if the function element PTOC1 has operated. Op is of the common data class ACT (Protection activation information) and has three mandatory attributes:

- general is a data attribute of the basic type BOOLEAN indicating the operation of the function element when the value is true, regardless of which is the faulted phase.
- q is a data attribute of type Quality and contains data that describe the quality of the data from the server.
- t is a data attribute of type Timestamp providing a timestamp of the last change of the value in any of the phase currents or the neutral current, as well as the quality.

ACT also includes optional data attributes phsA, phsB, phsC, and neut that can be used when we need to know which is the faulted phase.

The following is an example of another data object StrVal and some of its attributes in the protection time overcurrent logical node PTOC representing its pickup setting that belongs to a setting group (FC SG). It is of the common data class ASG (Analog setting).

PTOC1.SG.StrVal.setMag

This data object is also the same in many protection logical nodes and also plays a very important role in the modeling of different functions because it defines the value above which the protection function element will start. The value of the pickup setting of a protection function typically depends on the selected setting group, that is why it is with functional constraint SG. StrVal has one mandatory attribute:

setMag is a data attribute of the type Analog value control (AnalogueValueCtl) providing the value of the pickup setting.

There are additional commonly used optional data attributes for StrVal:

units, in SI, in this case, amps

minVal is a data attribute of the type Analog value and represents the minimum setting for setMag.

maxVal is a data attribute of the type Analog value and represents the maximum setting for setMag.

stepSize is a data attribute of the type Analog value and represents the step between the values of setMag in the range defined by minVal and maxVal.

The naming of all date objects and data attributes is defined in Part 7-4 of the standard and is based on using or combining abbreviated names from a list of abbreviations included in this part.

Considering the importance of a common understanding of the meaning of each name Part 74 also includes a table with detailed semantical description of each of the date objects or attributes. For example, setMag is defined by combining Set (setting) and Mag (magnitude).

9.6 The IEC 61850 Transformer Protection IED Model

As discussed earlier, in case of protective relays with more complex functional hierarchy, it might be necessary to group together several logical nodes in a functional group such as overcurrent protection. The fact that a logical node belongs to a functional group of logical nodes can be represented by containing them in a logical device. If the device has a very complex functional hierarchy, we need to use nested logical devices as described earlier. This is especially true for transformer protection IEDs.

When there are different functions and certain functional elements are grouped together (for example, for enabling the supervision of a group of functional elements), the modeling needs to be done using the available object hierarchy and the naming conventions for the data objects defined in IEC 61850. The model in this case will include multiple logical devices as shown in Figure 9.6.

IEC 61850 Edition 2 clearly defines the model hierarchy that can be used for different multifunctional IEDs. It does not specify how exactly they should be used for grouping of functional elements. This provides a lot of flexibility of the model. However, it creates problems with interoperability in the sense that the development of third-party tools that rely on a standardized functional model is not possible. That is why the industry needs to reach a common understanding of the principles of object modeling of complex hierarchical functions.

A further improvement will be the standardization of the commonly used logical devices' naming. If the grouping of functional elements represented by logical nodes in standardized logical devices is accepted by the industry in a way similar to

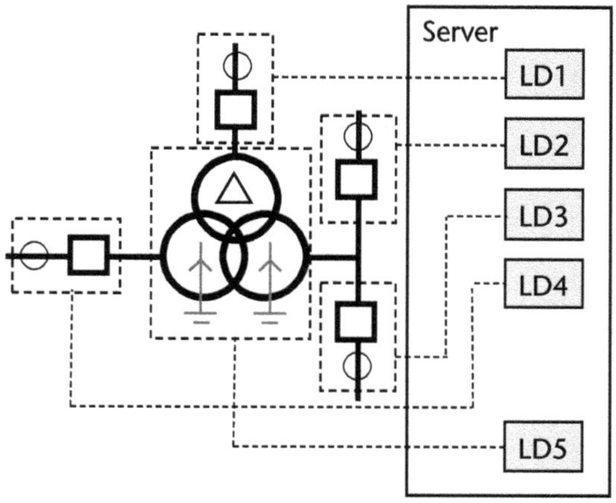

Figure 9.6 Server abstract model with multiple logical devices.

the standardization of logical nodes, it will create the foundation for the development of many applications with a standard interface as required by both users and vendors.

The model of such a device in IEC 61850 should reflect the functionality of more complex devices and can be done by mapping the different functions supported by the relay to different logical devices. This task is easier to achieve for distribution protection IEDs due to the fact that they are typically associated with a single breaker (i.e., they interface with a single set of current and voltage inputs).

One logical device (PROT) will represent the protection functions. Another will define the Control function (CTRL), and a third will define the Measuring function (MEAS). A fault locator and a circuit breaker monitor (if available) will be modeled with additional logical devices. A simplified block diagram of this model corresponding to its functional hierarchy is shown in Figure 9.7.

In the case of the transformer protection IED from Figure 9.6, each logical device, from LD1 to LD4, will have a nested functional hierarchy similar to the one shown in Figure 9.7.

If we go further down in the functional hierarchy from Figure 9.7, the protection logical device will include multiple protection functions. Each of these protection functions can be enabled or disabled. For example, the phase and ground overcurrent protection functions are typically enabled, while the negative sequence might be disabled. When a protection function is disabled, it means that all functional elements (logical nodes) included in it become disabled as well. This is one of the reasons that require the functional grouping of multiple logical nodes as described above and can be achieved by changing the Mod setting of LLN0 at the logical device level of the functional hierarchy.

The model in Figure 9.7 shows the hierarchy of the overcurrent protection in a transformer protection IED. The grouping of several instantaneous and time-delayed functional elements represented by the logical nodes PIOC (instantaneous overcurrent) and PTOC shows that they all can use the same directional function element represented by logical node RDIR. Each of these logical nodes has data object hierarchy as defined in IEC 61850.

The use of the nesting of logical devices as defined in IEC 61850 Edition 2 and corresponding to the object model from Figure 9.7 is shown in Figure 9.8. It is

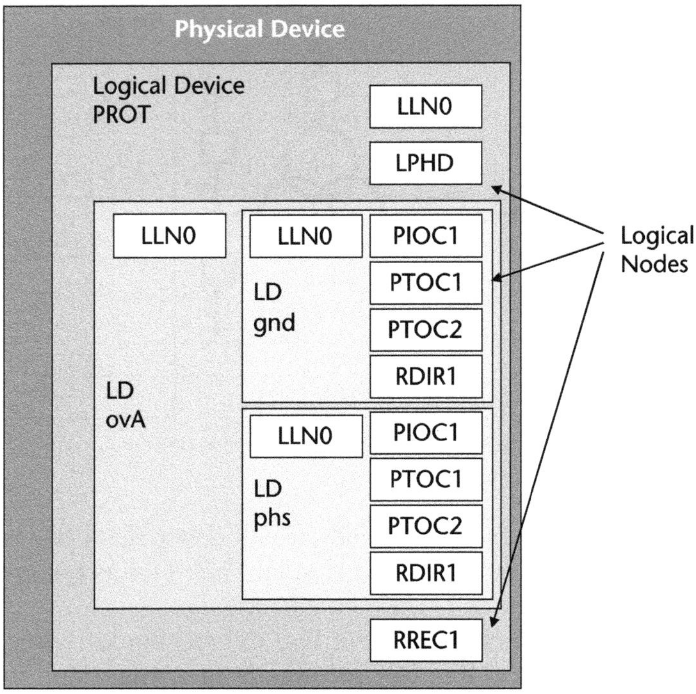

Figure 9.7 Logical device hierarchy: simplified object model.

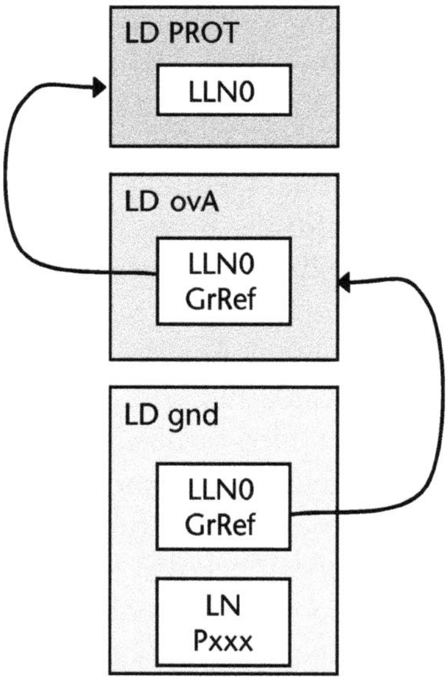

Figure 9.8 IEC 61850 LD management hierarchy example.

based on the group reference GrRef defined in IEC 61850 7-4 Edition 2 as "Reference to a higher-level logical device (LD). The Mod of this higher-level (referenced) LD influences the Beh of the LD and the Beh of all LNs in the LD where the GrRef is contained ..."

LLN0 of the nested logical device ovA in Figure 9.7 contains the setting data object GrRef whose value is PROT, meaning that the logical device function refers to the functional group represented by the parent logical device PROT. Similarly, LLN0 of logical device gnd refers to the functional group represented by the logical device ovA, indicating that the logical device gnd is a subfunction of logical device ovA.

One of the main advantages of the nested logical device hierarchy is that it allows the efficient control of the behavior of the logical devices and logical nodes in the functional hierarchy. If, for example, we set the mode of LLN0 of a logical device at a specific level of the hierarchy to off, this will disable all subfunctions and their function elements regardless of their original Mod setting values.

References

[1] IEC 61850-7-2:2020 Communication Networks and Systems for Power Utility Automation—Part 7-2: Basic Information and Communication Structure—Abstract Communication Service Interface (ACSI), Edition 2, Amendment 1, 2020.

[2] IEC 61850-7-3:2020 Communication Networks and Systems for Power Utility Automation—Part 7-3: Basic Communication Structure for Substation and Feeder Equipment—Common Data Classes, Edition 2, Amendment 1, 2020.

[3] IEC 61850-7-4:2020 Communication Networks and Systems for Power Utility Automation—Part 7-4: Basic Communication Structure—Compatible Logical Node Classes and Data Object Classes, Edition 2, Amendment 1, 2020.

CHAPTER 10

GOOSE Communications and Their Applications

10.1 Introduction

GOOSE messages are one of the most widely used components of the IEC 61850 standard and play a very important role in the digitization of today's PAC systems. The electric power system protection and control specialists are some of the most conservative because they need to ensure that the system will remain stable in case of short-circuit faults or other abnormal conditions. For more than a century, we have been using hardwired protection systems (Figure 10.1) and many people may be asking the question of why we should replace the wires with communication messages.

To implement a fairly simple hardwired protection scheme in the substation, for example, a breaker failure protection, it is necessary to:

1. Wire the relay output of a protection relay (for example, a distance protection) to the terminal block of the panel.
2. Wire the terminal block of the panel of a protection relay, initiating the breaker failure to the terminal block of the panel with the relay performing the breaker failure protection function.
3. Wire the terminal block of the panel of the relay performing the breaker failure protection function to its opto-input terminals.
4. Wire the relay output of the breaker failure protection relay to the terminal block of its panel to do the retrip.
5. Wire the terminal block of the panel of the breaker failure protection relay to the terminal block of the panel of the breaker to do the retrip.
6. Wire the relay output of the breaker failure protection relay to the terminal block of its panel to do the breaker failure trip (for example, through the bus differential protection).
7. Wire the terminal block of the panel of the breaker failure protection relay to the terminal block of the panel of the bus differential protection.
8. Wire the terminal block of the panel of the bus differential protection to its opto-input terminals.

Figure 10.1 Hardwired interfaces.

For each of the above steps, it is necessary to measure the pair of wires to be used and prepare them for connection with screws to the terminals of the relay outputs or inputs, as well as the terminals of the panels, and then to attach them to the terminals and make sure that there is a good connection.

The amount of wiring is multiplied in the case of a single pole trip and reclosing, breaker-and-a-half of ring bus schemes, requiring huge amounts of copper cables a lot of time to complete the work.

To accomplish all of the above with GOOSE, it is necessary to measure and prepare a fiber optic cable and connect it to the IED communications port and the communications port of the switch so that it can send GOOSE messages and interact with the substation clients.

The performance, reliability and availability, flexibility, or maintenance testing can all be improved using GOOSE. To take full advantage of the benefits that they provide, it is very important to understand the fundamentals behind its development, the structure of the messages, and the content of the data sets in order to support maintenance testing and cybersecurity.

This chapter describes the development of the GOOSE concept as part of the UCA 2 project and its evolution in the IEC 61850 standard.

It is very important to highlight that GOOSE messages are not commands. They represent unsolicited reporting of the change of state of a certain system parameter, and the decision of what action to take when received by a subscribing device depends on the application that is using the information contained in it.

Initially, GOOSE was intended to be used within the substation over the local area network, but, with the growing understanding of its benefits, there was a push to make it available outside of the substation to meet the needs of various smart grid-related applications. This led to the development of the routable GOOSE, which is described later in the chapter.

Once we understand the GOOSE concept and models, the remaining sections of this chapter focus on its applications both inside and outside of the substation.

10.2 GOOSE in UCA 2.0

The introduction of microprocessor-based protection relays and their evolution into multifunctional IEDs with communication interfaces resulted in significant efforts by the PAC community to improve the efficiency of their integration in substation automation systems. This led to the Electric Power Research Institute (EPRI)-sponsored development of UCA 2.0, and the Generic Object Models for Substation and Feeder Equipment (GOMSFE). This document not only introduced many of the modelling concepts later further developed in IEC 61850, but also proposed a solution to the communications requirements for protection applications.

We can consider the starting point of the GOOSE development the meeting of the "Chicago Seven + One" in May 1998 at the Courtyard by Marriott hotel, Wood Dale, Illinois. The participants (listed alphabetically by last name: Alex Apostolov, Kay Clinard, Herb Falk, George Schimmel, Mark Simon, Charles Sufana, John Tengdin, and Jim Whatley (via teleconference)) defined the main GOOSE requirements, data type, message structure, and communication principle that became part of the GOMSFE.

The definition of the GOOSE was based on the understanding of the critical role that protection devices play in maintaining electric power system stability when a short-circuit or other abnormal condition occurs. This means that any system needs to meet two key requirements: speed and reliability.

Considering that, at the time, the LANs that could be used in a substation were using hubs, which are exposed to collisions that may result in a failure to deliver a message when a fault occurs, it was necessary to find a way to ensure that the message will reach the recipients. To achieve this, a repetition mechanism (Figure 10.2) was introduced with high-speed retransmission (a few milliseconds) immediately following the event and with the retransmission interval increasing until it reaches a user-defined maximum value that can be multiple seconds.

To support this communication method, the UCA GOOSE message contains a state number that is incremented by 1 each time an IED sends information that is changed, indicating that the value of at least one bit pair has changed. This number and an associated time stamp can be used to identify an event and when it occurred.

Each GOOSE message also includes a sequence number, which is incremented by 1 each time a message is sent. Both the state and the sequence number roll over after the maximum count is reached.

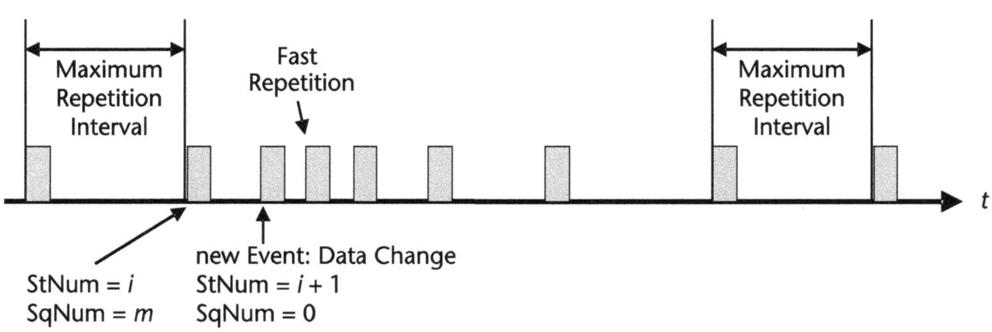

Figure 10.2 Repetition mechanism.

Because in many cases it may be necessary to send a message to multiple recipients, it was decided to use a multicast address as the destination address. The message includes a sending IED identifier that uniquely names the device reporting the GOOSE. When an IED is powered up, it is required to send its current status as an initial GOOSE message indicating to all other devices in the substation that it is available. The retransmission of the messages by an IED can be used to monitor its health.

The UCA GOOSE also contains the time since the last status change (back time), which can be used by a receiving IED to better implement some distributed protection and control functions. There is also a hold time used to ensure that the status conditions within the message remains valid and that a retransmission of the message has occurred before it expired.

As P2P communications are connectionless and there is no confirmation that the message has been received, during the development of the UCA GOOSE, there were a lot of discussions about how we know that the transmission was successful. The argument supporting this approach is that the sending device can monitor the change of state expected as the result of the transmission. For example, if a protection function has operated and a message was sent indicating that the breaker should be opened, the device monitoring the state of the breaker will send a GOOSE message that this change has occurred, thus confirming that the original message has been successfully received. It is clear that this approach requires development of programmable logic that can be used to verify that the transmission was successful, but this is the price that we pay when we are using high-speed communications that meet the requirements of protection applications.

At the same time, the requirements for the GOOSE message included a transfer time from the sending device application to the receiving devices applications of 4 ms. Because the fault-clearing time has an impact on the stability of the electric power system, it was not going to be acceptable to have a performance that is degraded in comparison to the existing protection systems. Some protection systems were using a high-speed auxiliary tripping relay introduced by Allmänna Svenska Elektriska Aktiebolaget (ASEA) in the mid-1970s, the RXMS, which was available with an operating time of just 4 ms and had contacts suitable for trip circuit duty. This is where the value of the required GOOSE transfer time came from.

Multifunctional protection devices at the time had multiple relay outputs providing signal exchange with other components of the protection system with each output being configured and wired to perform a specific task such as tripping or closing a circuit breaker, initiating auto reclosing or disturbance recording and many others.

In order to improve the efficiency of the communications interface between devices, the GOOSE message was designed to contain multiple data represented by bit pairs, with each one of them indicating a specific change of the state of certain signals. Some of them were predefined and grouped together in what was defined as DNA and included 32-bit pairs, with the first shown as an example in Table 10.1 [1].

Another group of 128-bit pairs was defined as user status and was used as needed by the user or manufacturer to provide additional information as required by the application.

Table 10.1 DNA Example for Protection Messages

		Bit Order	*00*	*01*	*10*	*11*
	Bit	*Value*	*0*	*1*	*2*	*3*
Bit #	*Pair*	*Definition*	*State*	*State*	*State*	*State*
0, 1	1	OperDev	Normal	Trip	Close	Invalid
2, 3	2	Lock Out	Invalid	Normal	LO	Invalid
4, 5	3	Initiate Reclosing	Normal	Cancel	Auto Reclosing	Invalid
6, 7	4	Block Reclosing	Normal	Cancel	Block	Invalid
8, 9	5	Breaker Failure Initiate	Normal	Cancel	Initiate	Invalid
10, 11	6	Send Transfer Trip	Normal	Cancel	Set	Invalid

One of the most amazing things in the development of the UCA GOOSE was the fact that both manufacturers and some users realized the huge potential of this technology and took a very active role by dedicating resources and participating at numerous meetings and plugfests before it was defined as a standard. The development of devices that supported GOOSE by different manufacturers and their willingness to participate in interoperability testing in the presence of utility representatives as witnesses brought significant benefits, because it allowed identifying issues with the GOOSE development that could be discovered only during real-life testing.

Many of the lessons learned during these initial stages of development were later used in the definitions of the IEC 61850 GOOSE.

10.3 IEC 61850 GOOSE

Because several key members of IEC TC 57 Working Groups 10 and 11 were also actively involved in the development of the GOMSFE, it was logical to expect that the GOOSE is going to find its place in the IEC 61850 standard. In the meantime, there have already been implementations of the UCA GOOSE in multifunctional IEDs by several manufacturers, as well as a few substation installations using it.

This experience was very useful in the further evolution of the GOOSE communications as part of IEC 61850. When IEC 61850-7-2 was published in 2003 [2], it included the Generic Substation Event (GSE) class model, which, in principle, was based on the UCA GOOSE model of P2P communications between publishers and subscribers (Figure 10.3).

However, it included two control classes and the structure of two messages:

- The Generic Substation State Event (GSSE) was actually the UCA GOOSE and was included to provide backward compatibility with existing devices and systems based on the UCA GOOSE bit pairs to represent the change of state.

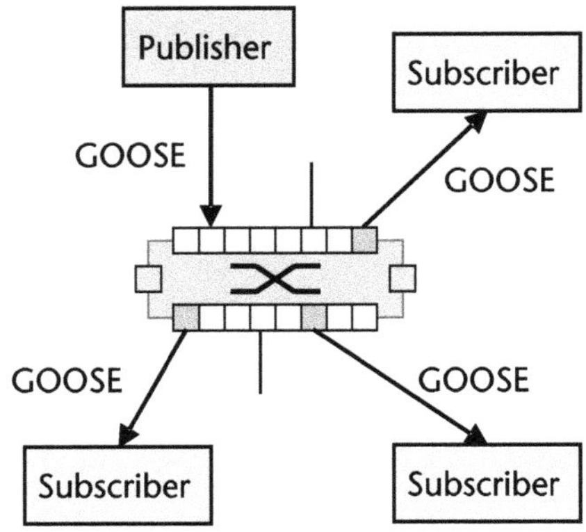

Figure 10.3 Publisher-subscriber communications.

- The second one, the GOOSE, supports the exchange of a wide range of different types of data available from the logical nodes that is organized in a user or manufacturer data set.

The decision to call the UCA GOOSE GSSE was very controversial, and we had many discussions in the working group if this was the right thing to do. However, this change actually makes a lot of sense. The IEC GOOSE is really object-oriented and may contain as members of the data set different data objects or their attributes to meet the requirements of many different applications.

Figure 10.4 shows a simplified block diagram of the GOOSE information exchange model using the publisher-subscriber mechanism. The publisher writes in a local buffer the values for the data objects and attributes defined in the data set to be sent in the GOOSE message over the substation network. The control of GOOSE communications is at the logical device level by LLN0, which contains the

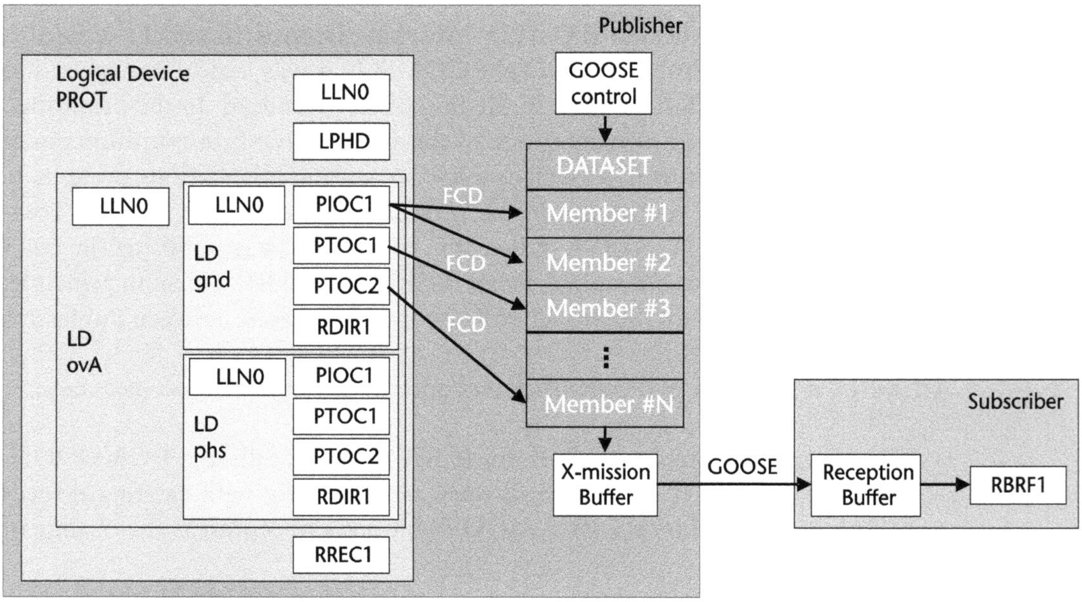

Figure 10.4 IEC 61850 GOOSE model.

GOOSE Control Block GoCB. The GOOSE Control Block class includes attributes that define the behavior of the P2P communications:

- *GoCBName* (GOOSE control name) identifies a GoCB within the scope of a GoCBRef (GOOSE control reference), a unique path-name of a GoCB within LLN0:
    - LDName/LLN0.GoCBName
- *GoEna* (GOOSE enable) indicates that the GoCB is enabled (if set to TRUE) to send GOOSE messages. If set to FALSE, it shall stop sending GOOSE messages.
- *AppID* is an application identification represented by a visible string that represents a logical device in which the GoCB is located.
- *DatSet* is the reference of the data set whose values of members shall be transmitted.
- *ConfRev* is the configuration revision indicating the number of times that the configuration of the data set referenced by DatSet has been changed. The counter is incremented every time when the configuration changes. This allows a subscriber to check that the expected information is being sent.
- *NdsCom* (needs commissioning) is TRUE if the attribute DatSet has a value of NULL and is used to indicate that the GoCB requires configuration.

Edition 2 of the standard [3] includes in GoCB also an attribute *DstAddress*, which provides SCSM-specific addressing information such as media access address, priority, and other information.

If the value of at least one of the data attributes in the data set has changed, the transmission buffer of the publisher is updated with the local service "publish" and the values are transmitted with a GOOSE message.

Once a new value of a date attributed has resulted in the multicasting of a new GOOSE message, the repetition mechanism ensures that the message is sent with a changing time interval between the repeated messages until a new change event occurs. As shown in Figure 10.2, at the beginning after a change, the interval is very short, a few milliseconds, which later increases until it reaches a value of a few seconds. This method achieves several important tasks:

- Ensures that a loss of a single message is not going to affect the functionality of the system;
- Allows any new device to inform all subscribing devices about its state;
- Allows any new device to learn the state of all publishing devices to which it subscribes;
- New subscribers can be added to the system without the need for total system reconfiguration.

The content of the GOOSE message allows the receiving devices to perform processing of the data in order to execute required actions. Some of the attributes in the GOOSE message that help perform the functions are:

- *StNum* indicates the current state number, a counter that increments every time a GOOSE message (including a changed value) has been sent for the first time. The initial value is 1.
- *T* is the time stamp representing the time at which the attribute StNum was incremented.
- *SqNum* is the sequence number, the value of a counter that increments each time a GOOSE message with the same values has been sent. The initial value is 1.
- *Test* is a parameter that indicates that the GOOSE message is used for test purposes (if the value is TRUE) and that the values of the message shall not be used for operational purposes.

The parameter *Test* in Edition 1 was replaced in Edition 2 by a parameter *Simulation*, which, when set to the value TRUE, indicates that the GOOSE message and its values have been published by a simulation unit for testing purposes. The GOOSE subscriber will report the value of the simulated message to its application instead of the real message depending on the setting of the receiving IED. The allowance for an IED to switch from acceptance of real messages to simulated messages is specified by a data object defined in IEC 61850-7-4.

The basic concept described above applies also to the GSSE model and is similar, but there are a couple of major differences:

- GOOSE provides flexibility in the definition of a data set with different data types, while GSSE provides only a simple list of status information represented by bit pairs.
- While the mapping of GOOSE to IEC 61850 8-1 supports VLAN and priority tagging, these are not available in GSSE messages.

A good example of the importance of the ability to include different data types as members of a GOOSE data set is what is known as an analog GOOSE. There are many protection and control applications that require monitoring and reporting of changes in an analog system parameter, such as the loading of a distribution feeders. Because such a value will be needed by an IED, which may not have client capabilities, it may subcribe to an analog GOOSE message, which will deliver a new value of the monitored parameter. A new analog GOOSE (with an incremented state number) is sent every time when the value of the parameter gets out of the deadband shown in Figure 10.5.

The deadbanded value calculation comes from the instantaneous value of the parameter (modeled as *instMag*) as shown. The value of *mag* is updated to the current instantaneous value when the value has changed outside of the deadband according to the configuration parameter *db*.

If db = 0, the value of mag is identical to the value of *instMag*. This is definitely not a good idea, especially for fast changing system parameters, because it will result in the continuous publishing of new GOOSE messages.

Specific communication services in the subscribers update the content of their reception buffers and processes them based on the values of the parameters de-

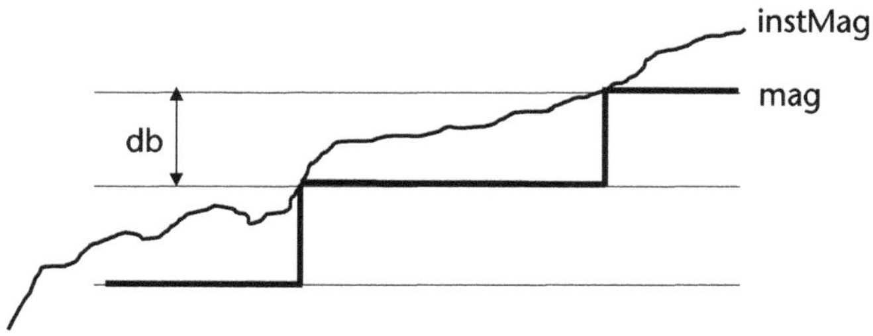

Figure 10.5 Deadband for analog GOOSE.

scribed above. If the state number has been incremented, it indicates that there are new values received to the related applications.

The Ethernet is used for substation communications based on the IEC 61850 standard. The Ethernet in this case is not the hub-based collision detection protocol from the time when the UCA GOOSE was invented; it is the advanced, high-speed protocol of today. Perhaps the most important advancement in Ethernet networks is the use of full-duplex switched Ethernet that replaced the shared medium of legacy Ethernet with a dedicated segment for each IED.

The performance of protection and other distributed functions is further improved through the availability of priority tagging defined, while improvements in security are the result of using a virtual LAN (VLAN).

Because priority code point (PCP) is a 3-bit field, which refers to the IEEE 802.1p priority, it indicates the frame priority level from 0 (lowest) to 7 (highest) and can be used to prioritize different classes of traffic such as GOOSE.

The Ethertype 2 byte code indicating protocol type in an Ethernet packet uses specific values related to IEC 61850 GSE as follows:

- IEC 61850 8-1 GOOSE: 88-B8;
- IEC 61850 8-1 GSE Management: 88-B9.

The use of switched Ethernet and priority tagging allows the transfer time for the messages to meet the high-speed requirements of time-critical protection functions.

The IEC 61850 transfer time definition [4] is based on Figure 10.6, where:

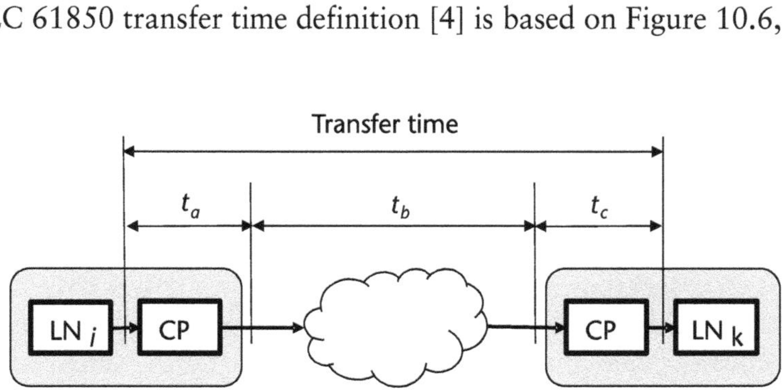

LN: Logical Node
CP: Communication Processor

Figure 10.6 Transfer time definition.

- ta: The time from the moment the sending IED puts the data content on top of its transmission stack until the message is sent on the network;
- tb: The time over the network;
- tc: The time from the moment the receiving IED gets the message from the network until the moment that it extracts the data from its transmission stack.

The overall performance requirements also depend on the message type. Type 1 is defined in the standard as Fast Messages. Because Trip (Type 1A) is the most important fast message in the substation, it has more demanding requirements of 3 ms compared to all other fast messages. The same performance may be requested for interlocking, intertrips, and logic discrimination between protection functions.

The success of using GOOSE messages for substation protection applications makes it attractive for use between IEDs in wide area protection systems. These are applications that impose different requirements on the P2P communications. Although GOOSE messages have already been used in protection and automation applications outside of the substation, the fact that they are transmitted over wide area networks makes them vulnerable to cyberattacks.

In order to address such concerns, the IEC TC 57 Working Group 10 developed the technical report IEC 61850 90-5 [5], which defined the routable communications of synchrophasors and GOOSE (R-GOOSE) over wide area networks. The communications are based on the full 7-layer OSI stack and use UDP multicast. The document also describes the use of end-to-end cybersecurity based on the definitions in the IEC 62351 standard.

10.4 GOOSE Applications to Adaptive Distribution Protection

Requirements for reduction in the duration of short-circuit faults on distribution feeders and substation equipment are based on significant increase in the numbers of DERs and customers with loads sensitive to voltage variations. High-speed fault clearing for different faults on distribution feeders and substation can be achieved by advanced communications-based protection schemes, which adapt to changes in their environment in order to improve their sensitivity and performance.

10.4.1 Adapting to Changes in Substation Configuration

There are many cases when the performance of the protection and control IED is affected by the substation or power system configuration. With conventional relays, for example, it is very difficult to adjust to changes in the grounding of the system at the relay location.

If the relay (feeder PIED in Figure 10.7) is located in a substation with two transformers, one grounded on the high side and the other on the low side, taking one of the transformers out of service will significantly affect the levels of zero sequence current seen by the ground overcurrent protection functions. If this is the transformer grounded on the distribution side (T1), the distribution system will

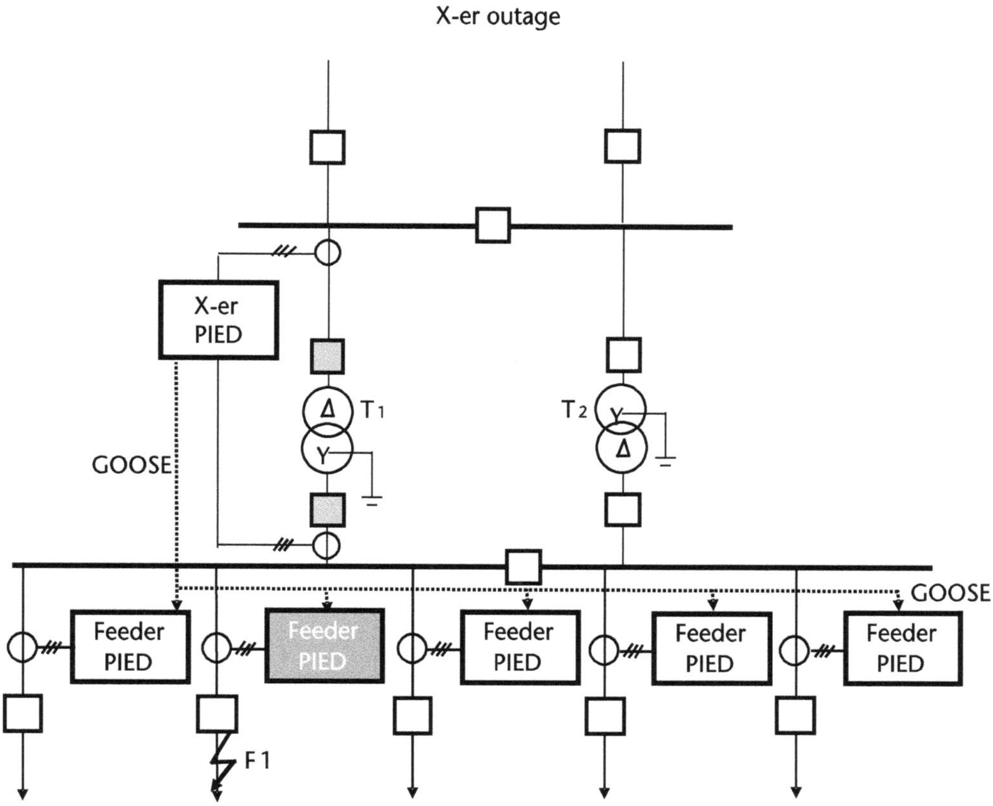

Figure 10.7 Protection IED with transformers with Δ/Y and Y/Δ transformers behind it.

change from solidly grounded to isolated, which will require significant changes in the protection.

The levels of fault currents for phase faults will also change with one of the transformers out of service. This requires monitoring of the state of the transformers and adapting to the new conditions as soon as one of the transformers is taken out of service. The changes will be different as a function of which transformer has been taken out of service.

To adapt the settings of all the distribution feeder relays to the changes in the system topology, the transformer protection IED will detect the tripping of transformer T1 and will send a GOOSE message indicating that this has occurred. All the feeder protection IEDs subscribe to this message and when it is received with values indicating the change, they will switch to the appropriate setting group, which will ensure their optimal sensitivity under the new conditions.

10.4.2 Adapting to Faults on Adjacent Feeders

The changes of fault conditions in the distribution system impact not only the sensitive loads, but also, depending on the load, may lead to the operation of protection elements of multifunctional relays on healthy feeders.

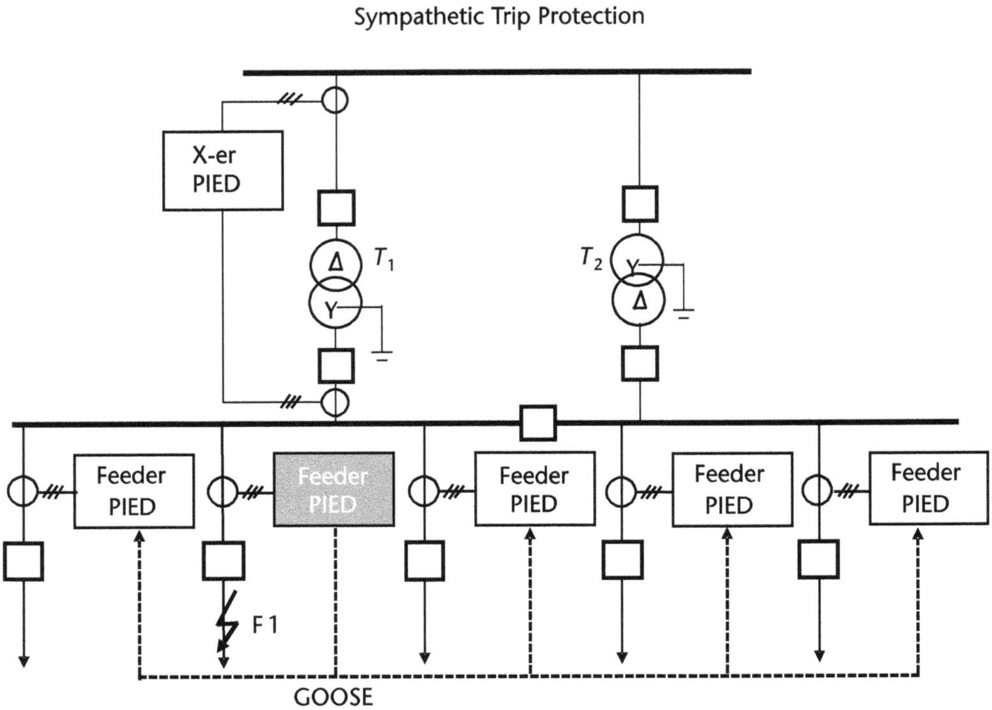

Figure 10.8 Sympathetic trip protection.

Detecting the operation of a relay on an adjacent feeder can be used to adjust the sensitive settings of the relays on the healthy feeders for the duration of an inrush condition following the clearing of a fault in a distribution system with a significant number of motor loads. This is known as a sympathetic trip scheme. As soon as a relay detects a fault on the feeder that it is protecting, it sends a signal to all other relays informing them to expect an inrush as a result of the voltage recovery following the clearing of the fault.

Each of the relays on the healthy feeders then adapts its settings for the period of time that the expected inrush condition is going to last. Two options are usually available:

- Block the sensitive overcurrent setting.
- Reduce the sensitivity by increasing the pickup setting for the duration of the inrush.

The challenge with implementing this scheme in conventional hardwired substations is that, depending on the number of feeders, there may not be enough relay outputs available in the IED on the faulted feeder to send the signal to all adjacent relays. This is where the GOOSE message provides a significant benefit because, at the moment when the IED on the faulted feeder detects the fault condition, it will send the GOOSE message indicating the fault detection and all the adjacent relays will subscribe to this message and adapt their behavior accordingly.

10.4.3 Adapting to the Loss of Protection IED

The common approach that many utilities have taken is to use a single protection IED on a feeder. In the case of failure of this relay, faults on the line are cleared by the backup overcurrent protection on the transformer or sectionalizing breaker.

The problem with this approach is the long fault-clearing time that may affect sensitive loads fed by the distribution substation. A solution that significantly reduces the duration of the fault is based on the adjustment that the backup relay can make in its decision to trip based on the knowledge that a specific IED has failed.

This adaptive form of protection uses the normally closed contacts of the feeder relays that close when the relay is not healthy. When the transformer of sectionalizing breaker relay sees a fault and does not get any blocking signal from any of the feeder relays, it knows that there are two possible cases:

- The fault is on the feeder with the failed relay.
- The fault is on the distribution bus.

The operation of this adaptive protection scheme is shown in Figure 10.9. Because the probability for a fault on a distribution feeder is much higher than the probability for a distribution bus fault, the relay first sends a signal (1) to trip the breaker of the failed relay. If this does not clear the fault, then it is clear that the fault is on the bus and it is cleared by tripping the source breakers using signal (2).

This scheme can be implemented in conventional hardwired substations with a limited number of feeders. However, in substations with a larger number of feeders,

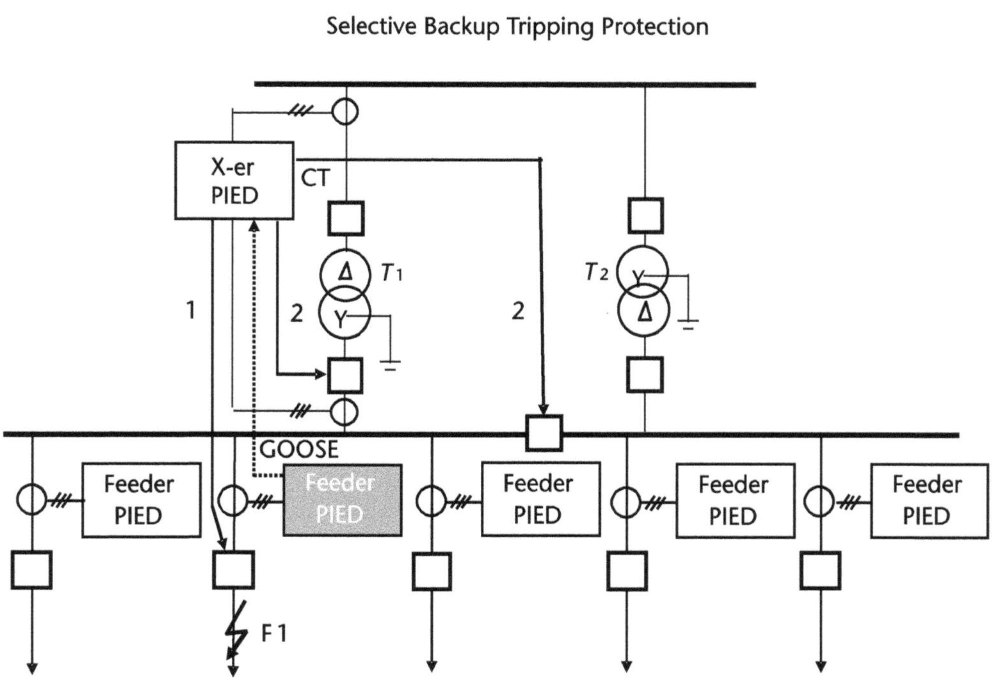

Figure 10.9 Selective backup tripping.

this will become practically impossible because the transformer protection device needs to have multiple input dedicated to the signals from each of the feeder relays in order to be able to understand which one is the one that has failed.

This is where again GOOSE messages can help. The transformer protection device will have to subscribe to GOOSE messages from each of the feeder relays and, when it detects that it is not receiving messages from a specific feeder relay and there is no indication of any communications failure, there is a high probability that the loss of messages is due to the failure of the publishing IED. Based on this knowledge, the transformer protection device can implement the above-described logic even in a substation with a very large number of feeders.

10.5 GOOSE Applications to Transmission Line Protection

Traditionally accelerated protection schemes have been used on transmission lines where longer fault clearing times may have an impact on the dynamic stability of the electric power grid. One of the challenges with their implementation has been the cost of the communications equipment, limiting their use only to critical locations.

Short-circuit faults at the transmission level of the electric power system result in voltage drops that are also experienced at the distribution level and may impact the ride through capability of the DERs connected to distribution feeders. That is why today we need protection schemes that will reduce the fault duration at the transmission level in a large number of cases. This can be accomplished by implementing accelerated protection schemes based on communications between the relays at both ends of the protected line, while at the same time solving the problem with the communications channel cost.

We can consider as an example of an accelerated transmission line protection scheme a permissive directional comparison scheme commonly used to accelerate the clearing of all kinds of faults, including high-resistance faults that are not seen by the distance elements of the transmission line protection relays or line differential relays.

The channel for a permission directional comparison scheme is keyed by operation of the forward-looking elements of the relay represented in the IEC 61860 model by the logical node RDIR.

If the fault direction seen by it is forward:

RDIR.Dir.dirGeneral = forward

A GOOSE message with the values for the data set containing data object *RDIR.Dir* or the data attribute *RDIR.Dir. dirGeneral* is sent to the substation at the other end of the transmission line. If the remote relay has also detected a forward fault upon receipt of this signal, the protection scheme represented by PDIR will operate. Such schemes offer some significant advantages, especially when high-speed directional detection methods based on superimposed current and voltage components are used.

Permissive schemes tend to be more secure than blocking schemes because forward directional decisions must be made at both ends of the line before tripping is allowed. Failure of the signaling channel will not result in unwanted tripping, because no signal is going to be received and the relay does not trip based on a forward directional detection only.

If the source at either end of the line is weak as in systems with large penetration of DERs, the permission directional comparison scheme uses weak infeed logic.

Current reversal guard logic is used to prevent healthy line protection maloperation for the high-speed current reversals experienced in double circuit lines, caused by sequential opening of circuit breakers.

If the signaling channel fails, the basic distance scheme tripping will be usually available.

The challenge for the implementation of accelerated transmission line protection schemes is that they require a communications channel, which, if it is a dedicated one, will add to the costs.

IEC 61850 GOOSE messages are the technology that can help us significantly reduce the fault clearing time and the accelerated transmission line protection scheme cost by using R-GOOSE (Figure 10.10).

Because the message goes through a wide area, the transfer time will be nondeterministic. However, the results of numerous measurements have shown that this time is in the range of 15 to 40 ms, which is comparable to the times of many traditional communication channels used in the past. Even if this delay adds 50 to 60 ms to the fault-clearing time, it is still much better than the time delay of zone 2, which is typically 300 to 400 ms, without the additional costs of a dedicated communications channel.

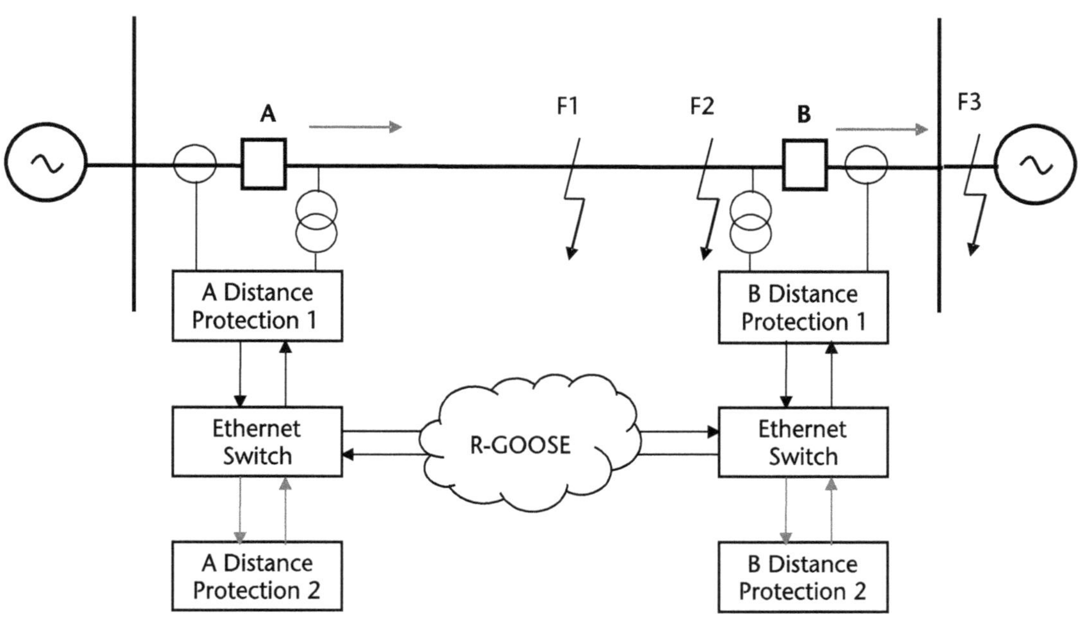

Figure 10.10 Accelerated protection scheme using R-GOOSE.

10.6 GOOSE Applications to System Integrity Protection Schemes

System integrity protection schemes (SIPS) are distributed applications based on exchange of information and control signals between IEDs located in different substations or feeders throughout the electric power system.

SIPS can be considered as systems that have three main types of functional elements:

- System monitoring elements;
- Protection elements;
- Execution elements.

The function of the system monitoring elements is to detect a change in the electric power system topology, system load, or generation.

Any information about changes is delivered between the components of the distributed SIPS using R-GOOSE messages containing status information or analog data.

Load-shedding is one of the main methods used by SIPS to prevent the spread of a disturbance over a wide area. Centralized load shedding can be used in substations when it is executed as the result of a message recieved from the regional SIPS or from the activation of the thermal or other transformer overload protection scheme as shown in Figure 10.11. This is done in this case in an attempt to prevent the tripping of the transformer due to the overload condition and creating a blackout in the distribution system.

The load-shedding is based on the coordinated operation of the two main components of such load-shedding scheme:

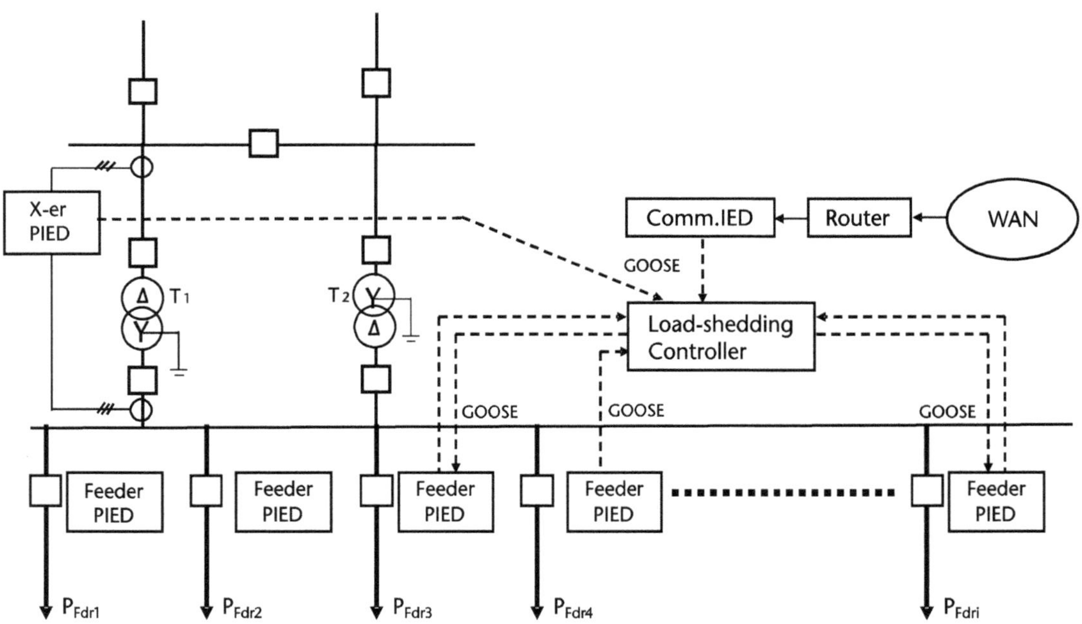

Figure 10.11 Adaptive load-shedding.

- Transformer protection IED;
- Load-shedding controller.

The role of the transformer protection IED is to determine the level of overloading of the transformer and calculate the amount of load that needs to be shed in order to bring the transformer temperature down to an acceptable level. Once the required amount of load to be shed is determined, the transformer protection IED sends a GOOSE message to the load-shedding controller responsible for its execution.

The load-shedding controller receives GOOSE messages from the distribution feeder relays every time there is a change in the load outside of the user-defined deadband setting. Based on the definitions in IEC 61850-7-3 [6] and IEC 61850-7-4 [7], the deadband value of the arithmetic average of the magnitude of active power of the three phases (Wa, Wb, Wc) *MMXU1.AvWPhs.mag* can be used.

The load-shedding controller uses this information to determine an optimal combination of loads to be shed based on the optimization criteria.

In the case of a sharp increase in the loading of the power transformer, the transformer protection IED sends a GOOSE message over the substation LAN to the substation controller to shed a specific amount of load. This amount of load is matched with the possible groups of feeder loads and a GOOSE message is sent to the network indicating which feeders need to be tripped in order to shed an amount of load as close as possible to the required value.

Because a centralized load-shedding system is subject to the failure of the load-shedding device, a second backup IED may be required to ensure the reliability of operation in the case of a substation event requiring the load-shedding.

10.7 GOOSE Benefits

The GOOSE model described above allows the development of protection and control systems that offer some significant advantages compared with conventional hardwired systems:

- Ability to exchange analog information between multifunctional IEDs using analog GOOSE messages;
- Support for the implementation of accelerated protection schemes using routable GOOSE messages over wide area networks instead of dedicated communication channels;
- Reduced wiring costs due to the replacement of numerous control cables with a limited number of fiber optic cables;
- Reduced commissioning costs due to the replacement of physical test switches with virtual isolation;
- Improved flexibility of the system due to the use of virtual signals described in a standard substation configuration language format;

- Reduced maintenance due to the fact that the state of the different components of the system and the interfaces between them can be continuously monitored;
- Improved interoperability due to the use of standard high-speed communications between devices of different manufacturers over a standard communications interface;
- Ability to test a subset of functions and their elements while keeping the rest of the system in operation;
- Support for the implementation of remote testing of GOOSE-based protection and control functions.

All of the above can provide good arguments to support the case for digitization of PAC systems that should convince even the most conservative specialists in the field.

References

[1] IEEE—SA TR 1550—1999 Technical Report on Utility Communications Architecture Version 2.0—Part 4: UCA Generic Object Models for Substation and Feeder Equipment (GOMSFE), 1999.

[2] IEC 61850-7-2:2003 Communication Networks and Systems in Substations—Part 7-2: Basic Communication Structure for Substation and Feeder Equipment—Abstract Communication Service Interface (ACSI), 2003.

[3] IEC 61850-7-2:2010 Communication Networks and Systems for Power Utility Automation—Part 7-2: Basic Information and Communication Structure—Abstract Communication Service Interface (ACSI), Edition 2, 2010.

[4] IEC 61850-5:2003 Communication Networks and Systems in Substations—Part 5: Communication Requirements for Functions and Device Models, 2003.

[5] IEC TR 61850-90-5:2012 Using IEC 61850 to Transmit Synchrophasor Information According to IEEE C37.118, 2012.

[6] IEC 61850-7-3:2010 Communication Networks and Systems for Power Utility Automation—Part 7-3: Basic Communication Structure for Substation and Feeder Equipment—Common Data Classes, Edition 2, 2010.

[7] IEC 61850-7-4:2010 Communication Networks and Systems for Power Utility Automation—Part 7-4: Basic Communication Structure—Compatible Logical Node Classes and Data Object Classes, Edition 2, 2010.

CHAPTER 11

Sampled Value Communications and Their Applications

11.1 Introduction

Sampled value messages are another major component of the IEC 61850 standard that makes it one of the main tools available to the industry for the digitization of today's PAC systems.

The replacement of the copper cables connecting the secondary side of the instrument transformers to the protection and control panels in the substation control house with optical cables carrying communications messages is also difficult to accept for many people that have no experience using such innovative technology. So again, it is a good idea to have a quick look at why we should consider making this change.

Similar to the hardwired binary signal interfaces, we can think about what we need to do to bring secondary current and voltage signals to the analog input of a multifunctional IED in a conventional substation and what are the potential issues resulting from it.

- First, we need to wire the secondary of the instrument transformers to the terminal block of the panel of a protection IED in many cases covering tens or even hundreds of meters with copper cables, as shown in Figure 11.1.
- Then we need to wire the terminal block of the panel with the analog input of the IED that needs these signals.
- If multiple devices need the same analog signal, for example, the bus voltage, we need to wire the terminal block of the panel that directly connected to the secondary of the instrument transformer to the terminal blocks of all the panels with IEDs that need that signal.
- Then we need to wire the terminal blocks of all these panels with the analog inputs of the IEDs that need these signals.

We should not forget that on each of the panels we also have test switches that need to be wired as well.

Figure 11.1 Substation yard to control house cables.

For each of the above steps, we need to measure the cables between the secondary of the instrument transformers and the control house, as well as the pairs of wires that we are going to use between panel terminal blocks, test switches, and IEDs. We have to prepare them so that we can connect them with screws to the terminals of the IED input, as well as the terminals of the panels. Then we need to attach them to the terminals and make sure that we have a good connection.

Now let us look at what it will take to accomplish all of the above with sampled values:

- We need to measure and prepare a fiber optic cable and connect it to the merging unit in the substation yard and the communication switch in the control house.
- We also need to measure and prepare the communication cables and connect the IED communications port and the communications port of the switch so we can send sampled value messages.

It is clear that installing a sampled value-based system is much more efficient than the traditional wiring.

The hardwired current circuits, in many cases, are challenging to substation and protection applications for other reasons that highlight the need for their digitization. First, we need to consider one of the main safety-related problems: an open current circuit condition. If this occurs while the primary winding is energized, the induced secondary electromagnetic field (e.m.f.) under these circumstances can be high enough to present a danger to people's health or even life.

One of the biggest challenges to some protection applications is current transformer (CT) saturation. Traditionally, the CT knee-point voltage is a function of the resistance of the different components of the current circuit:

$$V_K = f(R_{CT}, R_L, R_{RP})$$

where:

V_K = required CT knee-point voltage (volts);

R_{CT} = resistance of the current transformer secondary winding (ohms);

R_L = resistance of a single lead from relay to current transformer (ohms);

R_{RP} = impedance of a relay phase current input.

R_L plays a key role in determining the CT requirements, especially in large substations. If it is eliminated by digitizing the current signal by connecting directly to the secondary of the CT, we can practically eliminate CT saturation.

Because current circuits cannot be easily switched due to open-circuit concerns, the application of bus differential protection, as well as some backup protection schemes, becomes more complicated, requiring taking an outage, which may have economical and reliability implications.

Open CT circuit is also an issue during testing, so it requires the proper use of test switches when connecting test equipment to a test object in an energized substation.

Last, but not least, we need to think about reliability. If there is a problem with the CT circuit, it will have an impact on the protection function's performance, which will require their disabling.

All of the above demonstrate the challenges that conventional CT circuits present to PAC systems and how they can disrupt grid operations.

To take full advantage of the benefits that digitizing the analog circuits provides, it is very important to understand the fundamentals behind its development, the structure of the messages, and the content of the data sets in order to support efficient protection operation, maintenance testing, and cybersecurity.

In the following sections, we first describe the development of the sampled values concept as part of the IEC 60044-8 and its evolution in the IEC 61850 and IEC 61869 standards.

Initially, sampled values were intended to be used within the substation over the LAN, but with the growing understanding of its benefits, there was a push to make it available outside of the substation to meet the needs of various smart grid-related applications. This led to the development of the routable sampled values, which are described later in the chapter.

Once we understand the sampled value communications concept and models, the remaining sections of this chapter focus on their applications for protection and disturbance recording.

11.2 How Sampled Values Were Developed

We can consider the beginning of the digitization of the electric power system the groundbreaking paper by George Rockefeller, "Fault Protection with a Digital Computer," published in April 1969 in the *IEEE Transactions on Power Apparatus and Systems*. This introduced the concept of sampling voltages and currents and performing mathematical calculations on the individual sampled values to perform different protection functions in a single substation computer for all protection. This was followed more than a half-century ago by the installation in February 1971 of a Prodar 70-based system installation and service in the PG&E Tesla Substation 230-kV control house, connected to protect the Tesla-Bellota 230-kV line.

The evolution of microprocessor technology resulted in the development of digital relays by several global protection manufacturers in the early 1980s and their acceptance by utilities all over the world.

To better understand the principles of sampled value communications in digital substations, we need to look at the digitization of the analog signals in multifunctional IEDs.

Microprocessor-based relays are the typical protection devices in substations around the world. They are connected with hard wires to the voltage and current transformers and the auxiliary contacts and trip coils of the breakers.

The analog signals go through several transformations before they can be processed by the different elements in the protection module.

Once a fault or other abnormal condition has been detected and a decision to trip has been made, the protection module sends a command for the relay output module to close the required contacts and trip the breaker to clear the fault. Figure 11.2 shows a very simplified block diagram of a multifunctional protection IED connected to the substation primary equipment.

The analog input module provides the digitization interface between the protection IED processor board(s) and the voltage and current signals coming into the relay. This input module may consist of one or more boards. The number of current and voltage inputs depends on the primary protection function of the device. The input transformers are used both to step down the currents and voltages to levels appropriate to the relay's electronic circuitry and to provide effective isolation between the device and the power system. The connection arrangements of both

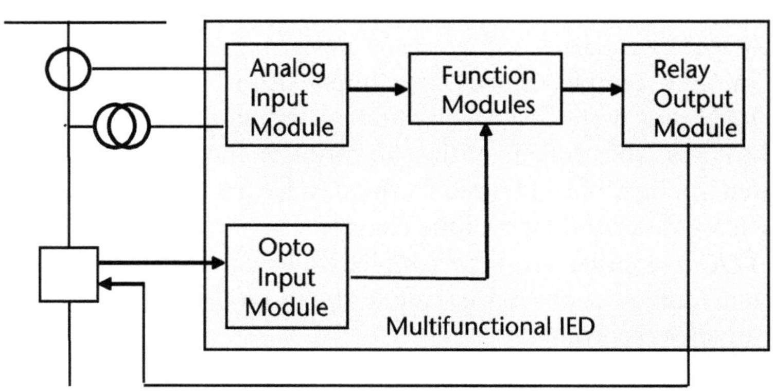

Figure 11.2 Simplified protection IED block diagram.

the current and voltage transformer secondary circuits provide differential input signals to the input board to reduce noise.

An analog input board is shown as a simplified block diagram in Figure 11.3. It provides the circuitry for the A/D conversion for the analog signals. Hence, it takes the differential analog signals from the current and voltage input transformers and converts them into digital sampled values and transmits the samples to the protection and other function modules via the digital data bus. On the input board, the analog signals are passed through an antialiasing filter before being multiplexed into an A/D converter chip. The A/D converter provides a sampled data stream output.

The sampling is typically based on a sample-hold mechanism that ensures that the samples on all channels are taken at the same moment in time. The signal multiplexing arrangement depends on the number of analog signals and provides for multiple analog channels to be sampled. The sampling rate for protection applications is usually maintained at a fixed number of samples per cycle of the power waveform by a logic control circuit which is driven by the frequency tracking function of the device. A calibration memory (usually an E2PROM) holds calibration coefficients that are used by the processor board to correct for any amplitude or phase error introduced by the transformers and analog circuitry.

This is one of the key differences that needs to be understood when analyzing the IEC 61850 communications: the fact that the analog signal's digitization is controlled by each individual IED with a specific sampling rate and in many cases with frequency tracking.

The challenges with hardwired current circuits mentioned in the previous section resulted in the development of nonconventional current transformers later in the last century. Several new methods of transforming the primary measured

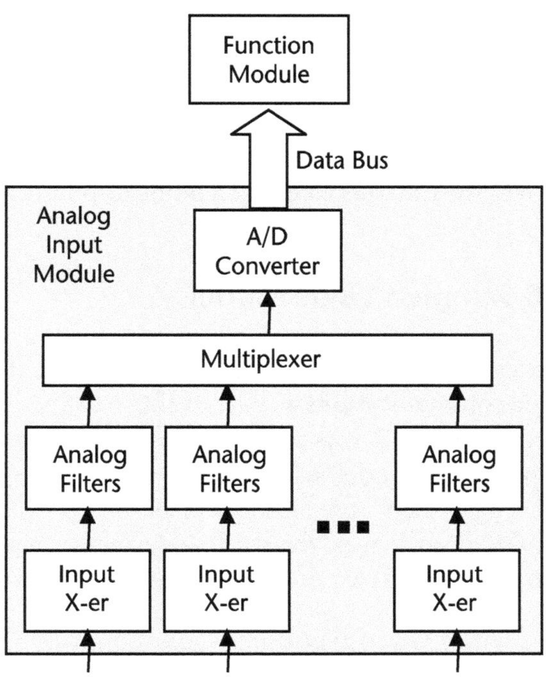

Figure 11.3 Simplified block diagram of an analog input module.

quantity are using optical and other methods and offer some significant advantages. Nonconventional optical and other transducers are usually smaller, lighter devices with an overall size and power rating of the unit that does not have any significant impact on the size and the complexity of the sensor. Small, lightweight insulator structures allow optical sensing devices to become an integral part of the insulator. One of the main advantages of this new sensing technology is that it provides an accurate representation of the primary currents and voltages over a wide range, thus meeting the requirements of both metering and protection applications.

The fundamental difference between an instrument transducer and a conventional instrument transformer is the electronic interface needed for its operation. This interface is required both for the sensing function and for adapting the new sensor technology to that of the secondary output currents and voltages. This introduces some problems due to the multiple A/D and digital-to-analog conversions. The introduction and widespread use of microprocessor-based protection devices, combined with the advancements in nonconventional instrument transformers, resulted in the development of digital interface between the censors and the IEDs. Digital interface in a point-to-point communications scheme was defined by the IEC in the IEC 60044-8 standard [1] by the IEC Technical Committee 38. The development of the concept of merging units that convert the optical signal into a digital message containing sampled values and protection devices with a digital interface that perform multiple protection functions resulted in demonstration projects that showed the advantages of this technology and laid the foundation for the development of the sampled value communications in the IEC 61850 standard.

The development of the sampled value communications in IEC 61850 was the task of TC 57 Working Group 12 Communication Standards for Substations: Communication Within and Between Process and Unit Levels. In Edition 1 of the standard, it developed two different versions for sampled value communications in Part 9-1 [2] and Part 9-2 [3].

IEC 61850-9-1 defined the specific communication service mappings for the communication between bay and process level on a serial unidirectional multidrop point-to-point link in accordance with IEC 60044-8. The frame format from IEC 60044-8 was reused, but the new standard defined the transmission of sampled analog values over the Ethernet in either a point-to-point (unicast) or multicast mode.

11.3 IEC 61850 Sampled Value Model

This experience with IEC 60044-8 was very useful in the further evolution of the sampled value communications as part of IEC 61850.

When IEC 61850-7-2 was published in 2003 [4], it included the transmission of sampled value class model, which, in principle, is also based on P2P communications between publishers and subscribers as shown in Figure 11.3, but instead of publishing GOOSE messages, the sending device is streaming sampled values.

However, it included two control classes and the structure of two messages:

- Transmission of sampled values using multicast;
- Transmission of sampled values using unicast.

Figure 11.4 shows a simplified block diagram of the sampled value information exchange model using the publisher/subscriber mechanism. The publisher writes in a local buffer the values for the data attributes defined in the data set to be sent in the sampled value message over the substation network. The control of sampled value communications is at the logical device level by LLN0, which contains the multicast sampled value control block (MSVCB). It includes attributes that define the behavior of the P2P communications:

- *MsvCBNam* (Instance name of MSVCB) identifies a MSVCB within the scope of a *MsvCBRef* (SV control reference), a unique path-name of a MS-VCB within LLN0:
  - LDName/LLN0.MsvCBName
- *SvEna* (SV enable) indicates that the MSVCB is enabled (if set to TRUE) to send SV messages. If set to FALSE, it shall stop sending sampled value messages.
- *DatSet* is the reference of the data set whose values of members shall be transmitted.
- *ConfRev* is the configuration revision indicating the number of times that the configuration of the data set referenced by DatSet has been changed. The counter is incremented every time when the configuration changes.
- *SmpRate* specifies the sampling rate in samples per cycle at the nominal frequency.

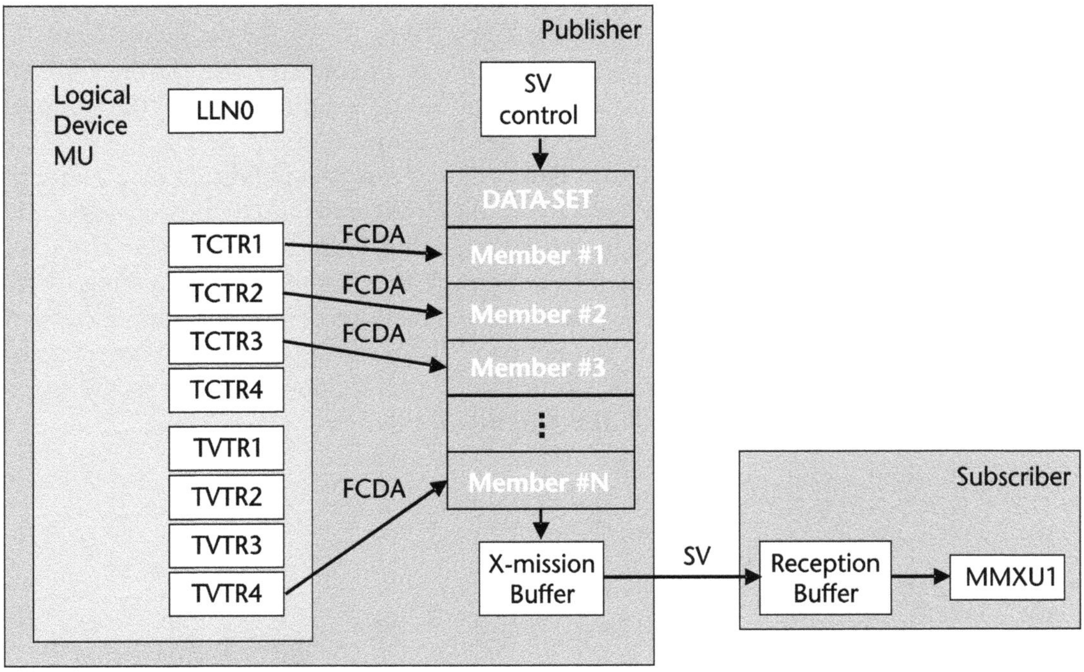

Figure 11.4 IEC 61850 sampled value model.

Edition 2 of the standard [5] includes in MSVCB also an enumerated attribute *SmpMod* defining the following options:

- Samples per nominal period (DEFAULT);
- Samples per second;
- Seconds per sample.

The standard also allows the transmission of sampled values using unicast based on a two-party application association established by a subscriber and a producer. After that, the subscriber may configure the instance of the control block class and enable the transmission using *SvEna* until the association is released. This method is not commonly used for sampled value communications.

The sampled value control model ensures their transmission in a time-controlled and organized manner. It represents the streaming of samples with a fixed interval between the messages as shown in Figure 11.5.

The content of the sampled value message allows the receiving devices to perform processing of the samples in order to execute the required actions. Some of the attributes in the message that help perform the functions are as follows.

Parameters *Sample* contain the values of the member of the data set sampled at a given time. They are of the common data classes sampled analog value (SAV) as defined in IEC 61850-7-3 [6].

The parameter *SmpCnt* is a counter, which is incremented each time a new sample of the analog value is taken. The counter is typically used to indicate time consistency of various sampled values and is reset by an external synchronization event, for example, the beginning of a new second.

An optional parameter *SmpSynch* indicates if the samples are synchronized by clock signals.

While in IEC 61850-7-2 Edition 1 there was a parameter *Test* for the GOOSE message indicating that it is used for test purposes, it did not exist in the sampled value message creating a significant testing challenge.

The GOOSE parameter *Test* in Edition 1 was replaced in Edition 2 by a parameter *Simulation*, which, when set to the value TRUE, indicates that the message and its values have been published by a simulation unit for testing purposes. The same parameter *Simulation* was added to the sampled values message indicating to the subscriber to report the values of the simulated message to its application instead of the real message depending on the setting of the receiving IED. The allowance for an IED to switch from acceptance of real messages to simulated messages is specified by a data object defined in IEC 61850-7-4 [7].

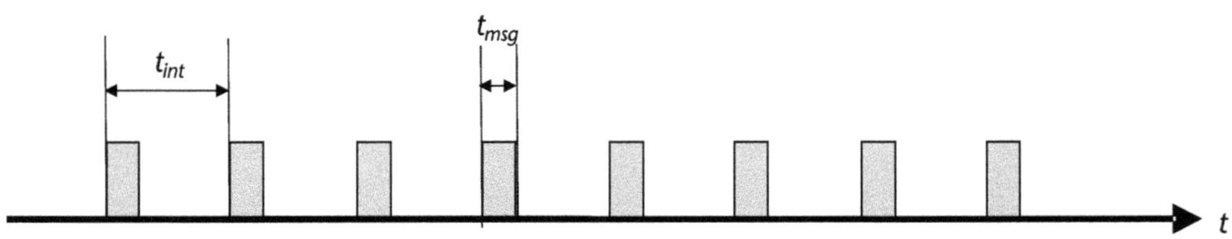

Figure 11.5 Streaming sampled values.

Specific communication services in the subscribers update the content of their reception buffers and processes them based on the values of the parameters described above.

The Ethertype 2-byte code indicating the protocol type in an Ethernet packet uses specific values related to the IEC 61850 sampled value 88-BA.

The performance requirements for the transfer time of sampled value messages are similar to GOOSE: 3 ms.

IEC 61850-9-2 [8] specified the mapping of sampled values, which provides the capability to concatenate more than one sample (Figure 11.6) represented by an Application–Service Data Unit (ASDU) into one message represented by the Application–Protocol Data Unit (APDU) before the APDU is posted into the transmission buffer. The numbers of samples that will be concatenated into one message are configurable and related to the sampling rate. The concatenation of ASDUs is not dynamically changeable to improve interoperability and reduce the implementation complexity. When concatenating several ASDUs into one frame, the ASDU with the oldest samples is the first one in the frame. The Application–Protocol Control Information (APCI) portion of the APDU contains a field with the number of samples in the message.

The success of using sampled value messages for substation protection applications makes it attractive for use between IEDs in transmission line protection systems. However, the streaming of the sampled value method is also attractive for the transmission of synchrophasor measurements for a different wide area protection, monitoring, and because they are used outside of the substation over wide area networks makes them vulnerable to cyberattacks.

To address such concerns IEC TC 57 Working Group 10 developed the technical report IEC 61850 90-5 [9], which defined the routable communications of synchrophasors (R-SV) over wide area networks. The communications are based on the full 7-layer OSI stack and use UDP multicast. The document also describes the use of end-to-end cybersecurity based on the definitions in the IEC 62351 standard.

From all of the above, it is clear that the standard defines a sampled value transmission model that can cover many different applications. It is obvious that it provides significant flexibility by supporting different sampling rates, datasets, and communication message structures. The price that we pay to have such flexibility is the impact that it has on interoperability. That is why something had to be done

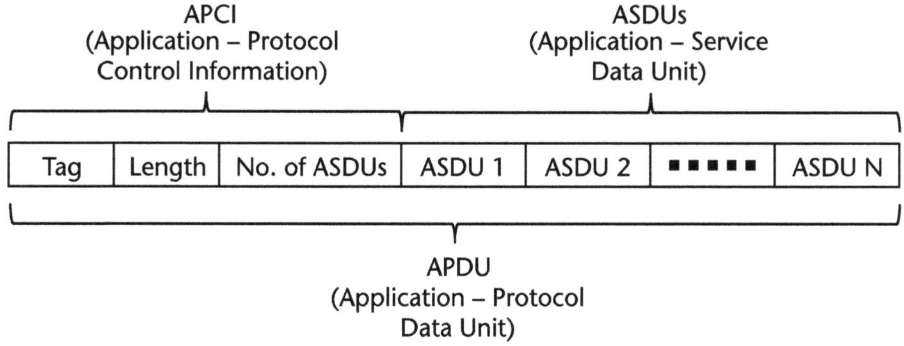

Figure 11.6 Combining multiple samples in a single frame.

to address this concern and this is what we know commonly in the industry as IEC 61850 9-2 LE.

11.4 Implementation Agreement IEC 61850 9-2 LE

During the development of IEC 61850-9-2, the international experts involved in it realized that it will be very important to establish some guidelines to help the different manufacturers with the implementation of the sampled value transmission because it was creating a completely new environment for a digital interface with the instrument transformers in the substation. The same way that we have limited choices for connecting to the secondary of a current transformer like having only two possible nominal currents of 1A and 5A, it was necessary to define a limited number of ways to implement sampled value communications. This required gathering a group of experts representing the global industry and coming up with an agreement of how exactly to implement what is in the standard to ensure interoperability between the devices digitizing the analog signals and the PAC equipment in the substation.

This task was taken by the UCA International Users Group, which immediately started work and, as a result, published the "Implementation Guideline for Digital Interface to Instrument Transformers Using IEC 61850-9-2" [10], published in 2004. This long title was too much to use on an everyday basis, so the document later received the nickname IEC 61850-9-2 LE.

Some people who opposed the implementation of the standard complained that this document was not a standard and that was why they were not interested in using it. This is not a valid argument because it represents a profile of the standard that defines specific interpretations accepted by all participating manufacturers to support interoperability. It is important to highlight that the document only supports the services related to sending sampled value messages and, as a result, it only defines unidirectional communication from the digitizing device to the subscribers of the sampled value stream. The global acceptance of IEC 61850-9-2 LE contributed to the development of digital substations all over the world and making it the de facto standard for many years.

From the modeling point of view, it accepted the IEC 60044-8 concept of a merging unit and defined a logical device merging unit containing a number of specific sensor logical nodes, the sampling rates, and the dataset used for the transmission.

The logical device merging unit can be part of any process interface device that is performing the digitization of specific analog signals. In the case of this implementation agreement, this is limited to the current and voltage signals from conventional or low-power instrument transformers.

Figure 11.7 shows the simplified block diagram of such a process interface device, which makes it clear that it performs the same functionality as the analog interface of a multifunctional IED, but, instead of sending the digitized samples over the internal digital data bus of the IED, it is publishing them over the substation LAN.

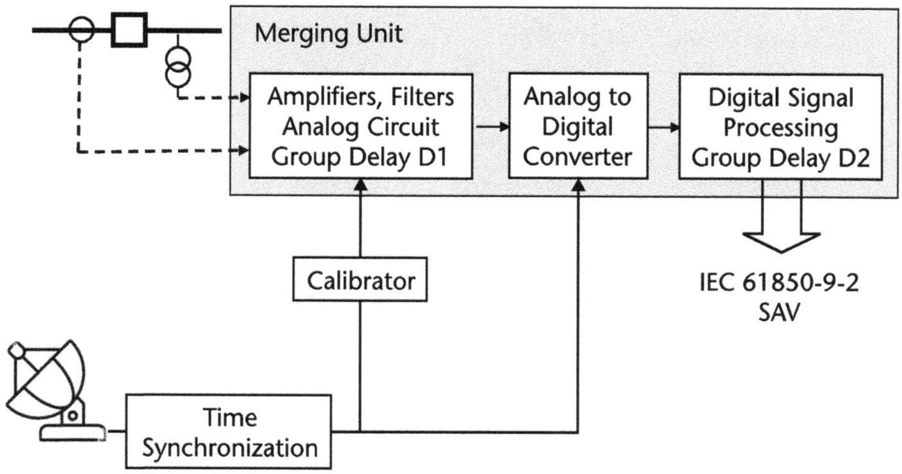

Figure 11.7 Merging unit.

Because specific devices may need to subscribe to sampled values from different merging units and the IEDs do not control the sampling, the synchronization of the sampling is based on a time signal typically from a GPS source.

It is recommended to use as a physical interface fiber optic 100Base-FX full duplex with ST connectors. The 100Base-FX with MT-RJ connectors or electrical transmission using 100Base-TX full duplex with RJ-45 connectors are the only allowed alternatives.

The quality attribute as usual plays a very important role and is extended with an additional Boolean attribute identifying if the value is produced by a real sensor or is being calculated, for example, if there is no direct measurement of the neutral current and it is calculated based on the phase currents, the *derived* attribute value will be set to TRUE.

IEC 61850 does not specify logical devices and this is left to the manufacturers. To improve the interoperability between merging units and IEDs from different manufacturers, the profile defines a specific logical device as a merging unit shown in Figure 11.8.

Each instance of the current transformer (TCTR) and voltage transformer (TVTR) logical nodes digitizes one phase or neutral current or voltage. The primary current and voltage samples values are members of the dataset PhsMeas1 used in the ASDU.

The profile defines two optional control blocks:

- MSVCB01 with a sampling rate of 80 samples/cycle and a single sample (ASDU)/message;
- MSVCB02 with a sampling rate of 256 samples/cycle and 8 samples (ASDU)/message.

The merging unit's mode is normally on, but also can be off or test. In the last case, the test bit in the quality attribute is used to indicate it.

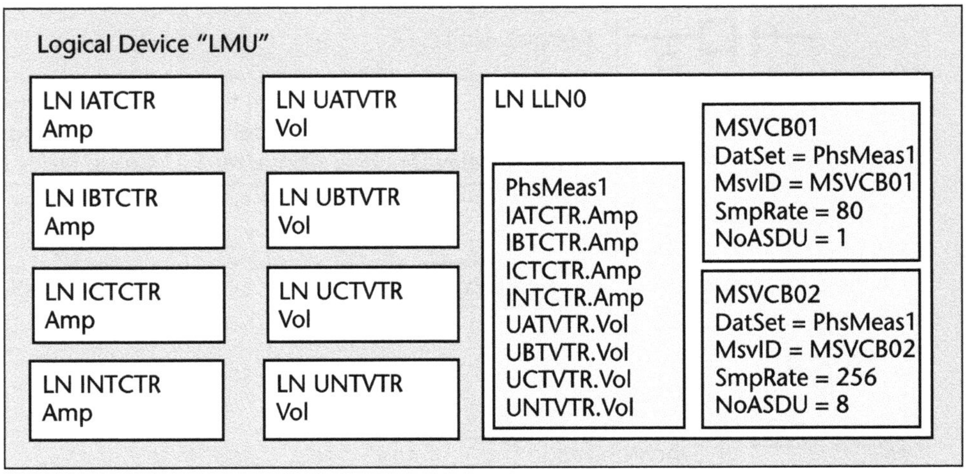

Figure 11.8 Logical device merging unit.

At the time of the development of the profile, it was decided to use 1 pulse per second (pps) as the timing event to synchronize the sampling of merging units. It supports the required accuracy of 1 μs, but the drawback is that it requires a dedicated time synchronization network. Today, because of the introduction of the IEEE 1588 (PTP) profile IEC 61850-9-3, it is used in many applications of merging units as the time synchronization signal over the Ethernet.

The profile specifies two nominal frequencies, 50 Hz and 60 Hz. This means that the merging unit manufacturer needs to design devices that will publish 4,000 samples/second when the nominal frequency is 50 Hz and 4,800 samples/second when the nominal frequency is 60 Hz with 80 samples/cycle at both nominal frequencies.

The publication of this implementation guideline had a tremendous impact on the acceptance of the digitized process interfaces in the substation leading to the development of many fully digital substations based on the IEC 61850 standard that we have all over the world today.

11.5 IEC 61869-9

In the previous section, we talked about the importance of the IEC 61850-9-2 LE profile defining an implementation of the standard and the fact that some people complained that it is actually not a standard by itself. IEC 61850 is focused on defining the digitalization of the current and voltage signals by defining the object models and communication services. However, IEC TC 38 is the one responsible for instrument transformers and that is why, based on the experience with the implementation of IEC 61850 and the interest in development of fully digital substations due to the significant benefits that they bring, it took the task to develop an international standard for the digital interface for instrument transformers.

With the support of some of the most active members of IEC TC 57 Working Group 10 and based on the IEC 61850 principles, the new standard IEC 61869-9:2016 Instrument Transformers—Part 9: Digital Interface for Instrument

Transformers was published in 2016. As the title indicates, it defines requirements for digital communications of instrument transformer measurements. It replaced IEC 60044-8 and is based on the IEC 61850 series, the "Implementation Guideline for Digital Interface to Instrument Transformers Using IEC 61850-9-2," published by the UCA International Users' Group and the relevant parts of IEC 60044-8. It contains specific requirements for electronic low-power instrument transformers (LPITs) (see Figure 11.9) having a digital output, as well as additional improvements such as time synchronization based on the IEC 61588 precision time protocol.

IEC 61869-9 contains many similarities and considering the large base of existing devices and systems is backward-compatible with IEC 61850-9-2 LE communications. However, there are some very significant differences that have an impact on the design and implementation of protection and control systems based on the new standard compared to the implementation guideline.

Probably the most important is the fact that IEC 61850-9-2 LE specifies the sampling rates of the merging unit as a number of samples per cycle at the nominal frequency while IEC 61869-9 specifies the sampling rate in hertz (i.e., the number of samples per second regardless of the nominal frequency).

This means that the same merging unit can be used in a system with either 50- or 60-Hz nominal frequency. However, the processing of the samples will be different because it will correspond to a different number of samples per cycle at a different nominal frequency.

If we take as an example the case with 80 samples per cycle at the 60-Hz system, it will correspond to 4,800-Hz sampling rate, which is one of the default values in IEC 61869-9, so no change needs to be made in the design of the merging units or the IEDs with this specific sampling rate selected.

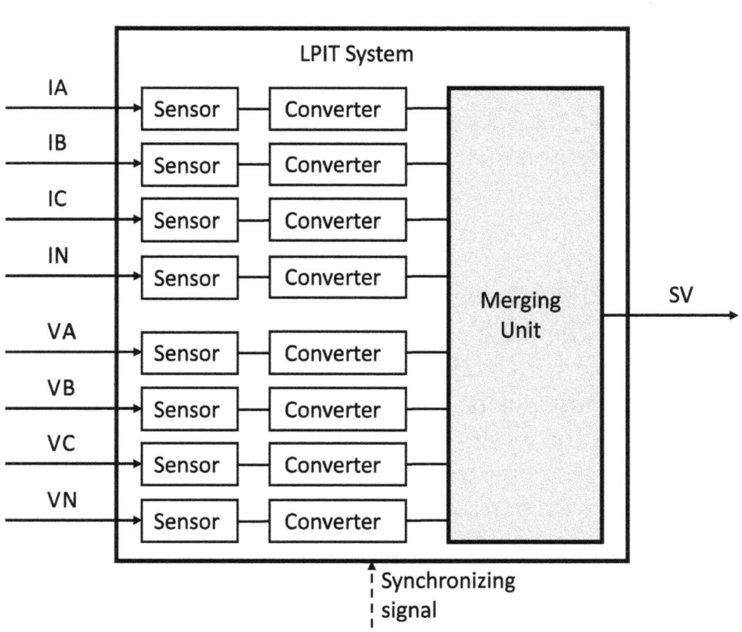

Figure 11.9 LPIT with sampled values.

However, if the same merging unit is used in a 50-Hz system, the number of samples per cycle will be 96, which will require different processing from what was used in the IEC 61850-9-2 LE-based solution.

Another significant difference is in the number of samples per message. While for protection applications we have 1 sample per message and 8 samples per message for power quality-related applications in the case of IEC 61869-9 for protection applications, we have 2 samples per message and for power quality applications with a sampling rate of 14,400 Hz we have 6 samples per message. A simple calculation shows that, for both sampling rates, we have the same number of messages per second: 2,400. This will lead to a significant reduction on the loading of the communications network because the number of messages used for sampled values will be cut in half.

IEC 61869-9 introduces delay time (td), which is measured as the difference between the time encoded by the *SmpCnt* and the time that the message timestamp point appears at the digital output. For protection, the requirement is 2 ms, while, for power quality, it is 10 ms.

To ensure better interoperability, only a limited variability is allowed for naming, message structure, sampling rate, analog signal content, and scaling.

The supported variants are included in the device nameplate using the following notation in a human readable text format:

F*f*S*s*I*i*U*u*

where

f is the digital output sample rate expressed in samples per second.

s is the number of ASDUs (samples) contained in a sampled value message.

I is the number of current quantities contained in each ASDU (maximum is 24).

u is the number of voltage quantities contained in each ASDU (maximum is 24).

For example, the backward-compatible to the 9-2 LE MSVCB01 sampled values with a 50-Hz nominal system frequency variant is described as

F4000S1I4U4

Although 61869-9 documents this as backward-compatible with 9-2 LE, it is only backward-compatible for sampling rate and number of ASDUs. To be backward-compatible with 9-2LE, the SyncSrchID/GMID field that was added in 9-2 LE Edition 2 must not be transmitted. The IEC 61850-6 sampled value control block configuration allows this to be accomplished.

The backward-compatible to the 9-2 LE MSVCB02 sampled values with the 50-Hz nominal system frequency variant is described as:

F12800S8I4U4

For future applications, the standard defines the following preferred rates:

- For protection and general measurements: 4,800 Hz with two ASDU (samples) per message;
- For quality measurements and transient recording: 14,400 Hz with six ASDU (samples) per message;
- For high bandwidth direct current (DC) control applications: 96,000 Hz with one ASDU (sample) per message.

DC applications may require point-to-point connection and Gigabit Ethernet links.

F4800S2I4U4 describes sampled values with 4,800 samples per second, two ASDU (samples) per message, 4 currents, and 4 voltages.

The IEC 61869-9 standard also defines the modeling of the merging unit (Figure 11.9) and specifies naming conventions for the individual instances that are used in the substation section of the SCL as part of the engineering process. For example, the names of instances of TCTR are based on:

I *nn* *p* TCTR *n*

where

nn is the instance number of the current measurement point (01-99) and should be unique within the bay.

p is the phase identification of the primary current, either A, B, C, or N for alternating current (AC) instrument transformers.

n is the attribute "inst" of the element LN in the substation and IED sections of the ICD file.

An example of a TCTR instance name is:

I04ATCTR7

The same principles are applied to the naming of instances of TVTR.

The IEC 61869-9 standard also defines rules for naming datasets according to:

PhsMeas*x*

where *x* is the instance number of the dataset (1 to 99) that makes the dataset name unique within LLN0.

Another step in supporting interoperability based on the experience with IEC 61850-9-2 LE is specifying that the dataset name PhsMeas1 may only be used for a dataset whose members in order are:

InnATCTR1.AmpSv.instMag.i_

InnATCTR1.AmpSv.q

InnBTCTR2.AmpSv.instMag.i

InnBTCTR2.AmpSv.q

InnCTCTR3.AmpSv.instMag.i

InnCTCTR3.AmpSv.q
InnNTCTR4.AmpSv.instMag.i
InnNTCTR4.AmpSv.q
UnnATVTR1.VolSv.instMag.i
UnnATVTR1.VolSv.q
UnnBTVTR2.VolSv.instMag.i
UnnBTVTR2.VolSv.q
UnnCTVTR3.VolSv.instMag.i
UnnCTVTR3.VolSv.q
UnnNTVTR4.VolSv.instMag.i
UnnNTVTR4.VolSv.q

Looking at this data set, it is very important to note that each current or voltage sample attribute is followed immediately by the corresponding quality attribute: AmpSv.q or VolSv.q. To ensure the correct processing of the received values, this should be a mandatory requirement for any dataset whose members are data attributes.

11.6 Using Sampled Values for High-Voltage Bus Protection

One of the best examples of the benefits of using sampled values for protection applications is the busbar differential protection. While bus differential protection was centralized in the past, based on a summation of currents from all current transformers defining the zone of protection, the introduction of sampled values combined with advanced high-speed substation communications has resulted in the development and implementation of a new generation of bus differential protection systems with distributed architecture.

This technology and its implementations to bus protection offer some significant advantages. The classical bus differential schemes have more or less a fixed zone of protection because they have a single current input that is the sum of all current circuits wired into the relay. IEC 61850-based distributed schemes can easily change the zone of protection based on the knowledge of the substation topology and the state of breakers and disconnecting switches at any moment in time.

The bus differential protection system is based on using sampled analog values published over the substation process bus. The architecture of such a system is shown in Figure 11.10.

The distributed bus differential protection system architecture includes four types of devices:

- Central unit performing the differential, and all other related functions;
- Merging units, providing the analog signal processing functions;
- Ethernet switch (process bus);
- Time synchronization device.

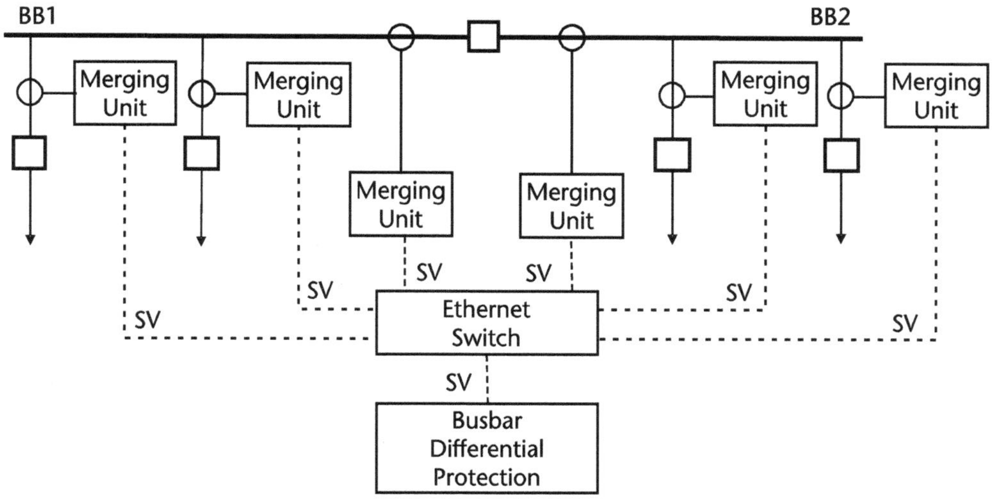

Figure 11.10 Sampled value-based bus differential protection.

The merging units have the task to digitize the analog signals from conventional or nonconventional sensors to sampled values that can be transmitted over the fiber cables to the device that performs the differential current calculation and makes a decision if there is a fault within the zone of protection.

Interoperability between merging units and protection, control, monitoring, or recording devices is ensured through the implementation guidelines or IEC 61869-9 as described earlier in this chapter. A sufficient sampling rate for bus differential applications is 80 samples/cycle.

The transmission of sampled values for bus differential protection requires special attention with regard to the time constraints. The model provides transmission of sampled values in an organized and time-controlled way so that the combined jitter of sampling and transmission is minimized to a degree that an unambiguous allocation of the samples, times, and sequence is provided based on the sequence number of the received samples. This minimizes the impact of the different transfer times between the individual merging units and the central unit performing the bus differential function.

By monitoring the status of the switchgear by subscribing to GOOSE messages from the switchgear interface units by the central unit, it has continuous knowledge of the actual substation topology, which allows it to automatically adapt its zone of differential protection. This is a significant benefit compared to conventional bus differential protection schemes that require an outage to change the zone of protection by switching current circuits while avoiding creating an open CT circuit condition representing a safety hazard.

The status of the breakers and disconnecting switches in the substation is modeled using the circuit breaker (XCBR) and circuit switch (XSWI) logical nodes. It will provide information on the three phases or single-phase status of the switching device, as well as the normally open or closed auxiliary contacts to be used to determine the topology of the substation and zones of bus protection.

PDIF is the logical node representing the bus differential element in the central unit. The calculation of the differential current sample can be performed simply by summing the values of the current samples from the different merging units with the same sequence number and is not dependent on the actual time when the message containing the sample was received. Which currents will be used for the differential current calculation depend on the zone of busbar protection defined by the substation topology determined by the status of the circuit breakers and disconnecting switches at any moment in time.

Another benefit of the use of IEC 61850 is that it provides a standard format for the description of the substation topology. It is based on the substation configuration description (SCD) file.

The synchronization device in early implementations was using a 1-pps signal through an RS485 network to all interface devices included in the system. Time-synchronization accuracy better than 1 μs is achieved by this solution, but it required a dedicated synchronization network. Today it is based on the IEC 61850-9-3 profile of the IEEE 1588 standard (PTP) over Ethernet.

Each merging unit is connected to an Ethernet switch that, in this case, is dedicated to the process bus. The central unit receives from the switch all Ethernet messages from the merging units included in the system. Considering the size of the Ethernet frames and the number of lines and transformers connected to the bus, the number of Ethernet ports may need to be increased.

Another alternative solution for a large number of merging units in a big substation is to use a central unit with a 1-Gbps Ethernet port connected to a 1-Gbps Ethernet switch with 100-Mbps ports connected to the merging units.

In the case of large substations, it will be beneficial to use IEC 61869-9 with 2 samples/messages, which will double the throughput of the communication links.

In the case of conventional hardwired differential protection systems, any problem with a current transformer will result in some differential current, which may lead to an undesired operation. In the sampled value-based implementation, the loss of samples or issues with the sensors will have a similar impact. However, the sampled value communications provide an opportunity for continuous monitoring of the sensors and their digital interfaces, while at the same time, by using redundant communications based on PRP or HSR, significantly improve the reliability and security of the differential protection scheme.

Another challenge for hardwired differential protection systems is CT saturation, which may lead to an undesired operation of the protection scheme for external faults. The use of sampled values for differential protection practically eliminates this problem if the standalone merging units are connected directly to the secondary of the current transformers and due to the fact that they have a very small impedance themselves. It is even better if the differential scheme is using samples from low-power instrument transformers (LPITs), which do not saturate under fault conditions.

11.7 Using Sampled Values for Disturbance Recording

IEC 61850-based sampled value transmission is a technology that brings significant benefits to the development and implementation of disturbance recording systems. We can concentrate on two main types:

- Transient recordings;
- Disturbance recordings.

Transient or waveform recording in conventional substations is available in many multifunctional protection and control IEDs. It captures the individual samples of the currents and voltages measured by the IED with a sampling rate that may be in the hundreds of samples per cycle for high-end monitoring and recording IEDs.

In IEC 61850-based digital substations, the transient recording is based on the sampled values available from merging units and in the existing implementations using the UCA international implementation agreement known as IEC 61850 9-2 LE.

The user typically has options to define the triggering criteria, the pre-trigger or post-trigger intervals, and if extended recording should be available in cases of evolving faults or other changing system conditions. The capture of several cycles of pre-fault data, as well as the ability to record the waveform over a period of several seconds, will result in better use of the record.

Transient recording also may be performed by dedicated disturbance recording devices with multiple hardwired analog and binary inputs. In IEC 61850-based digital substations, this is implemented by a centralized disturbance recording function connected to the substation local area network and subscribing to multiple streams of sampled values and GOOSE messages.

The trigger for waveform recording can be local or external. An internal trigger can be based on measured or instantaneous values and, in some cases, may use superimposed components of the current and voltage signals.

An external trigger in IEC 61850-based digital substations is typically a GOOSE message indicating the start or operation of a P-class logical node.

The transient records are typically available as COMTRADE files.

High-speed or low-speed disturbance recording is intended for capturing events such as local or wide area disturbances following short-circuit faults on the transmission or distribution system or fault induced delayed voltage recovery (FIDVR).

The disturbance recording device stores the values of a user-defined set of parameters for a period of time that is too long for capturing using sampled values. The setting range is dependent on the available memory in the device. If the sampling rate is more than one cycle per sample, the user should be able to select the recording of minimum, maximum, and average values through the specified sampling interval.

The recorded values can be RMS values, phasors, or synchrophasors, depending on the capabilities of the recording device and the requirements of the application.

An option to trigger high-speed disturbance recording when a waveform capture is triggered is achieved by using the same trigger with different recording modes. The combination of waveform capture and high-speed or low-speed disturbance recording triggered by the same power system event allows the recording of long events, while, at the same time, the details of the transitions from one state to another are recorded in the waveform sampled value-based capture. This allows the use of the same event record for the analysis of relay operation or verification of the system models used by different analysis tools.

Recently, the IEEE PES Power System Relaying and Control (PSRC) Committee defined a schema that allows the use of COMTRADE files for synchrophasor-based disturbance recording.

The process interfaces related to the recording of short-circuit faults or other electric power system events depend on the substation design and equipment used.

The communications architecture of centralized IEC 61850-based disturbance recording systems is quite simple and very similar to the bus differential example in Figure 11.10. At the bottom of the hierarchy are the merging units or process interface devices (PIDs).

The PID contains two logical devices:

- *Merging unit (MU):* It provides the current and voltage interface to the process based on the available interface options described earlier and publishing sampled values over the logical process bus to the disturbance recorder or any other substation function.
- *Switchgear interface unit (SIU):* It provides the GOOSE-based interface with the circuit breakers and the disconnector switches as required by the different substation functions, including the disturbance recorder.

The PID may also include local protection or measurement functions, such as an MMXU. Because the PID is accurately time-synchronized, it will calculate M or P class synchrophasor measurements and will publish them over the substation LAN for disturbance recording purposes.

Today the communications of the sampled values are typically based on the IEC 61850 9-2 LE profile, but, in the future, this will be replaced by IEC 61869-9. As a result of the transition, the traffic on the communications network will be reduced in half, but the size of the frames will increase.

Centralized disturbance recording systems provide many benefits and at the same time face some challenges. The benefits include, but are not limited to:

- The principles allow the implementation in substations of any configuration and at any voltage level.
- The efficiency of the engineering of the system is significantly improved based on the use of system configuration description (SCD) files.
- It does not require the installation, commissioning, and maintenance of dedicated disturbance recording devices.
- It has high reliability and availability due to the high level of integration and the availability of redundant process interfaces.
- The sampled value communication services can be used both for waveform recording using samples and disturbance recording using synchrophasors.

The main challenge with centralized disturbance recording in IEC 61850 digital substations is the throughput of communication links. Typically, it will not be a problem in substations with start topology with the exception of the link to the substation computer hosting the disturbance recording function, which has to be 1 Gbps.

In substation communications with ring topology, the links have to be 1 Gbps.

In very large substations with many PIDs or MUs, while considering the maintenance testing of bus differential protection schemes that doubles the sampled values traffic, it may be necessary to use 10 Gbps.

References

[1] IEC 60044-8:2002 Instrument Transformers—Part 8: Electronic Current Transformers, 2002.

[2] IEC 61850-9-1:2003 Communication Networks and Systems in Substations—Part 9-1: Specific Communication Service Mapping (SCSM)—Sampled Values over Serial Unidirectional Multidrop Point to Point Link, 2003.

[3] IEC 61850-9-2: 2004 Communication Networks and Systems in Substations—Part 9-2: Specific Communication Service Mapping (SCSM)—Sampled Values over ISO/IEC 8802-3, 2004.

[4] IEC 61850-7-2:2003 Communication Networks and Systems in Substations—Part 7-2: Basic Communication Structure for Substation and Feeder Equipment—Abstract Communication Service Interface (ACSI) Implementation Guideline for Digital Interface to Instrument Transformers Using IEC 61850-9-2, UCA International Users Group, 2003.

[5] IEC 61850-7-2:2010 Communication Networks and Systems for Power Utility Automation—Part 7-2: Basic Information and Communication Structure—Abstract Communication Service Interface (ACSI), Edition 2, 2010.

[6] IEC 61850-7-3:2003 Communication Networks and Systems in Substations—Part 7-3: Basic Communication Structure for Substation and Feeder Equipment—Common Data Classes, 2003.

[7] IEC 61850-7-4:2003 Communication Networks and Systems in Substations—Part 7-4: Basic Communication Structure for Substation and Feeder Equipment—Compatible Logical Node Classes and Data Classes, 2003.

[8] IEC 61850-9-2: 2004 Communication Networks and Systems in Substations—Part 9-2: Specific Communication Service Mapping (SCSM)—Sampled Values over ISO/IEC 8802-3, 2004.

[9] IEC TR 61850-90-5:2012 Using IEC 61850 to Transmit Synchrophasor Information According to IEEE C37.118, 2012.

[10] Implementation Guideline for Digital Interface to Instrument Transformers Using IEC 61850-9-2, UCA International Users Group, 2004.

[11] IEC 61869-9:2016 Instrument Transformers—Part 9: Digital Interface for Instrument Transformers, 2016.

CHAPTER 12

Standards-Based Engineering

12.1 Introduction

The requirements for improvements in the efficiency and quality of PAC systems at all levels of the electric power systems highlights the need for development of new methods and tools that can help the industry to achieve these goals. That is why it is very important to analyze the opportunities that exist to develop a set of engineering tools based on the creation and experience with the use of utility standards and the IEC 61850 substation configuration language.

We need to consider the benefits of an object-oriented engineering process for the development and implementation of PAC systems based on five stages in a utility's standardization process:

- Definition of PAC system philosophy;
- Definition of PAC system types;
- Selection of approved PAC IEDs;
- Design of the PAC system;
- Instantiation of the standardized PAC system for specific substations.

Traditionally, the engineering process has been based on manually created drawings and textual description of the requirement specifications as well as the use of spreadsheets to identify the signal flow. This is a very time-consuming process that is prone to human errors that may have a significant impact of the quality of the PAC systems and undesired operations that may result in local or wide area disturbances.

Digitalizing the engineering process based on the advancements of technology and the development of the IEC 61850 standard offers an opportunity to improve all aspects of the engineering process. This requires the development of methods and tools that allow us to model all components of an electric power substation and the interactions between them. Such methods need to cover the different components of the model, such as:

- Substation topology;

- Transmission bays;
- Distribution bays;
- Power transformer bays;
- Generator bays;
- Transmission lines;
- Distribution feeders;
- Communications infrastructure;
- PAC, energy management, and other functions;
- Multifunctional IEDs;
- Their associations with the primary equipment and the communication system.

The benefits of digitalization of the engineering process come from the ability to develop tools that can automatically process the computer-readable files describing all primary and secondary components of an electrical substation and the relationships between them.

A 4-stage standardization process defined by CIGRE Working Group B5.27 and published in June 2014 as Technical Brochure 584 [1] envisioned the use of such formalized description to improve the efficiency and quality of the engineering process.

Standards-based engineering offers some significant advantages that are described at the end of the chapter. At the same time, this approach may require some changes in the organizations and the methods and tools used for engineering. It will also need some initial investment, but the long-term benefits will result in significant savings in time and money, as well as improved quality of the schemes and the reliability and security of the electric power system.

The availability of a standardized description of the PAC system's functionality can also help with improving the efficiency of commissioning and maintenance testing, as well as the cybersecurity of the system.

12.2 Object-Oriented Standards-Based Engineering of Protection Systems

IEDs (microprocessor-based) for data acquisition, protection, measurements, and control have gained widespread acceptance and are recognized as essential to the efficient operation and management of substations. Their integration in hierarchical substation protection and controls systems over a substation local area network allows significant improvement in the functionality of the system without any increase in the cost. This integration process in substations using IEC 61850 as the communications protocol is based on object models that require the use of appropriate tools to represent the complex architecture of the substation, the communication system, and the multiple functions in the IEDs themselves. A major part of the engineering of a substation automation system is related to the architecture and configuration of the secondary equipment in the substation. This requires the development of a formalized format that allows the description of all different elements and their relationships. IEC 61850 defines the object models of the different types of primary and secondary equipment, as well as their functionality in the substation.

The object-oriented approach to the engineering of the substation PAC system is based on its hierarchy and contains nested objects with different levels of complexity that can be defined as part of the standardization process.

At the top of the hierarchy is the substation PAC system (SPACS) that contains multiple instances of bay PAC schemes (BPACS), each defined as a complex object: SPACSO or BPACSO (see Figure 12.1).

Each BPACS contains multifunctional IEDs, defined in the object-oriented design process as a PAC object (PACO) with scheme specific functionality. For example, we may have a Main 1 PACO and a Main 2 PACO.

Each PACO contains multiple function objects (FOs) with specific functionality:

- Protection;
- Automation;
- Control;
- Measurements;
- Monitoring;
- Recording;
- Analysis;
- Others.

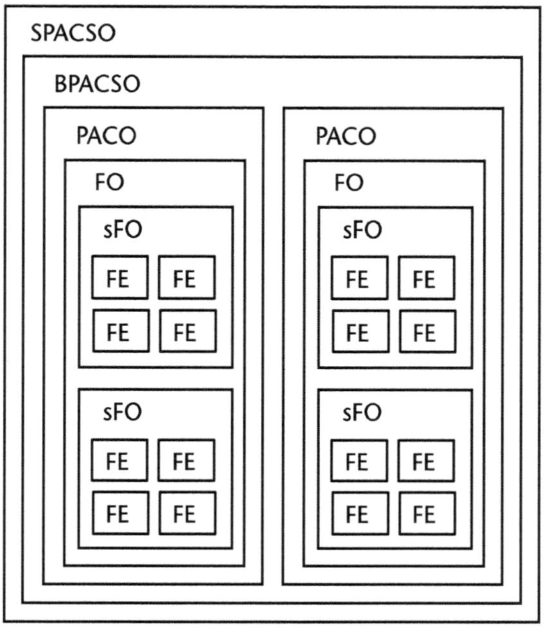

Figure 12.1 Substation object model hierarchy.

Each FO can contain one to many subfunction objects (sFOs). For example, we may have a protection FO that contains an overcurrent protection sFO, containing phase, ground, and negative sequence sFOs.

The sFO at the bottom of the protection system/scheme hierarchy contains the function elements (FE), the smallest functional objects that are represented by logical nodes in the IEC 61850 model. For the overcurrent protection sFOs, we may have instantaneous, definite time-delayed, or inverse time function elements.

A substation protection and automation system also includes different tools for visualization and control of the primary and secondary substation equipment, the substation HMI. The user can navigate through the multiple views of the substation's one-line or communications diagrams or check the status or settings of a specific IED. The development of the HMI and the mapping of the multiple analog and binary signals from the IEDs is a very labor-intensive process that also can be subject to errors at different stages of the engineering process.

The standardization process typically defines bay-level objects, but more utilities are going in the direction of using standard substations, especially at the distribution level of the electric power system.

It is important to understand that standardization, like everything else, has benefits and drawbacks. The analysis of both clearly shows that the benefits are much more than the drawbacks, especially if we consider the long-term benefits against the short-term drawbacks. Even though it will impose an initial cost and resource burden, in the long run, it will lead to significant cost savings and improvement in the quality of the secondary systems. The benefits of such an approach can be further improved if the standardization applies not only to the protection schemes' engineering, but also to IED configurations, settings, and logic.

Although some nonmonetary benefits might be achieved in the short term, the standardized designs should be applied for a period of time in order to realize the anticipated full benefits, but this period should not be so long that the technology becomes obsolete or too far out of date compared with the latest available technology.

The development of standard secondary schemes is based on:

- Utility standards;
- Utility best practice;
- National standards;
- International standards:
  - IEC;
  - IEEE.
- Industry best practice:
  - CIGRE reports;
  - IEEE Power System Relaying Committee reports.

Detailed analysis of the standardization of protection and control schemes, the definition of a standardization process, and the benefits and challenges of this approach based on the experience and practices of many utilities from around the world is available in the CIGRE Technical Brochure 584, "Implications and

Benefits of Standardized Protection and Control Schemes," prepared by Working Group B5.27.

The contributions of this work, combined with the best practices from the established standardization process within a utility, provide the foundation for the standardization strategy described later in this chapter.

12.2.1 Standard Bays

The efficiency of the standardization process can be significantly improved if the design of the substations is based on standard bays.

Standard bay design includes the following elements:

- Bay scheme;
- Bay layout;
- Bay primary equipment;
- Instrument transformer(s) location;
- Instrument transformers.

The following bay types are commonly used in a utility's power system and included in the standardization process:

- HV breaker-and-a-half;
- HV single breaker;
- MV single breaker;
- HV transformer;
- MV transformer;
- MV feeder.

The functional and performance requirements in standard secondary systems are defined by the philosophy and criticality of the application.

The overall standard scheme design will be based on the combination of the bay type and voltage level or criticality factor.

12.2.2 Standard Substations

The highest level of efficiency of the standardization process can be achieved using standard substations. It is possible to design standard distribution substations and it is already a common practice in some utilities.

Such a strategy will offer significant benefits, as it will support the design of standard container style control houses that can be produced, configured, and commissioned in a factory environment. This will result in a significant reduction of the amount of work that needs to be done at the site, especially if standardized interface between the substation process and the control house is designed and implemented.

Figure 12.2 shows a simplified one-line diagram of a small standard distribution substation.

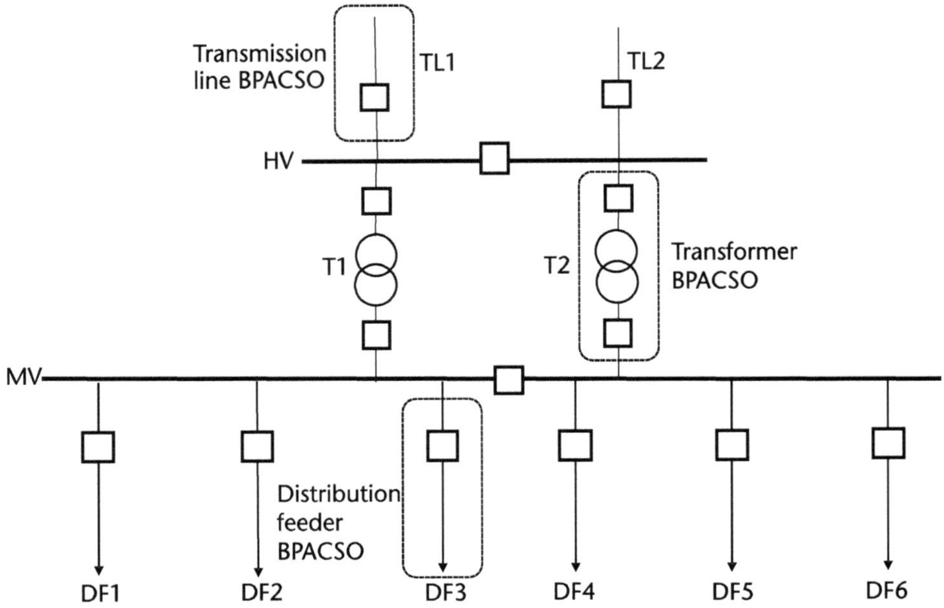

Figure 12.2 Simplified one-line diagram of a small standard distribution substation.

The engineering of such a standard substation with two transmission lines, two power transformers, and six distribution feeders connected to a sectionalized distribution bus is based on the instantiation of the standard bay PAC scheme objects such as the transmission line BPACSO, the transformer BPACSO, and distribution feeder BPACSO shown in the figure.

Because each of these standard schemes has been extensively tested during development, they do not have to be tested during commissioning other than to verify that the settings used for a specific installation are correct. This approach significantly improves the efficiency of the engineering process and limits the amount of work that needs to be performed at the substation only to verify the connections to the primary substation equipment.

12.3 IEC 61850 Substation Configuration Language

The object-oriented approach to the engineering of substation PAC systems requires the definition of a formal model that meets a set of requirements. It should be capable of:

- Describing the substation's one-line diagram and the functions associated with the different primary substation components;
- Describing in detail the functionality of PAC IEDs;
- Describing in detail the functionality of PAC objects at the higher levels of the hierarchy;

- Configuring all aspects of PAC objects at the different levels of the hierarchy, including the complete substation;
- Completely configuring the data flow between all components of the system;
- Supporting modifications and extensions to the system.

The development of the object and data models and the definition of various services described in different parts of the IEC 61850 standard create a foundation on which a substation configuration language meeting the above requirements can be built.

Many people think of the IEC 61850 standard as defining just another communications protocol. It is true that it defines a protocol, but, with the development of the substation configuration language [2], it is much more than that because it creates a brand-new engineering environment that can be used to support all stages of the life cycle of a substation PAC system.

The substation configuration language was developed by a group of members of IEC TC 57 Working Group 10 and defined a complete substation model that includes:

- A description of all the primary equipment in the substations such as breakers and their disconnecting switches, power transformers, and other devices, as well as how they are connected (based on the principles defined in the standard IEC 81346-1 [3]);
- Description of the substation communication architecture, including information about how the individual IEDs are connected to the different networks or subnetworks and to which communication port they are connected;
- The structure of the individual IEDs including a description of the logical devices configured on the IED, as well as the logical nodes belonging to each logical device with class and type;
- Definitions of the common logical nodes data classes instantiated in the specific IED, identifying which of the optional data objects or attributes are supported;
- The preconfigured associations and which data shall be reported or logged, exchanged, or subscribed to.

As it is impossible for any standard to cover all possible requirements of different users from all over the world, another requirement for the substation configuration language (SCL) is based on the rules of IEC 61850-7-1 [4] to support using namespaces in the specification of completely user-defined logical nodes and data objects as an extension of standard LN classes.

An SCL file describes an instance of the model engineered to represent a specific substation and its PAC system. To do that, the object model includes three main parts:

- Substation;
- Product;
- Communication.

Another section specifies the data type templates indicating which data objects and attributes really exist in the IED, thus creating a template for an instantiable logical node.

12.3.1 The Substation Model

The substation model appears in the left part of Figure 12.3 and shows the hierarchical structure of the substation. As can be seen from Figure 12.3, the substation model has a hierarchy as follows.

At the top is a substation object representing a complete substation. It contains at least one voltage level with primary substation equipment operating at the same voltage. Each voltage level contains bays with different complexity depending on the substation design.

Bays contain primary equipment such as circuit breakers and disconnectors, current and voltage transformers, and power transformer windings. There are also connectivity nodes that allow the modeling of how the primary equipment that belongs to the bay is connected together.

The primary equipment may contain subequipment that is usually used to model individual phases of single-phase high-voltage equipment that has single pole control in the 3-phase system.

Typically, each primary equipment device will have two terminals that will be connected to connectivity nodes in order to model the electrical connections between them and describe the substation topology. A typical example of a connectivity node is a busbar connecting several bays at the same voltage level. Each primary device contains at its terminals references to the connectivity nodes to which it is connected. This approach can be used to model the one-line diagram of the substation.

The substation structure also includes functions and subfunction related to equipment. As can be seen from the figure, functions can be at the substation, voltage level, or bay level.

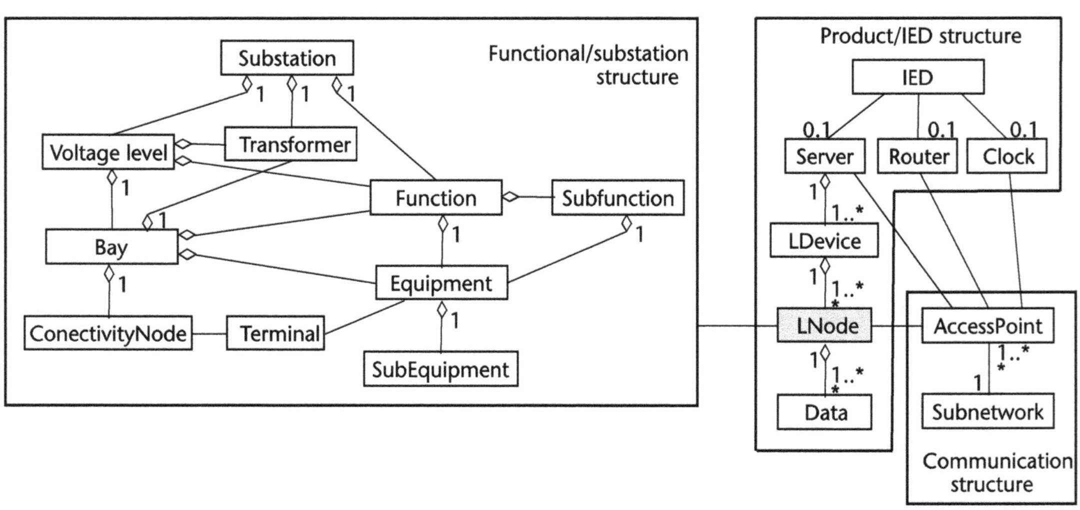

Figure 12.3 SCL object model.

Another component of the substation structure is the power transformer. Hierarchically, it is at a level below the substation, voltage level, or a bay because it contains windings that are each connected to a bay at a different voltage level.

This hierarchical structure can be used for functional designations by appropriate structured names. The SCL syntax requires the use of at least one character as name on each level.

Because logical nodes represent function elements that can be related to any substation component or function, the logical node in the center of Figure 12.3 is shown as attached to the substation part of the model.

For example, an XCBR logical node is attached to a circuit breaker, while an undervoltage logical node under voltage protection (PTUV) is attached to a voltage level. A transformer differential protection logical node PDIF is attached to a power transformer, while the underfrequency protection logical node under frequency protection (PTUF) can be at the substation level of the hierarchy.

As the substation model is also used to relate logical nodes functionality to subequipment, instances of XCBR are used for the single-phase breakers at high voltage substations.

12.3.2 The Product (IED) Model

Within the context of the substation configuration language products are considered multifunctional IEDs that use software modules to implement different protection, automation, monitoring, recording, and other functions required in the substation. The IED model is restricted to such devices and the term product is not used for any of the primary equipment in the substation.

As we have seen earlier in this book, IEDs operate using a software architecture based on the integration of functional elements modeled in IEC 61850 as logical nodes that interact with each other over communication links.

The IED model in SCL is based on the object models defined in IEC 61850-5 [5] and the different parts of IEC 61850-7.

As can be seen from Figure 12.3, the IED model has a hierarchy as follows.

On top of the hierarchy is a server that can communicate the data available from the different logical nodes using the access point providing interface to the communication system. The server contains at least one logical device and the logical devices contain logical nodes representing its functionality.

Each logical node contains data that is exchanged with other logical nodes in the device or the system to perform the different local or distributed functions. SCL supports the modeling of the data flow between logical nodes by describing both the offered and needed data objects.

12.3.3 The Communication System Model

The main difference between the communication model and the substation and product models is that it is not hierarchical. It is used to describe how different products are connected to the substation communications network based on the access points of each server or other system components that need communications interface.

We can think of the access point playing the role of a terminal from the substation model while the role of the connectivity node in the communications system model is performed by the subnetwork. We have to note that the subnetwork is not necessarily a physical structure. An access point may be a physical port, but also a logical address of an IED.

In the model, subnetworks are used for direct (link layer) communication between access points. They are a logical concept and several logical subnetworks using different protocols can be used on the same physical bus.

As can be seen in Figure 12.3, an IED may also have a router function that can be connected with two different access points to two different subnetworks and allow TCP-based messages to reach IEDs connected to another subnetwork. However, Layer-2 GOOSE and sampled values messages can be delivered only to IEDs connected to the same subnetwork as the publisher.

Time synchronization meeting the precision requirements for merging units and phasor measurement units also requires careful consideration and may need a dedicated clock connected to a subnetwork in order to meet this requirement.

In the SCL model, subnetworks in principle model logical connections, but it is also possible to build a representation of the physical structure based on the appropriate naming of the access points in subnetworks and their relation to physical connection points. In this case, the access points play the role of transition objects of the abstract communication model and its physical implementation in the substation communication system.

Another addition to the model is the clock function, also shown in in Figure 12.3, indicating where the subnetwork clock is located.

A special IED type SWITCH is used to model switched Ethernet networks and typically includes only access points to their IP subnetwork.

12.3.4 Data Flow Modeling

Modeling the data flow between the different components of the substation PAC system is another important part of the engineering process. Because the logical node is the building block of the system that is a source of data and a user of data, modeling the exchange of data is commonly done at the logical node level. This applies especially to client/server communications when one of the logical nodes acts as a client requesting data that it needs and another acts as a server providing this data. This exchange requires an association channel that is assigned to different components of the model hierarchy, all the way down to a logical node.

For the modeling of GOOSE or sample value communications at the channel level, the publisher and the subscribers are whole IEDs connected to the same subnetwork. As we remember, the publisher in this communication uses a multicast address, thus making the whole receiving IEDs the subscribers through their access points.

The granularity of the data flow model is much higher at the data object level. In this case, we can use the data inputs of the logical nodes by specifying which data they should use. The efficiency of the dataflow modeling can be significantly improved if, instead of the mapping to a data input of the individual logical nodes, we feed the data at an LLN0 at the logical device level that contains all logical nodes using the same data as an input. For example, if we have a phase overcurrent protection function represented by a nested logical device containing an instantaneous and several time over current logical nodes all using the phase current as an input, instead of feeding each one of them separately, we can map them to an input of LLN0.

This data flow modeling approach is much easier to understand by protection and control engineers with traditional training in hardwired systems because it represents the direct mapping of signals into input.

The signal naming is based on the rules defined in IEC 61850-7-2 and is shown in Figure 12.4.

The logical device name LDName shown in Figure 12.4 can be defined based on product-related naming or function-related naming. In the first case, it is the name of the IED in the IED (product) section, concatenated with the IED relative LD Instance identification. Function-related naming at the communication level is enabled by free setting of the LD name based on the decision of the IED manufacturer to support it with his or her tools.

An example of function-related naming of LD name is the name of the switch yard function or function type, to which the LN relates. Figure 12.5(a) shows the IEC 81346-1 object structure that can be used for the logical device name, as can be seen in Figure 12.5(b). In this example, E1 is a voltage level that contains bays Q1 and Q2. Bay Q1 contains a circuit breaker QA1 and an IED SB1. The IED contains three logical devices LD1, LD2, and LD3. Based on this example, a logical device can be identified as E1Q1SB1LD2.

12.3.5 Modeling of Redundancy

The reliability and security of protection and control systems are key requirements that become even more important during the transition to a new technology. One of the main tools available that can help us achieve this goal is redundancy.

Redundancy can be implemented in various forms with the different components of the protection and control system. We can start at the IED level where we can implement different protection algorithms to protect a power system component. For example, for a transmission line, we can use distance protection as the main protection function and directional ground overcurrent as a backup protection function. For the overcurrent, we can have phase, ground, and negative sequence elements that provide significant levels of redundancy for different types of faults.

Because the process interface has a significant impact on the performance of the system, we may use redundant process interface devices and redundant communications. We typically also have, at least at the transmission level, redundant IEDs that act, for example, as the Main 1 in Main 2 protection IEDs.

LDName	LNName			DataName	DataAttributeName
	LN Prefix	LN Class	LN Instance		

Figure 12.4 Signal name.

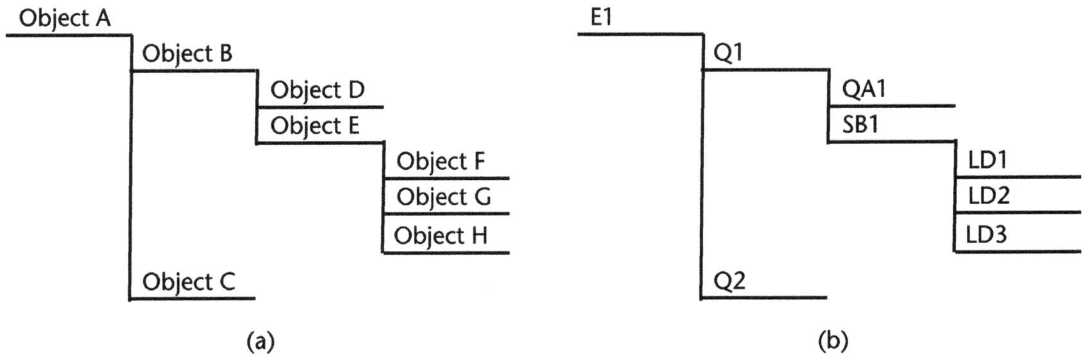

Figure 12.5 IEC 81346-1 object structure tree and its application: (a) object structure, and (b) SCL application.

When we are looking at the use of the SCL as part of the engineering of the protection and control system, it is clear that the redundancy has to be properly modeled in the process.

The redundant functional elements in an IED are naturally modeled by the inclusion of the logical nodes representing their functionality in the model. Redundant IEDs are simply included in the model as instances with unique identifiers within the bay to which they belong. Redundant process interface devices are modeled using a similar approach.

Redundancy in the communication system can be described in SCL at the level of physical connections of an access point.

12.4 SCL Files

The idea behind the substation configuration language is to support the engineering of the system by allowing the exchange of configuration information between different tools from various manufacturers. IEC 61850 defines several different types of files required to support the intended engineering process that can be recognized by their file extension. Because the system and its configuration may evolve during the engineering process and following the substation energization, it is required to provide means for adequate version control. This is achieved by the use of version and revision numbers for each SCL file.

In order for an IED or a system solution by a manufacturer to be compliant with the standard, they have to support the use of the files described below directly from the IEDs or through tools delivered by the supplier with the system.

12.4.1 IED Specification Description

This is a new SCL file that was proposed several years ago and is already used in some of the available tools. It represents the required functionality of an IED in the substation configuration language. It is similar in structure to the ICD file, but, instead of capabilities, is describing specifications for an IED. It is used for data exchange from the IED specification tool (IST) to the system specification tool (SST).

It contains exactly one IED section for the IED whose specifications it represents. Because it describes the specified functionality (i.e., not an existing device), the IED name in this file is SPEC. The file also includes the different logical node types with the required by the user data objects. The file extension shall be .ISD for IED Specification Description (ISD).

IEC 61850 does not yet specify where the ISD file comes from. The engineering tools used for the development of utility standards designed for IEC 61850 environment should have the capability of producing this XML file. The ISD file can also be used to specify the requirements for an IED as part of the procurement process to improve its quality and efficiency.

12.4.2 IED Capability Description

The default functionality of an IED in the substation configuration language is represented by the IED Capability Description (ICD) file. It is used for data exchange from the IED configuration tool (ICT) to the system configuration tool (SCT).

This ICD file describes the capabilities of an IED. It contains exactly one IED section for the IED whose capabilities it represents. As it describes the default functionality (i.e., before it has been configured), the IED name in this file is TEMPLATE. The file also includes the different logical node types as they are instantiated in the device. The file extension shall be .ICD for IED Capability Description.

IEC 61850 does not specify where the ICD file comes from. In IEDs designed for IEC 61850 environment and with large memory, this XML file may be available from the device itself. For IEDs that are based on existing platforms that were adapted to support the standard, the manufacturer is required to provide tools that output ICD files.

During the procurement process, an ISD file provided by the user will be directly compared with the ICD files of IEDs produced by the manufacturer to find the ones that contain all required logical nodes and data objects.

The ICD files approved by the user IEDs for application in standard protection and control schemes can be also available during the engineering process from a database containing a library of the files for all approved devices.

12.4.3 Instantiated IED Description

Considering the large number of components of the IED model that are optional, in order to improve the efficiency of the engineering process as part of the development of a utility standard that is implemented using specific devices from various manufacturers, in many cases, the IEDs may be delivered to the user as partially configured, thus leading to the creation of a file that represents the functionality of an instantiated ICD file, which is used instead of the ICD file. Obviously, the IED name is not going to be TEMPLATE anymore, but rather, for example, something representing the role of the IED in a standard scheme.

This file is also used during the life cycle of the substation PAC system following modification on it using the IED configuration tool. The file extension shall be .IID for Instantiated IED Description (IID).

12.4.4 System Specification Description

The description of the system is the first step in the engineering process and until now has not been based on any standardized approach. The IEC 61850 engineering process envisions the use of substation specification tools that allow the user to describe the substation design and associated functional requirements for the substation protection and automation systems.

The data exchange from such a system specification tool and other tools utilized in the process should be based on the System Specification Description (SSD) files defined in the standard; they have an SSD extension.

The SSD file describes the one-line diagram of the substation and the functional requirements related to the different primary substation devices represented by logical nodes. The logical nodes can be abstract in the sense that they are not allocated to specific IEDs. However, the ISD file allows the user to allocate logical nodes to abstract specification IEDs that can be used to engineer in a device independent from the substation specification, including the data flow.

12.4.5 Substation Configuration Description

The configuration of the engineered system is represented by the Substation Configuration Description (SCD) file. It contains a substation description section, a communication configuration section, and all IEDs.

The IEDs in the SCD file are not in their default configuration any more, but they are configured to operate within the substation protection and automation system. These files are then used to configure the individual IEDs in the system.

This file is typically used for data exchange between the system configuration tool and the IED configuration tool during the engineering of the system. However, the information that it contains about the overall configuration of the system can be very helpful in improving cybersecurity or the quality and efficiency of testing.

12.4.6 Configured IED Description

Once the overall engineering of a substation project is completed and saved as an SCD file, the IED-specific configuration information needs to be delivered to the device itself. This data exchange between the IED configuration tool and the IED uses a stripped-down version of the SCD file that contains only the substation and communications sections data related to the IED. The difference between the ICD file and the Configured IED Description (CID) file is that the second includes the substation-specific names and addresses instead of the default ones in the first.

We need to remember that the file that is actually downloaded to the IED may not be an XML file but one in a proprietary format. Also, the information in this file may not be everything that is needed for the complete configuration of the IED. For example, the device-specific programmable scheme logic may need to be added.

12.4.7 System Interface Exchange Description

With the change of focus of Edition 2 of IEC 61850 from substation to power delivery automation, as well as the use of GOOSE and sampled value messages for transmission line protection, the SCL files described above do not cover the data exchange between substations required by such applications. This resulted in the development of a new SCL file to be used for data exchange between system configuration tools for different substation projects.

This file describes the interfaces of one project to be used by the other project to establish intersubstation communications. It is important that it fixes already defined source object references in the involved IEDs and, as part of the engineering, specifies new interface connections between the projects. It also states the engineering rights and the owning project at each IED from the perspective of the importing project.

12.5 IEC 61850 SCL Engineering Process

The first step of the engineering of an IEC 61850-based substation PAC system is to define the functional specification according to the approved PAC concepts and user's standards. This is done using the substation one-line diagram and based on the established philosophy and practice defining:

- Protection functions required for each primary substation or system component;
- Measurements and status information needed;
- Controls to be used;
- Reporting requirements;
- Monitoring and recording requirements;
- Redundancy requirements;
- Communications architecture;
- Substation-level functions;
- Other as necessary.

The above should be produced by a system specification tool that provides as an output an SSD file.

If the user has a well-established standardization process that is formalized by defining the PAC requirements for a specific substation component in the form of a virtual IED represented by an ISD file, these files can be also used is an input to the system specification tool to define the data flow in the system specification description.

Once it is clear what PAC, monitoring, and recording functions are required, the system designer needs to select approved by the user IEC 61850-compliant IEDs. This means that they had successfully passed at least the conformance tests defined in Part 10 of the standard and by the UCA International Users Group Testing Subcommittee. Functional and interoperability testing of the IEDs as part of the acceptance process within the user's organization is recommended.

After the selection of the IEDs, their ICD files and the SSD file become inputs into the system configurator, the tool used to configure the substation PAC system. The key requirement for this tool is that it should support the import and export of the different types of files defined by the SCL. It is also important that such tools should be user-friendly to simplify the system engineering process.

The output of the system configurator is an SCD file that can be later used for many different applications (Figure 12.6).

The future success of IEC 61850 is, to a great extent, dependent on the completion of the object models with all function-specific settings, implementation agreements on the use of logical devices, and development of advanced tools that take full advantage of the SCD files that will then provide the complete configuration information for the substation PAC system.

The SCD file becomes an input to any of the IEDs configuration tools and is used to produce the CID file for each specific IED. Today this typically is limited to the communications section. In the future, this should include all settings of the IED as well.

The use of the CID file depends on the implementation of IEC 61850 in the IED. One option is to directly download the file in the device. Another is to convert it to a proprietary file format that is then downloaded in the IED.

If, during the engineering of the system or later after it has been energized, there are certain modifications required in one or more of the IEDs, after they have been performed by the IED configurator and updated, an IID file is created and imported to the system configurator to update the SCD file as shown in Figure 12.7.

The engineering process described above is based on the current implementation of the standard. It is just the first step in what may become a real revolution in the field of electric power systems' PAC. For this to happen, there is a need for some further development, including:

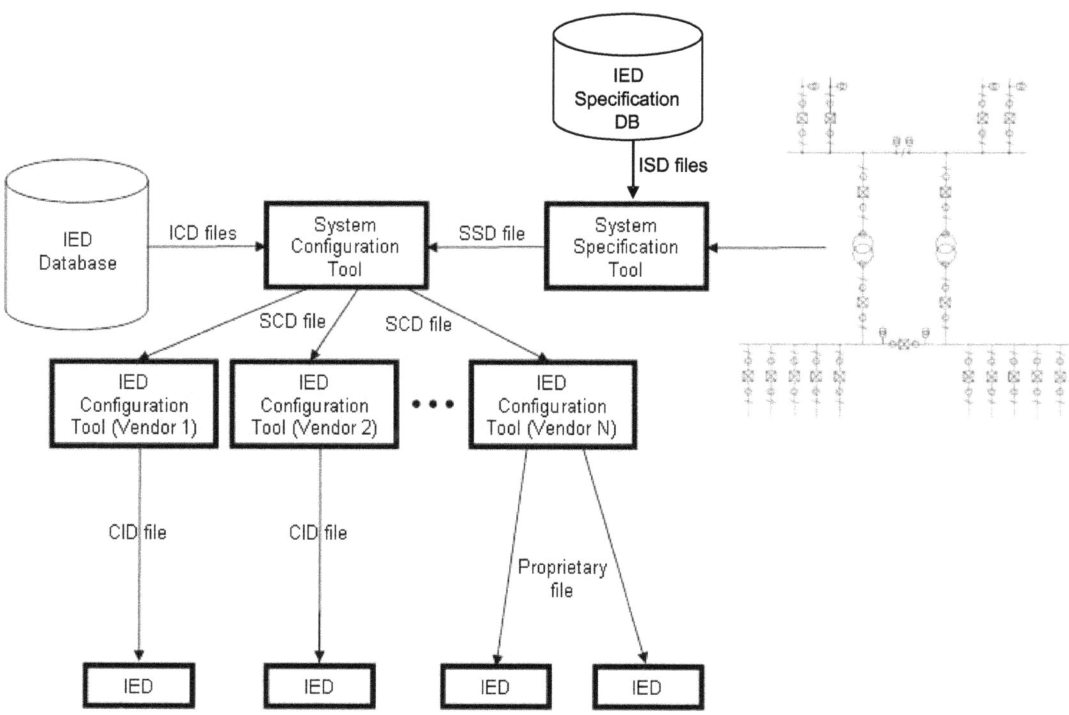

Figure 12.6 IEC 61850 SCL engineering process.

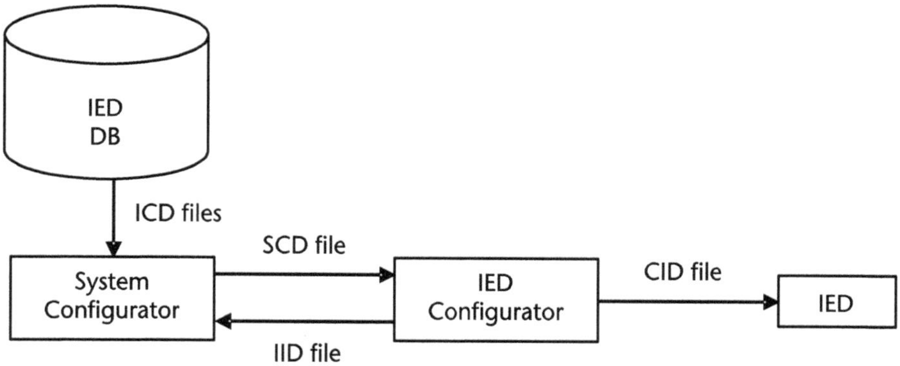

Figure 12.7 IEC 61850 SCL modification process.

- Completion of the setting parts of the logical node class models;
- Inclusion of the settings in the ICD and CID files;
- Harmonization between the common information model (CIM) model and the IEC 61850 model;
- Extensions in the system model to cover the needs of coordination and analysis tools;
- Addition of IED setting configuration functions in the system configuration tools;
- Addition of automatic test tool configuration capabilities.

As a result, it will be possible to extend the use of the SCD files for the following substation automation systems engineering tasks:

- Automatic creation of the graphical user interface from the SCD file, including the different screen layouts;
- Automatic mapping of the different measurements and status information from the IEDs to the substation HMI;
- Automatic configuration of the IED or the substation protection and control system testing process;
- Automatic substation event analysis.

The benefits from such developments will be quite significant and will not only reduce the costs for system design, factory, and site acceptance testing and maintenance, but also will improve the overall quality of the substation automation system.

12.6 SCL-Based Standardization Process

A standard secondary scheme is defined as a single set of multifunctional IEDs integrated using process and interdevice interfaces in order to provide all required by the application functions, such as protection, control, status monitoring, measurements (including synchrophasors when required), communication, condition monitoring, recording, event reporting, fault location, and power quality.

For each standard scheme, a 4-step standardization process based on the IEC 61850 SCL should be followed to achieve a high level of efficiency of the digitalization.

12.6.1 Standard Scheme Template

This is step 1 of the standardization process and covers the definition of the functional requirement specification for a standard secondary scheme based on a utility's philosophy and practice.

It is a conceptual description of the scheme (Figure 12.8). This is typically the formalized description of the application of protection and control philosophy to a specific type of bay as described above. The templates should include all of the necessary components of the documentation of each subsequent stage.

At this stage, the functional requirements and integration constraints need to be defined. These are detailed requirements associated with the bay topology, voltage level and criticality, and communications requirements, resulting in some interfaces and functions being defined.

This stage should also include primary single-line diagram, secondary functional diagrams, trip matrix, setting, and testing philosophies.

Items that are left generic at this stage are types of primary equipment or IEDs. They may be considered, but are not specified at this stage.

In addition, the description of the functional specification in the form of an IEC 61850 Substation Specification Description (SSD) file is recommended, thus allowing the automation of the procurement process based on exchange of such files between a utility and its suppliers.

Step 1 of the standardization process is performed by a utility's secondary core group of experts responsible for the engineering of standard schemes at the company level.

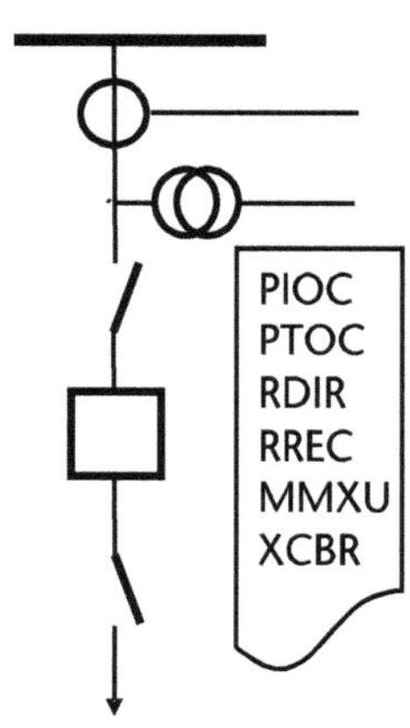

Figure 12.8 Standard scheme "template."

12.6.2 Defined Standard Scheme

This is step 2 of the standardization process and represents a development stage of a standard scheme that defines the primary plant and the hardware interfaces with the specific type of bay covered by the scheme. The circuit breaker, disconnectors, Earth switches, CT/VT and auxiliary interface specifications, and signals list or diagram (hardwired or communications based) are specified at this stage.

The defined standard scheme can be used for the same or similar types of new or existing installations without any changes in external wiring, signaling, and equipment.

The allocation of functions to generic (abstract) IEDs is also defined at this stage (Figure 12.9). The required functionality of individual IEDs can be described also using the newly defined IED Specification Description (ISD) file, thus allowing the automation of the procurement process based on exchange of such files between the utility and its suppliers. This will support automatic selection of the IEDs that meet the requirement specification for a specific standard scheme by comparing the ISD file with the existing ICD files.

The definition of the required interfaces (including quality) with the process at this stage also allows the definition of the scheme terminal blocks that are signal-specific, but not product-specific.

Step 2 of the standardization process is performed by the utility's experts responsible for the engineering of standard schemes at the company level.

12.6.3 Applied Standard Scheme

Step 3 of the standardization process is what is typically considered by the utility as a standard secondary scheme. This includes the use of approved specific IEDs or other secondary equipment (Figure 12.10). The IED selection should ensure that all functions and functional elements defined for the scheme template in stage 2 are available in the selected IEDs. This means that all logical nodes, data objects, and data attributes from the ISD file from step 2 should exist in the ICD file of the selected IED.

The IED HW, SW, and parameter-set versions, IED configuration tools, signal list (hardwired or communications based), wiring diagrams, cable lists, and

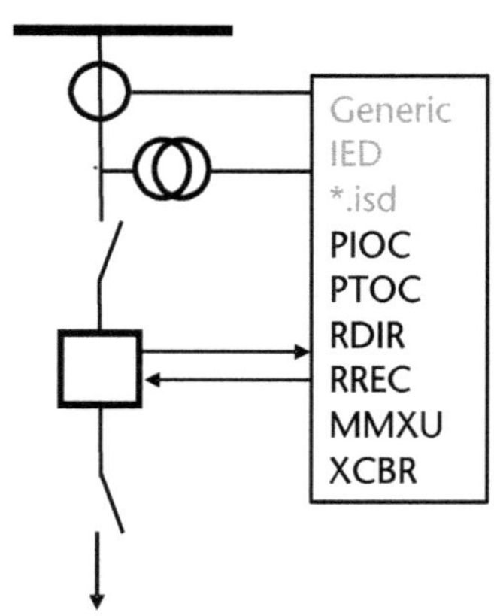

Figure 12.9 Standard scheme defined.

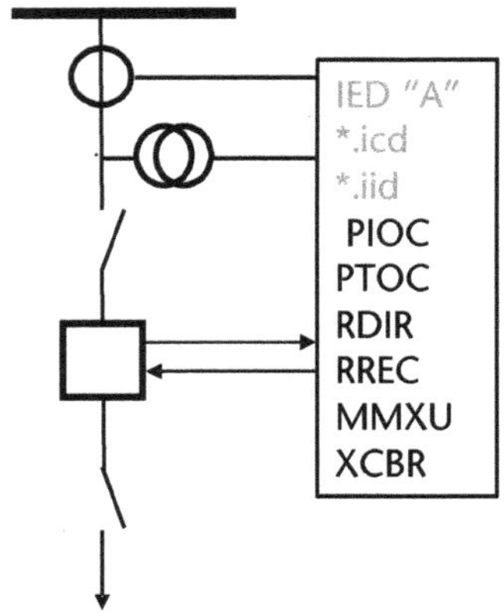

Figure 12.10 Standard scheme applied.

standard settings are specified at this stage. The global settings, including programmable scheme logic, of the IEDs are introduced at this level of standardization. In case the utility wants to use a different supplier for the same scheme, it will result in a different scheme template implementation that meets the requirements of the above definitions.

Because, at this stage, all IEDs and their interfaces are defined, the functionality of the standard scheme is configured using IEC 61850 engineering tools based on the ICD files of the individual IEDs and documented as an IID file that will be available at this stage to be used in the engineering of specific substations. However, at this stage, there are still no local settings or other site-specific configuration parameters.

Step 3 of the standardization process is performed by the scheme supplier based on the documentation produced in step 2. The development of the standard scheme should also involve at least two members of the utility's group of experts responsible for the engineering of standard schemes at the company level. The standard scheme should be subject to type testing before it is approved for use by authorized utility representatives.

12.6.4 Instantiated Standard Scheme

This is a site-specific implementation of the standard scheme (i.e., an instantiated standard scheme from stage 3). Site and application-specific settings are implemented at this stage, and all hardware is defined. Based on this information and the IID files for the individual IEDs, a substation configuration description (SCD) file is created using the system configuration tool (Figure 12.11). From this file the individual IED configuration tools extract the configured IED description (CID) files used to configure the IEDs for operation in the substation.

While instantiation excludes any modifications besides site-specific setting parameters and site-specific naming, specialization on stage 4 offers the opportunity to adapt the standard scheme typically to variations in primary HW-components when used in existing sites.

At this stage, the IED-specific calculations and setting files as well as specific commissioning, maintenance, and testing procedures are applied.

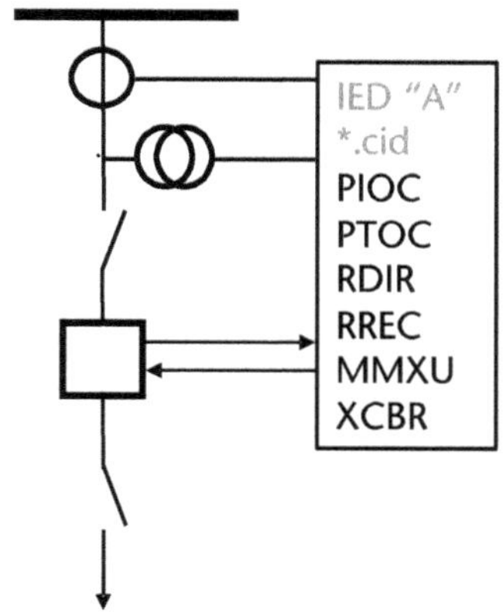

Figure 12.11 Standard scheme instantiated.

In special cases, other modifications can also be considered. This may appear as a deviation from the standardization process; however, it takes advantage of the developed standard scheme for a special application that may not justify going back to stage 3 of the process.

Step 4 is performed by the secondary scheme supplier based on setting files and procedures, as well as SSD files supplied by the utility. Test reports and other documentation produced during the production, configuration, and commissioning of the scheme should be reviewed and approved by authorized utility representatives before the scheme is delivered to the site.

References

[1] Implications and Benefits of Standardised Protection and Control Schemes, CIGRE TB 584, 2014.

[2] IEC 61850-6:2009 Communication Networks and Systems for Power Utility Automation—Part 6: Configuration Description Language for Communication in Electrical Substations Related to IEDs, Edition 2, 2009.

[3] IEC 81346-1 ED. 1.0 B:2009 Industrial Systems, Installations and Equipment and Industrial Products—Structuring Principles and Reference Designations—Part 1: Basic Rules. 2009.

[4] IEC 61850-5:2013 Communication Networks and Systems for Power Utility Automation—Part 5 Communication Requirements for Functions and Device Models, Edition 2, 2013.

[5] IEC 61850-7-2:2010 Communication Networks and Systems for Power Utility Automation—Part 7-2: Basic Information and Communication Structure—Abstract Communication Service Interface (ACSI), Edition 2, 2010.

CHAPTER 13

Time and Its Applications in Protection and Control Systems

13.1 Introduction

Time is something that is a fundamental part of our lives, something that no one of us can live without. To start with a definition of time, it is "the duration in which all things happen, or a precise instant that something happens" (https://www.yourdictionary.com/time). If we look at the world around us, we can see that most of it exists without time. All creatures, except most humans, do not have calendars and clocks. They live in the moment, based on their sensors that tell them what to do at that moment. Even humans have been living like this for thousands of years. If we reach a state of enlightenment and we are able to live in the moment, we do not need time, because the past and the future do not exist; only the present exists. So we do not need to worry about when something happened in the past, when it will happen in the future, or how long it will last.

However, with human evolution, we started developing as a very different kind of beings, observing the stars and the movements of the planets, asking questions about what the world around us is, growing plants, and raising animals for food. We started learning from the past and planning for the future, so we can improve the efficiency of our existence. This is why the ancient people started first developing calendars and introducing the concepts of time and methods and tools for how to measure it. From sundials and water clocks at the beginning, we are now using satellites' synchronized clocks to coordinate our daily activities.

Something different happened in the protection and control industry. It has existed for less than 200 years and started at a time when we already had a well-established concept about time and had clocks and watches. However, the electromechanical protection devices had no idea about what time it is and when an operation occurred. If the relay tripped during a fault, in the best case, we have a target indicating that it operated. Maybe it also showed which was the faulted phase, but we had no idea at what time the fault occurred. We did not have the technology and the power system was different.

The development of microprocessor relays changed all of that because now they had a clock. We set the time and when a fault occurs, we have a time stamp telling us when it happened. If it is a single relay operation, knowing the approximate time of the event is good enough. If we have multiple relays operating within the same period of time, how do we know if there was a misoperation?

This is when we started thinking about and requiring time synchronization. We decided that, for proper analysis of protection operation and wide area disturbances, the accuracy of the time synchronization of all devices needs to be at least 1 ms.

In this new century, we live in the world of smart grid, digitization, and IEC 61850. We have synchrophasor measurements and merging units that need to be synchronized with 1-μs accuracy based on Global Positioning System (GPS) and other satellite signals from atomic clocks. Time has become a critical component for the efficient management of the changing electric power grid.

That is the reason to focus this chapter on the different aspects of time and its impact on PAC systems: to talk about time settings and measurements, time synchronization standards and networks, time models and latency, and many other time-related issues, because in the PAC world we need to know when things happen.

13.2 Time in PAC Systems

PAC applications are some of the key components in the reliable and secure operation of the evolving grids of the twenty-first century. The complexity of today's electric power grid and its changing dynamic characteristics together with the rapid development of computer and communication technologies resulted in the transition from the conventional hardwired PAC systems to IEC 61850-based solutions in digital substations.

Time plays an important role in everything in our lives. It is used to:

- Establish when an event occurred;
- Measure the duration of an event;
- Specify the duration of an event;
- Measure the time interval between events;
- Schedule an event;
- Specify the time interval between events.

It is also required for several purposes:

- To synchronize multifunctional protection IEDs in order to be able to analyze their operation following fault detection and clearing, or wide area disturbances;
- To synchronize merging units or phasor measurement units (PMUs) in order to be able to use analog measurements from different locations;
- To synchronize transient or disturbance recording devices in order to be able to align the records for fault or disturbance analysis purposes.

That is why the development of the IEC 61850 standard had to consider the requirements for time in the data and object models, as well as in the communications service definitions.

At the same time, the standard has to adapt to the developments in time synchronization technology and how it can be implemented in the different PAC systems.

13.3 Time-Related Definitions

In the previous section, we used the term time in many different contexts, which underlines the fact that time is a very complex issue. That is why we need to first look at some of the definitions related to the term time and how it is used and measured.

If we do a search on the Internet, we will find different definitions, but one that is probably most suitable is the starting point that it represents the indefinite continued progress of existence and events in the past, present, and future regarded as a whole. At the same time, we can say that it is an instance of something happening, a point of time as measured in hours and minutes past midnight or noon.

The term second was used for the first time more than 1,000 years ago by the Persian scholar Abu Rayhan Muhammad ibn Ahmad al-Biruni, who subdivided the hour into minutes and seconds. However, developing the tools to measure the second took a few more hundreds of years with the first mechanical clocks to mark the second appearing in the 1500s. In 1644, the French priest and mathematician Marin Mersenne suggested using a pendulum to define the second for the first time, leading to the second pendulum clock built around 1673 by Christiaan Huygens and the international adoption of grandfather clocks by the end of the seventeenth century.

In 1967, according to the International System of Units (SI), the second became the basic unit of measuring time with the second representing 1⁄86,400 of a day or of the time that it takes the Earth to rotate once on its axis. This number is derived based on:

- 1 day has 24 hours;
- 1 hour has 60 minutes;
- 1 minute has 60 seconds.

The problem with this definition is that the rotation of the Earth is not constant, but varies depending on different factors, which means that the duration of a second will be different at different moments in time. This required the identification of a different way of measuring the second that is independent from the Earth's rotation.

A really precise measurement of the passing of time became possible with the discovery of atomic clocks, which work by measuring properties of certain atoms. The invention of the cesium atomic clock in 1955 laid the foundation for the definition of the SI second. In 1967, the SI second was officially linked to the cesium clock and defined as the duration of 9192631770 periods of the radiation corresponding to the transition between the two hyperfine levels of the ground state of the cesium 133. This solved the problem with accurately measuring the second but does not answer the question of what time it is.

Even though this looks like the simplest question to answer, in reality, it is quite complicated and starts with “It depends!” First, it depends on what time reference is being used by the clock measuring the time.

Greenwich Mean Time (GMT) is the first such reference. In 1884, the local mean solar time at the Royal Observatory in Greenwich near London was chosen as the international time reference to be used for defining the universal day. This was followed in 1928 with the introduction of the term Universal Time (UT), referring to the astronomical GMT time with the day starting at midnight.

From 1961, the Bureau International de l’Heure began coordinating internationally UT, which, in 1970, as the result of the work by an international advisory group of technical experts within the International Telecommunication Union (ITU), became the Coordinated Universal Time system. The acronym for this name would be CUT, while for the French “temps universel coordonné,” it will be TUC. To avoid issues related to favoritism towards any particular language, the abbreviation UTC was selected and is commonly used today as a nonlanguage-specific acronym, in English standing for Universal Time Coordinated. The modern term for this astronomical time is UT1.

Today the term GMT is more commonly used to refer to the time zone at the prime meridian at Greenwich (0° longitude), in which case it is being used as a local representation of UTC and not UT1. UTC is based on the International Atomic Time (TAI), but it is adjusted by leap seconds to account for the difference between the definition of the second and the rotation of Earth to always keep UTC within less than 1 second of UT1.

UT1, or, as it is sometimes called, mean solar time, is based on the rotation of Earth, which is irregular and leads to the requirement for the adjustment. The first leap second was inserted into the UTC time scale on June 30, 1972, and the last was inserted on December 31, 2016. These corrections keep UTC in conjunction with the apparent position of the Sun and the stars, and it is the standard used for all general timekeeping applications. This creates a lot of challenges for time-related PAC applications, as well as its relationship to other time scales in use.

International Atomic Time (TAI for Temps Atomique International in French) is an international time scale [8] that is computed by taking the weighted average of more than 300 atomic clocks located at more than 60 timing laboratories around the world. The accuracy of TAI is derived from data from primary frequency standards, which are clocks built at several national metrology institutes. TAI is computed by the International Bureau of Weights and Measures (BIPM) located near Paris, France.

The final implementation of the SI second for UTC took place on January 1, 1972, at 00:00:00 UTC defining an offset of exactly 10 seconds to TAI. From 1972 until now, a total of 27 leap seconds were inserted, resulting in a total offset of 37 seconds between TAI and UTC, with TAI being ahead of UTC as shown in Figure 13.1. Leap seconds can be positive or negative. The insertion of a negative leap second reduces the offset between TAI and UTC. When necessary, leap seconds are always added at the end of December or June.

The GPS is a constellation of satellites, each of them carrying atomic clocks. The time on each satellite is defined based on the time scale at the GPS Master

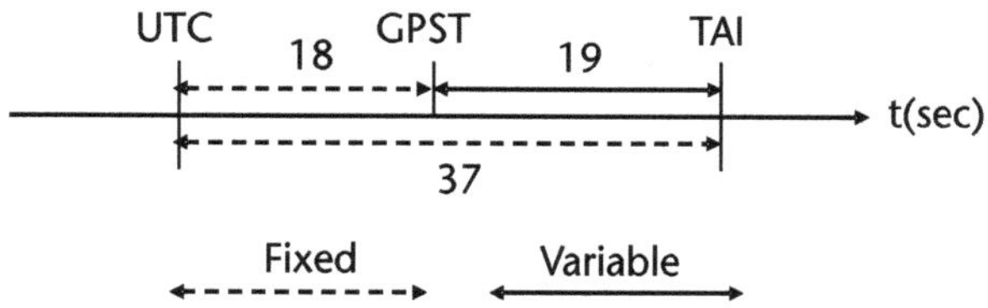

Figure 13.1 Relationship between time scales.

Control Station, which is monitored and compared to UTC(USNO). The GPS time scale (GPST) was started on January 6, 1980, 0:00:00 UTC. At that date, the offset between UTC and TAI was 19 seconds. The GPST is a linear time scale not taking into consideration leap seconds. Therefore, the offset between GPST and TAI will remain constant. The time offset of GPST from UTC is contained in the GPS broadcast message and is usually applied automatically by GPS receivers.

While UTC is typically used for many applications, including in PAC systems, many times, it is necessary to display it in local time. To do that requires taking into consideration other factors such as the time zone and if daylight savings time is enabled or not.

13.4 Time-Related Requirements

Time synchronization has been a requirement for electric power systems' PAC systems since the introduction of microprocessor based multifunctional PAC and recording devices.

The need was to be able to analyze the operation of multiple devices at different physical locations in response to an electric power system event. This was further extended based on the requirement to measure electric power system parameters in different physical locations.

The IEC 61850 standard uses the Piece of Information for Communication (PICOM) concept developed by CIGRE Working Group 34.03 and published in Technical Brochure B5 180 "Communication Requirements in Terms of Data Flow Within Substations." It is focused by definition on the exchanged data between two application functions or subfunctions within a substation. One of the components of the PICOM is the Time tag defined as an absolute time to identify the age of the data if applicable.

IEC 61850 defines different requirements for the development of the IEC 61850 standard [1] that also include a time-related requirement as follows:

- Accuracy: Depending on the application, different time accuracy is required.
- The time stamp shall be based on an existing time standard (UTC is generally accepted as the base time standard).
- The time model shall be able to track leap seconds and provide enough information to allow the user to perform delta time calculation for events crossing the leap second boundary.

- The time-stamp model shall contain sufficient information that would allow the client to compute a date and time without additional information such as the number of leap seconds from the beginning of time.
- The time stamp shall be easily derived from commercially available time sources.
- The overall time model shall include information to allow the computation of local time.
- The time model shall allow for half-hour offsets for local time.
- The time model shall indicate whether Daylight Savings is in effect or not.
- The format shall last at least 100 years.
- The time-stamp format shall be compact and easily machine-manipulated.

The industry has been looking at the topic of time stamping for many years in an effort to standardize the definitions based on a common understanding of the types of time stamps required. IEC 61850-5 [1] defined three different kinds of events, which need a dedicated time allocation procedure:

- If an event is defined as result of computation (internal or calculated event), allocation of time (time tagging) shall be done immediately within the time resolution of the clock. No special measures are needed.
- If an event is defined as change of a binary input, the delay of the debouncing procedure of the input contact has to be considered. The event time shall be locally corrected.
- If an event is defined as change of an analog input, the delay of the filtering procedure of the input circuit has to be considered. The event time shall be locally corrected.

This strong event time definition ensures that the processing of the time stamp becomes independent from the communications system latency and does not require correction by the receiving function.

Table 13.1 [1] shows the time performance classes for almost all event-related applications and depends on the supported functionality.

Table 13.1 Time Synchronization Classes for AC Applications Synchronization

Time Synchronization Class	*Accuracy [μs]*	*Phase Angle Accuracy for 50 Hz [°]*	*Phase Angle Accuracy for 60 Hz [°]*
T0	10,000	180	216
T1	1,000	18	21.6
T2	100	1.8	2.2
T3	25	0.5	0.5
T4	4	0.1	0.1
T5	1	0.02	0.02

With the introduction of synchrophasor measurements, the need for much more precise time synchronization arose. This is due to the fact that a 1-ms accuracy (T1 class in Table 13.1) results in a phase angle measurement error of 21.6° in a system with a nominal frequency of 60 Hz. To reduce this error, the time synchronized has to be in the microsecond range with different performance classes being defined in IEC 61850. T3 accuracy will be OK for most applications, but it is commonly accepted that PMUs need to be synchronized to meet T5 (i.e., with an accuracy of 1 μs).

With the introduction of optical current and voltage sensors or stand-alone merging units used for process interface in digital substations, a similar requirement was established for their time synchronization.

13.5 Time in the IEC 61850 Model

As described in the previous section, IEC 61850 introduces requirements for the time-tagging of events for use in different PAC applications. This requires also the definition of time-related data objects and attributes in the IEC 61850 model.

The different data attributes and data types used in the standard data model were defined in IEC 61850-7-3 [3].

One of the common attributes of many data objects is *t*, representing a time stamp related to the last change of the value of an attribute in the data object. A time stamp may also reflect a change in the quality attribute. For the different CDCs, *t* applies to different data attributes, such as:

- Status;
- Activation;
- Operation;
- Control;
- Measured value.

The TimeStamp type represents a UTC time with the epoch of midnight (00:00:00) of 1970-01-01 specified.

Another time-related attribute in the model is FractionOfSecond representing the fraction of the current second when the value of the TimeStamp has been determined.

The TimeQuality is an attribute in the object models that provides information about the time source of the sending device and includes information about:

- LeapSecondsKnown;
- ClockFailure;
- ClockNotSynchronized;
- TimeAccuracy.

The above information is used by the receiving device to properly process the data in the message.

13.6 Time Settings for Protection Functions

Many protection functions do not operate instantaneously after the parameter that they monitor exceeds the setting threshold. They are delayed by a time setting or a time characteristic. As we have already discussed, the names of data objects in the standard typically are based on combining abbreviations of specific words. The abbreviations related to time are as follows:

- Tm for time;
- Tmh for time in hours;
- Tmm for time in minutes;
- Tms for time in seconds;
- Tmms for time in milliseconds.

A simple example is the time delay of the operation of a distance element in a transmission line protection represented by the logical node PDIS.

It is represented by the data object OpDlTmms (Figure 13.2) and is defined in the IEC 61850 standard as "time delay before operating once operate conditions have been met" in milliseconds.

The time delay setting is of the CDC ING (Common Data Class Integer) status setting. The attributes of ING are:

- setVal: Value of the setting;
- minVal: Minimum setting for "setVal";
- maxVal: Maximum setting for "setVal";
- stepSize: The step size of the setting;
- units: The units of the setting.

In the case of the time delay setting OpDlTmms, it is in milliseconds. Similarly, RsDlTmms represents the setting of the reset time delay in milliseconds.

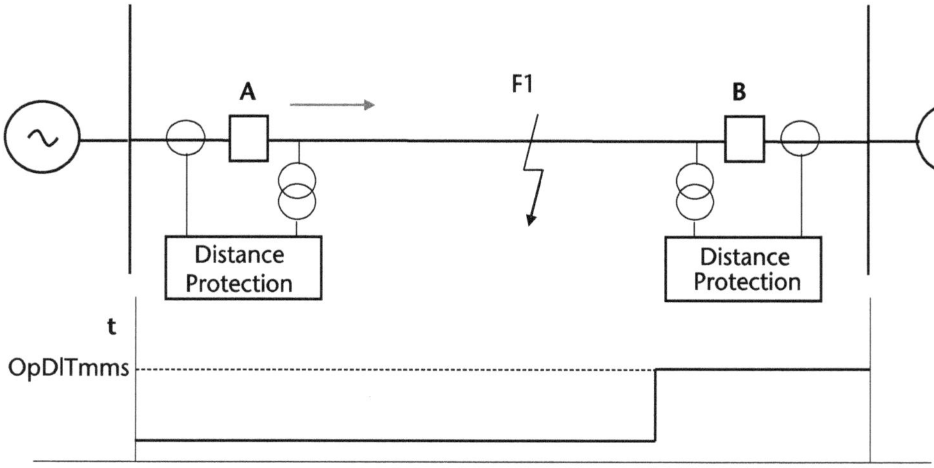

Figure 13.2 Distance Zone 2 time delay setting.

13.7 Time in the GOOSE Model

High-speed P2P in IEC 61850-based protection and control systems use a specific method designed to meet a variety of requirements. The GSE method can be considered as a mechanism for unsolicited reporting by a logical device. The achievement of speed performance, availability, and reliability depends on the implementation in any specific device.

The generic substation event model is discussed in detail in Chapter 10, so here we are just going to look at the time-related issues.

The data in the published GOOSE messages is a collection of values of data attributes defined as members of a data set. As many of the attributes that are included in the data set have an associated time stamp, if the subscribing device needs the information about the moment in time when the value of a specific attribute changed, it becomes critical to include the time-stamp attribute in the data set.

A GSE control class in the publisher is used to control the process. If the value of at least one of the DataAttributes has changed, the transmission buffer of the publisher is updated with the local service "publish" and the values are transmitted with a GOOSE message that includes the time when the value has changed. The receiver reads the values from a local buffer at the receiving side and, in time-related functions, processes the data not based on the time when the message was received but based on the time stamp in that message.

The publisher/subscriber mechanism allows the source IED to reach multiple receiving IEDs, thus significantly improving the efficiency of the communications interface. The time when a message will be received by the subscribers may vary depending on the latency affected by the communications architecture and the traffic on the network.

Because the GOOSE messages replace hardwired signals used for protection and control applications, IEC 61850 introduces mechanisms that ensure the delivery of the required information. Once a new value of a data attribute has resulted in the multicasting of a new GOOSE message, the repetition mechanism ensures that the message is sent with a changing time interval between the repeated messages until a new change event occurs.

One of the data objects in a GOOSE message is StNum, state number. It represents the value of a counter that increments each time a GOOSE message has been sent and a value change has been detected within the GOOSE dataset. The initial value for StNum upon a transition of GOOSE Enable (GoEna) to TRUE is 1.

Any change of value in the GOOSE data set will also result in a new value of the parameter T (TimeStamp) and StNum in the GOOSE message.

As shown in Figure 13.3, at the beginning after a change (newEvent in Figure 13.3), the parameter StNum is incremented by 1 and SqNum is set to a value of 0. It is a counter that increments each time a repeated GOOSE message has been sent, while StNum maintains the same value.

Following a change, the repetition interval is very short: a few milliseconds. This time is set by *MinTime*, an MMS component defining the sending delay on a data change between the first immediate sending of the change and the first repetition in milliseconds.

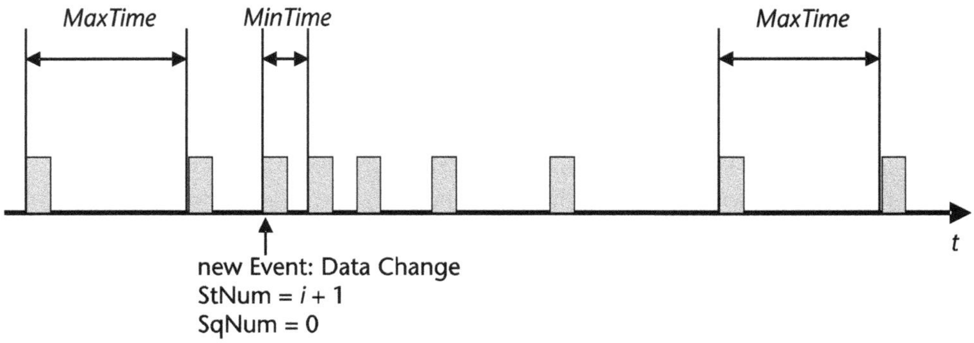

Figure 13.3 Repetition intervals.

The repetition interval gradually increases until it reaches a maximum value of one or more seconds. This time is set by *MaxTime*, the source supervision time in milliseconds (supervision heartbeat cycle time).

This method achieves several important tasks:

- Ensures that a loss of a single message is not going to affect the functionality of the system;
- Allows any new device to inform all subscribing devices about its state;
- Allows any new device to learn the state of all publishing devices to which it subscribes.

The GOOSE messages contain information that allows the receiving devices to know not only that a status has changed, but also the time of the last status change. This allows a receiving device to set local timers relating to a given event.

As we have seen earlier in the book, IEC 61850 defines what is the transfer time but does not describe how it can be measured or calculated. However, if we look at some of the time-related attributes defined in the standard, we can find a fairly straightforward method for calculating the transfer time based on time stamps.

If we consider two IEDs that are synchronized using Precision Time Protocol (PTP) and one of them operates, which makes the trip-conditioning logical node PTRC operate, the moment when this happens is time-stamped with the value available in the attribute PTRC1.Op.t. A GOOSE message is sent by the communication processor over the communication network and, for example, received by the device performing breaker failure protection, which will result in the start of the breaker failure logical node RBRF. This is also time-stamped with the value available in the attribute RBRF1.Str.t.

If we have an analysis tool that will capture GOOSE messages from the two IEDs containing these two data attributes in the datasets, we can easily calculate the transfer time as the difference between the two time stamps.

Transfer time = RBRF1.Str.t - PTRC1.Op.t

As can be seen from Figure 13.4, this time includes the time of the two communication processors plus the time over the communication network.

If we are interested in the time over the network, it can be measured by capturing Ethernet packets by PTP-synchronized devices at the locations of the publishing

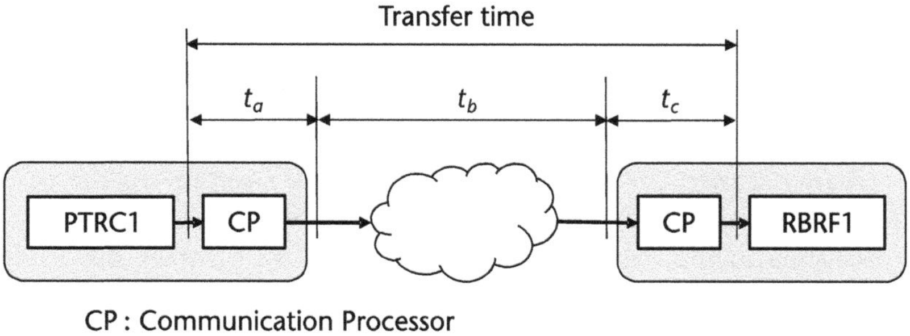

Figure 13.4 Transfer time calculation principle.

and subscribing devices. An analysis tool will then evaluate the differences from the time stamps of the same packet captured at the two locations. Considering the GOOSE repetition mechanism, a large number of packets will be available and can be used to perform statistical calculations.

These methods can be used to verify that the implementation meets the requirement according to the transfer time class [1] specified in Table 13.2. For example, for breaker failure initiation, the transfer time should be within T6, while, for a change of state of a disconnecting switch, T4 or even T3 will be sufficient.

13.8 Time in Sampled Value Communications

IEC 61850 uses merging units based on IEC 61850-9-2 or IEC 61869-9 for the analog interface in digital substations.

Most existing installations are based on IEC 61850-9-2 LE and use two modes of sending sampled values between a merging unit and a device that uses the data. For protection applications, the merging units send 80 samples/cycle in 80 messages/cycle (i.e., each Ethernet frame contains a single set of V and I samples). For waveform-recording applications, such a sampling rate may not be sufficient. That is why 256 samples/cycle can be sent in groups of 8 sets of samples per Ethernet frame sent 32 times/cycle.

When we think about time, it is clear that both the number of samples per cycle and the number of messages through which they are sent will have an impact on the performance of the system and have to be analyzed as part of its design. They

Table 13.2 Transfer Time Classes

Transfer Time Class	*Transfer Time [ms]*	*Transfer of:*
TT0	>1,000	Files, events, log control
TT1	1,000	Events, alarms
TT2	500	Operator commands
TT3	100	Slow automatic interaction
TT4	20	Fast automatic interaction
TT5	10	Releases, status changes
TT6	3	Trips, blockings

also should be measured as part of the commissioning and maintenance of the PAC system in digital substations.

The sampled analog values model in Edition 1 of the standard was intended for the communication of individual samples in a waveform with very short time intervals between the publishing of individual samples on the substation communications network. These time intervals t_{int} were in the range of a hundred microseconds in the case of the typically used 80 samples per cycle for protection applications.

With the change of focus of the standard from substation automation to power system automation, it was decided that the same streaming mechanism used for sampled values and described in detail in Chapter 11 can be used also for the transmission of synchrophasor measurements over wide area networks.

This resulted in some changes in the options available for the definition of the sampling mode *SmpMod* as follows:

- Samples per nominal period;
- Samples per second;
- Seconds per sample.

The default mode is samples per nominal period and it can be used for P class synchrophasors as well, as, in some cases, there are 4 calculations per period. The option samples per second is the one typically used for M class synchrophasors, while seconds per sample can be used for slowly changing parameters such as temperature.

The duration of the time intervals t_{int} between messages depends on the sampling rate and the number of samples in a message. As can be seen from Figure 13.5, with the same sampling rate of 80 samples per cycle, if we use IEC 61869-9, the time interval between messages will be twice as long as the time interval if we are using IEC 61850-9-2 LE. This is because with IEC 61869-9 we have two samples per message, while with IEC 61850-9-2 LE we have a single sample per

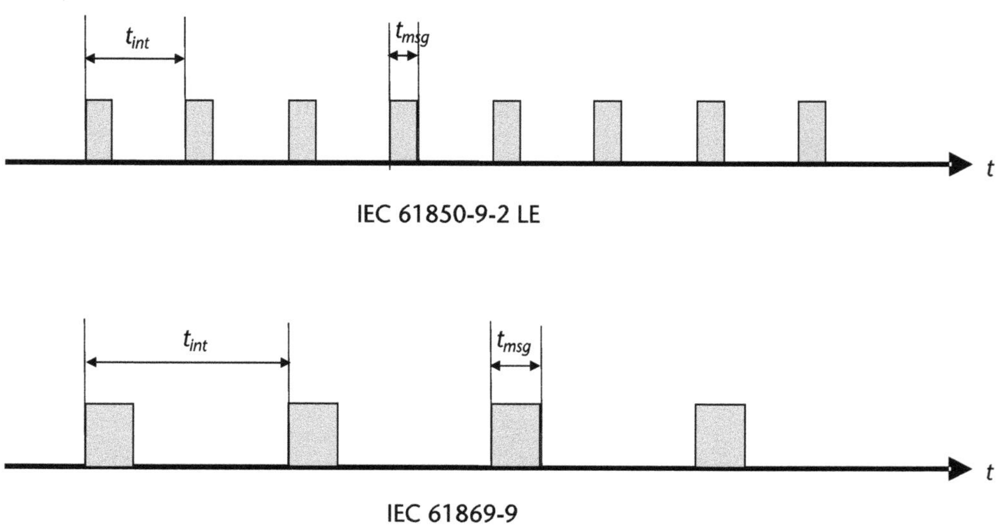

Figure 13.5 Sampled analog values publishing.

message. However, it is obvious that, with more samples, the time t_{msg} that it takes for the message to go through the switch will become longer. This time also will depend on the communication speed, but, for a high-speed Ethernet, this time will not be significant.

It is important to note that each sample comes with a time stamp, which, in reality, does not show the time when the sample was taken but instead indicates which consecutive sample it is within a second. When a synchronization pulse or message is received, this counter *SmpCnt* value is set to 0 and is incremented by 1 with every consecutive sample until the start of the next second. That is why, for protection applications, such as busbar differential protection, it is sufficient to use the samples from the different merging units that have the same sequence number. However, if we are using the sampled values for disturbance recording, the sequence number has to be converted to UTC time in order to be stored for processing and visualization by the different tools.

13.9 Time Protocols

Edition 1 of IEC 61850 defined the Simple Network Time Protocol (SNTP) as the protocol for time synchronization. It provides sufficient accuracy (1 ms) for the time-tagging of events but cannot be used for the synchronizing of merging units and PMUs that require time synchronization accuracy of 1 μs. That is why, in Edition 1-based systems, the accurate time synchronization is based on 1-pps signals.

The introduction of 1588-2002—IEEE Standard for a Precision Clock Synchronization Protocol for Networked Measurement and Control Systems [4] provided a much better solution for accurate time synchronization in digital substations. It is known as Precision Time Protocol (PTP) and is applicable to systems communicating by LANs supporting multicast messaging including but not limited to Ethernet. It enables heterogeneous systems that include clocks of various inherent precision, resolution, and stability to synchronize to a grandmaster clock and supports system-wide synchronization accuracy in the submicrosecond range with minimal network and local clock computing resources. PTP utilizes a continuous timescale based on TAI.

IEEE 1588 V2 [4] was published in 2008 to improve the accuracy, precision and robustness; however, it is not backward-compatible with the 2002 standard. The last edition of the standard "1588-2019—IEEE Standard for a Precision Clock Synchronization Protocol for Networked Measurement and Control Systems" [5] was officially published on June 16, 2019, and is also referred to as PTP v2.1. It supersedes both standards 1588-2002—IEEE Standard for a Precision Clock Synchronization Protocol for Networked Measurement and Control Systems and 1588-2008—IEEE Standard for a Precision Clock Synchronization Protocol for Networked Measurement and Control Systems.

It allows the operation of 1588-2008 (PTP v2.0) systems together with IEEE 1588-2019 (PTP v2.1) systems with certain limitations but discontinues the compatibility with 1588-2002 (PTP v1.0).

In the 2019 PTP v2.1 document includes many changes compared to PTP v2 such as a revision of the specifications of a transparent clock, replacement of the 2008 architectural model by a layered model, and definition of TLVs (type, length, value) that are used to extend a PTP message with some extra information for some optional feature, such as for cryptographic message and source authentication.

PTP is using a master-slave principle in which all slave clocks are synchronized to a single grandmaster clock of the system. The grandmaster clock itself is typically synchronized to a primary time reference, a GNSS receiver. The synchronization of the slaves is performed based on the exchange of data packets as shown in Figure 13.6 in a way similar to what protection engineers know as the ping-pong principle used in line differential protection. To ensure that the correct time offset between the master and the slave can be determined, it is necessary that the propagation delay for data packets through the network is equal in both directions. This is accomplished by using in the time synchronization system transparent clocks: PTP-capable Ethernet switches. They time-stamp PTP packets at ingress and egress. From these time stamps, the time of the packet inside the switch is calculated and added into a field of the respective data packet or a follow-up message.

Considering the many options available in IEEE 1588 V2 in order to adapt it for the needs of power system applications, IEEE and IEC developed profiles of the standard. The IEEE PSRC initially released IEEE C37.238-2011, the original power profile for PTP [8]. It defined clock accuracy requirements; additional type, length, and value structures; and settings limits. In 2016, IEC TC 57 Working Group 10 released their version of the power profile, IEC/IEEE 61850-9-3 Edition 1 [7]. This version adjusted the profile to include specific requirements deemed critical for use in power system timing by IEC.

IEEE updated the C37.238 standard (C37.238-2017) to align with the IEC standard and provide additional requirements. The most significant change was splitting this standard into two standards: the new joint standard IEC/IEEE 61850-9-3 and the now revised IEEE C37.238-2017. The reason for the change was that the two main sponsoring organizations IEC Technical Committee (TC) 57 Working Group (WG) 10 and IEEE Power and Energy Society (PES) Power Systems Relaying and Control Committee (PSRC) expressed significant differences in their expectations from this work. The result was that IEC/IEEE 61850-9-3 became the technical specification document, providing the baseline performance level required for all applications. Additional requirements supported by the revised IEEE C37.238 include real-time updates of the delivered time quality (inaccuracy), plus inclu-

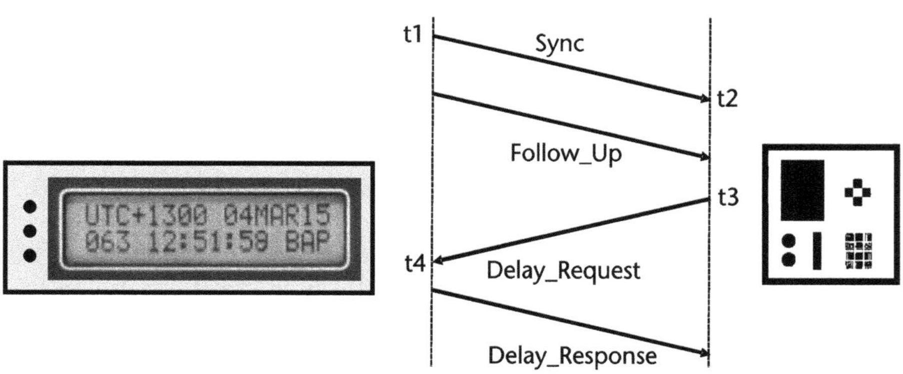

Figure 13.6 PTP synchronization.

sion of the data needed to support the migration of substation time-distribution networks from IRIG-B.

13.10 Time Synchronization Systems

Time synchronization in digital substations is achieved based on the interaction of multiple devices with different clock capabilities over the LAN.

IEC 61850-9-3 describes several clock types that play different roles in the time synchronization system shown in Figure 13.7:

- A transparent clock is an Ethernet switch with PTP capability. It time-stamps PTP packages at ingress and egress. From these time stamps, the residence time of the package inside the switch is calculated and written into a field of the respective data package or a follow-up message.
- A boundary clock is a clock that has ports in two or more domains. The boundary clock synchronizes to a grandmaster in one domain and acts as a grandmaster in all other domains. It is used to time-synchronize two or more separate networks infrastructures to one grandmaster without the need of bridging data packages between the networks.
- A slave-only clock has port(s) that are always in the slave state. It will lock itself to the grandmaster of the network. In the case that no grandmaster is present, it will remain in the slave state and never announce itself as a grandmaster.
- A grandmaster-only clock announces itself as a grandmaster. If it is the best clock in the network, it will become the grandmaster. Otherwise, it will

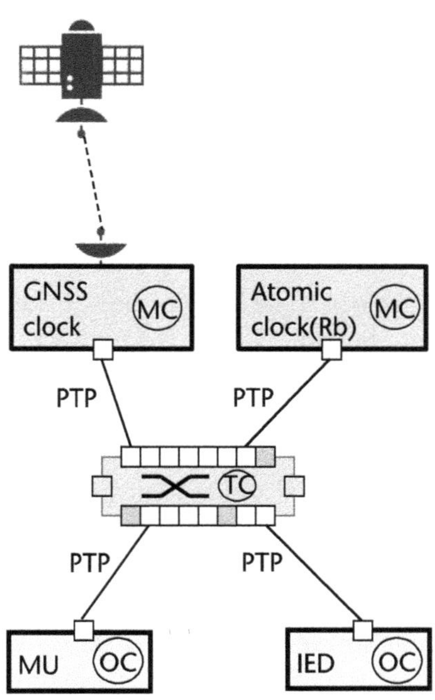

Figure 13.7 Time synchronization system components.

switch its ports to passive. A grandmaster-only clock will always lock only to its primary time reference (GPS or other GNSS) but never to another grandmaster in the network.

- A grandmaster-capable clock can switch its ports either to the master or the slave state. Further on, a grandmaster-capable clock does not necessarily require a primary time reference. In case that all grandmaster-only clocks in a network are malfunctioning or are switching to holdover, a grandmaster-capable clock equipped with an accurate internal oscillator can become grandmaster of the network.

The merging unit and the IED have ordinary clocks (OC) and are synchronized over PTP through a transparent clock (TC) with a grandmaster clock GNSS (MC). A Rubidium atomic clock can be used as a backup in case of loss or jamming of the radio signal.

To achieve the required values for the network time synchronization parameters, it is necessary to carefully analyze the characteristics of all involved devices and to consider the placement of masters and redundant masters and the impact of the possible network topology. This is especially important in larger networks and more demanding applications.

Considering the importance of precise time synchronization for the operation of merging units and phasor measurement units and related protection and control functions, the time synchronization system must meet the requirements for resilience, which means that it should be able to withstand any degradation in the system performance, failure of equipment, or cyberattacks.

To meet such requirements, a substation should rely on two redundant time servers. To avoid the loss of the reference signal for a longer period in challenging geographical locations, GNSS signal jamming or spoofing, and the impact of solar storms, the redundant time servers should be of different types and it is recommended to use an atomic clock as a backup (Figure 13.7). Substation hardened clocks capable of receiving multiple GNSS signals with built-in Rubidium oscillators are available and affordable.

13.11 Time Synchronization Sources

Time synchronization of the devices in the substation is based on a dedicated time server that receives the time signal from single or multiple GNSS (space-based global navigation satellite system): GPS, GLONASS, Galileo satellites (Figure 13.8). GNSS [9] provides reliable positioning, navigation, and timing services to users on a continuous worldwide basis freely available to all.

The GPS is a U.S.-owned utility that provides users with positioning, navigation, and timing (PNT) services. This system consists of three segments: the space segment, the control segment, and the user segment. The U.S. Air Force develops, maintains, and operates the space and control segments. The GPS space segment consists of a constellation of satellites transmitting radio signals to users. The United States is committed to maintaining the availability of at least 24 operational GPS satellites, 95% of the time.

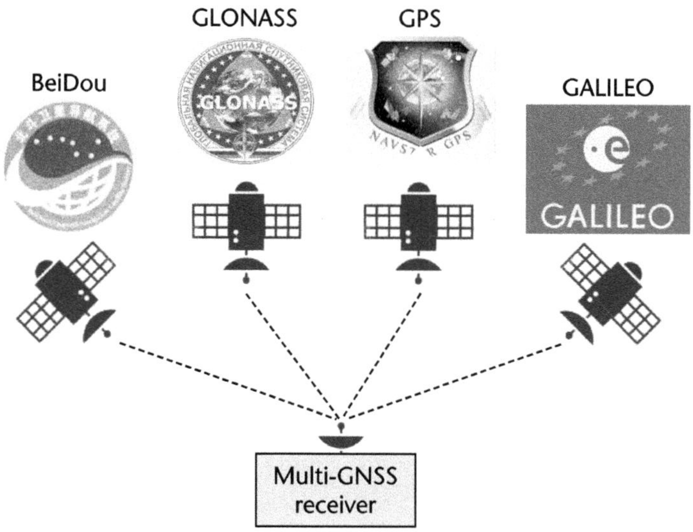

Figure 13.8 GNNS system.

Galileo is the European system that also requires 24 satellites to operate and provides better accuracy than GPS (20 cm for the encrypted version and 1m for the regular, compared to 3m for the regular GPS).

GLONASS is the Russian system that needs 24 satellites as well and has comparable accuracy to GPS.

The Chinese GNSS is called BeiDou Navigation Satellite System (BDS) and its first version was also known as Compass (providing limited coverage). The third generation of BeiDou was completed in 2020 and was expected to provide subcentimeter accuracy.

Each of these four main GNSS can be used separately or combined to provide time synchronization in digital substations. Utilizing more constellations, improves the reliability of the system by the increased number of satellites and the fact that different constellations use multiple frequency bands. This also reduces the likelihood of lost GNSS reception and spoofing.

Considering the criticality of the accurate time synchronization and the cybersecurity concerns, some advanced substation clocks combine GNNS signals with a built-in atomic clock utilizing a highly stable Rubidium reference module.

If jamming or spoofing is a concern, using two or more antennas with different locations within the substation would make them affected differently by the radiation and the system can select to use the antenna that is less affected.

References

[1] IEC 61850 Communication Networks and Systems for Power Utility Automation Part 5: Communication Requirements for Functions and Device Models, 2003.

[2] IEC 61850 Communication Networks and Systems for Power Utility Automation Part 7-2: Basic Communication Structure—Abstract Communication Service Interface (ACSI), 2003.

[3] IEC 61850 Communication Networks and Systems for Power Utility Automation Part 7-3: Basic Communication Structure—Common Data Classes, 2003.

[4] IEEE 1588-2008 IEEE Standard for a Precision Clock Synchronization Protocol for Networked Measurement and Control Systems, 2008.

[5] IEEE 1588-2019—IEEE Standard for a Precision Clock Synchronization Protocol for Networked Measurement and Control Systems, 2019.

[6] IEC 61850 Communication Networks and Systems for Power Utility Automation Part 90-4: Network Engineering Guidelines, Technical Report, 2013.

[7] IEC/IEEE 61850-9-3:2016 Communication Networks and Systems for Power Utility Automation—Part 9-3: Precision Time Protocol Profile for Power Utility Automation, 2016.

[8] IEEE C37.238-2011, IEEE Standard Profile for Use of IEEE 1588 Precision Time Protocol in Power System Applications, 2011.

[9] Time and Frequency Division | NIST, https://www.nist.gov/pml/time-and-frequency-division.

[10] GPS.gov: Other Global Navigation Satellite Systems (GNSS).

CHAPTER 14

Testing of IEC 61850-Based Devices and Systems

14.1 Introduction

For more than a century, PAC systems have been tested based on the knowledge of electric power systems with large and small power plants delivering power to the load centers of a vast network of transmission lines, the world of significant short-circuit currents and DC offset, the world of current transformers saturation and capacitor coupled voltage transformer (CCVT) transients, the world of hardwired electromechanical relays.

So we have developed everything that we needed to test the protection in this world:

- The models of the transient behavior of the electric power system components;
- The simulation tools;
- The test devices that can apply the simulated currents and voltages to the tested protection relay and detect and measure its response time;
- The test switches that can isolate the test object from the energized substation when doing maintenance testing in a live substation.

Based on all of that, there is an established testing process; every couple of years, a testing crew goes to the substation with their test equipment, and they isolate the relays that are scheduled for testing, connect the test equipment, and run the tests using the tools available to them. It is necessary, because nobody knew which characteristic has changed and which terminal screw has loosened.

In the last few decades, this world has been changing. Everything is changing, from the electric power grid to the protection and control systems and the testing technology. The large synchronous generators in nuclear or coal power stations are disappearing and being replaced by wind turbines and photovoltaic panels everywhere: at the transmission and distribution level in wind or solar farms and at the low-voltage level as rooftop solar panels. This is all good for the environment, but it creates a problem for the protection and control systems; these distributed energy resources are not synchronous machines. They are connected to the electric power grid through inverters that do not produce fault current in case of a short-circuit but behave in a way that depends on the algorithms implemented in the inverter controller by its design engineer. So many of the protection functions that we use may not work. The testing tools that we have developed are not able to accurately simulate the behavior of the inverters under fault conditions.

Another major change is in communications technology and its impact on protection and control systems. First, we are not talking anymore about the testing of protection relays, but about testing of protection functions, schemes, and distributed or centralized protection systems interfacing over IEC 61850 communications interfaces. They replace the copper wires with fiber, thus eliminating another familiar tool, the test switch. So we must come up with new methods and tools for functional isolation when doing maintenance testing in an energized substation.

The availability of optical and other nonconventional sensors also requires a different approach to the testing.

Considering that the use of test equipment in energized substations introduces a potential cybersecurity threat, IEC TC 57 Working Group 15 developed the IEC 62351-6:2020 standard [1], which specifies messages, procedures, and algorithms for securing the operation of all protocols based on or derived from the IEC 61850 series, including issues such as the processing of GOOSE messages during testing.

14.2 Requirements for Isolation During Testing

The requirements for isolation depend mainly on the purpose of the test and what is being tested. Considering that the testing might be performed in a lab environment, in the substation that is being commissioned, or in the energized substation, it is clear that the requirements are going to be very different and the tools available to meet them will vary as well. This is also related to who is doing the testing and whether it is part of the development of a new function or product by a manufacturer or it is part of its acceptance or maintenance by the user (Figure 14.1).

That is why first we are going to look at the different types of tests from the point of view of the issues listed above. What, when, why, and how it is being tested are the questions that we need to answer in the following sections.

The requirements for functional testing of devices and distributed functions also determine the methods for testing of both types of systems are proposed based on the following order of system components tests:

- Functional testing of individual function elements used in the IEDs;
- Functional testing of complex functions used in the IEDs;
- Functional testing of individual IEDs used in the scheme;
- Functional testing of distributed functions within a substation;
- Functional testing of distributed functions between substations.

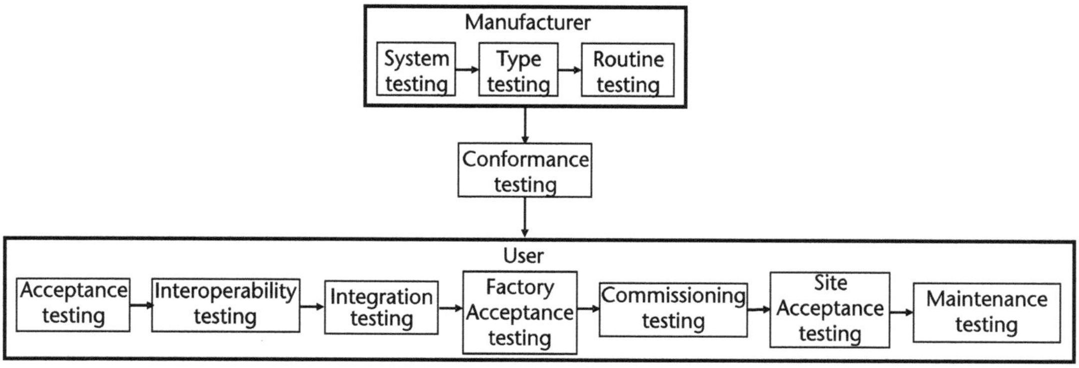

Figure 14.1 Testing use cases.

14.2.1 Device Development Tests

The process of development of any new product by a manufacturer is quite complex and requires a lot of testing to make sure that the product that will be brought to market is going to meet the functional specification. Because in today's digital world we do not have single function protection relays working in isolation but we have advanced multifunctional protection and control devices that do not operate in isolation but work with each other in communications-based protection schemes, the development process needs to ensure that they not only can perform their local protection monitoring and control functions but also can work successfully within a system.

Once it has been verified that the developed product successfully passes the system testing, the manufacturer can proceed with its type testing. It is performed on an approved hardware and firmware implementation to be used in production and is intended to verify the product can withstand the harsh environment of substations in different environmental conditions.

It is the most extensive testing performed by the manufacturer to ensure the completeness of the final product and its readiness to be delivered to the market. Once the product is considered ready for the market, routine testing of components and the finished devices is performed as part of the manufacturing process before delivery to the customers.

14.2.2 Conformance Test

To increase the probability that all devices will understand each other as part of protection and control schemes, Working Group 10 developed IEC 61850-10, which defines the conformance testing of the devices compliant with the standard.

The definition of specific test procedures and the tools to be used to perform them is the responsibility of the Testing Subcommittee of the UCA International Users Group. Its members are utility users, suppliers, integrators, universities, consulting companies, and individuals.

Like any other testing, the problem of conformance testing is the completeness of the tests. The number of all possible situations can be very large. It may be possible to cover all normal operating cases but challenging to cover all failure cases.

In the standard conformance test, only the application according to ACSI can be tested based on the documentation provided by the manufacturer. The conformance test establishes that the communication of the device under test (DUT) works according to the IEC 61850 series specification.

The conformance testing is performed using testing tools certified by the UCA International Users Group and by organizations that are approved to perform it. Even if the device passes the test, there is no guarantee that it will interoperate with any other device from the same or different manufacturer because of the differences in the supported object models and communication services that were declared in the documentation provided by the manufacturer. This documentation includes the following:

- The Model Implementation Conformance Statement (MICS) details the standard data object model elements supported by the system or device.
- The Protocol Implementation Conformance Statement (PICS) is a summary of the capabilities of the system to be tested.
- The SCL Implementation Conformance Statement (SICS) details the mandatory and optional features of system configuration tools and IED configuration tools.
- The Protocol Implementation eXtra Information for Testing (PIXIT) contains system-specific information regarding the capabilities of the system to be tested and which are outside the scope of the IEC 61850 standard. The PIXIT is not subject to standardization.

Conformance tests do not completely guarantee interoperability and that all functional and performance requirements are met. They verify the expected behavior of the DUT versus the conformance tests and document those results. The issued certificate is not a general declaration of conformance to all parts of IEC 61850, which is why the certificate "has not been proven to be non-conformant." When properly performed, conformance tests simply significantly reduce the risk of costly problems occurring during system integration in the factory and on site.

14.2.3 Device Acceptance Test

The process of acceptance of any new product by a user is known as a device acceptance test. Its goal is the verification of the correct behavior of the individual device to be used in substation PAC system.

It is done by use of the system testing tools under the substation, electric power system, and environmental test conditions corresponding with the technical specification of the tested devices.

The acceptance test is a precondition for making a product acceptable for use in the protection scheme.

Acceptance tests ensure that the device really meets all technical specifications listed in the device documentation that are of interest to the user. It should include any additional user tests that are needed by the system, including but not limited to exception/error testing. Acceptance testing also may require the use of some specific to the user real-life fault or disturbance records.

Depending on the acceptance testing philosophy of the user and the available testing facilities, the test may also cover environmental condition testing, including some functional tests under extreme environmental conditions specific to the user's territory. The functional part of the acceptance testing should be based on a set of test scenarios that, as realistically as possible, simulate the user's substation or electric power system conditions for which the tested function is design to operate.

Because acceptance testing is performed in laboratory conditions, from the point of view of the requirements for virtual isolation, most of the tests should be performed using the normal operating modes of all function elements in the test object. The only exception is when performing the tests to verify the virtual isolation capabilities of the IED under test. In this case, the different modes and data objects/attributes used for virtual isolation should be tested to verify that they are properly implemented in the IED. Acceptance tests should be performed on any new version of an already accepted device.

14.2.4 Device Interoperability Test

A device interoperability test is required as part of the product testing before it is accepted for use in the user's system. It is intended to check the correct behavior of any device when integrated as part of a system. This is to ensure that the device interoperates correctly with other devices approved by the user for application in their substations.

Because interoperability testing is performed in laboratory conditions, from the point of view of the requirements for virtual isolation, most of the tests should be performed using the normal operating modes of all function elements in the test object. The only exception is when performing the tests to verify the virtual isolation capabilities of the IED working together as a protection scheme. In this case, the different combinations of modes and data objects/attributes used for virtual isolation should be tested to verify that they are properly implemented in the IEDs.

14.2.5 Integration Test

Integration test is used to ensure that the individual components of the system not only interoperate correctly, but also meet the performance requirements according to the protection system development specification.

In this case, they will include testing of devices at two ends of a communications link. The methods and tools used are in the category of end-to-end and subsystem testing as described earlier.

One of the goals of the integration test is also to ensure that the performance of the distributed scheme meets the requirements of the application. While the interoperability test proves that the IEDs talk together, the integration test goes one step further and proves that they talk fast enough. There is no need for virtual isolation during integration testing.

14.2.6 Factory Acceptance Test

The factory acceptance test (FAT) is a customer-agreed functional tests of the specifically designed and implemented protection system. It is a subject of agreement between the final user and the system integrator and is highly recommended, as it allows the detection of any potential problems in an earlier stage of the project when it is less expensive and easier to fix them.

The FAT should be performed using a top-down approach based on a test plan including test scenarios defined as part of the design of the system.

Black box testing methods can be used until any failure of the system for a specific test occurs. White box testing will then be used to determine the reason for the test failure.

One of the main characteristics of factory acceptance testing is that not all components of the system are available. That requires from the test system the ability to simulate any device missing from the factory system, which is a part of the real protection system.

Another differentiating factor for the FAT is that all existing components of the system are configured and set according to the requirements of the real system application. The factory acceptance test thus should be based on configuration of all devices using the SCD file for the project.

Because the FAT is performed in a laboratory environment with IEDs that have passed acceptance testing that verified that they properly support virtual isolation, it is not necessary to use any isolation tests at this time.

14.2.7 Commissioning Test

The commissioning test of any device is intended to prove that it is properly configured for the specific application. While many of the tests described above can be performed with generic settings, the commissioning tests are performed after the device (the test object) has been configured with the specific settings for the equipment that it has to protect.

As it is possible to have errors in the downloaded setting file, the tests should be based on the parameters of the protected system component: transmission line, distribution feeder, and power transformer.

During the commissioning, it is important to verify that all functional elements that are required for the protection and control of the power system equipment in the zone of protection are in a normal operating mode. No isolation is required during these tests.

14.2.8 Site Acceptance Test

The site acceptance test (SAT), similar to the FAT, is a set of customer-agreed functional tests of the specifically manufactured substation PAC system, performed with the complete system as installed in the substation. It is also a subject of agreement between the final user and the system integrator from the point of view of the content of the test plan and the responsibilities of the involved parties. The SAT should be performed using a top-down approach based on a test plan including test scenarios defined as part of the design of the system.

Black box testing methods can be used until any failure of the system for a specific test occurs. White box testing will then be used to determine the reason for the test failure.

One of the main characteristics of the SAT is that all components of the system are available. That requires from the test system the ability to simulate all required analog, binary, or other signals required for the testing of any specific substation or electric power system condition that the real system is designed to handle.

The SAT should be based on the configuration of all devices using the SCD file for the project.

The final stage of the SAT should be performed as end-to-end testing to ensure that all the wiring between the process and the devices included in the substation PAC system are properly done. It also verifies that all components of the substation communication system are connected according to the system design.

To compare with the FAT, which is performed off site where the system was assembled and has to prove that the complete system fulfills the properties specified in the contract between the manufacturer and the user before it leaves the factory, the SAT concludes commissioning and proves that the system fulfills the contract before it goes into operation.

14.2.9 Maintenance Testing

Maintenance testing, in general, is the testing that is performed to diagnose and identify equipment problems or confirm that different actions taken to change settings, upgrade or repair the protection device, or another component of the fault-clearing system have been effective. The tests to be included in the maintenance test will depend on which of the listed above actions have been implemented.

Scheduled tests are part of preventive maintenance and are typically performed at predefined intervals (schedule) based on regulatory requirements or the recommendations of the manufacturers, as well as other established practices by a utility.

Because microprocessor-based relays have characteristics defined by equations, they do not change with time, and, because many of their components are continuously monitored, scheduled testing is not necessary. The only exception might be the relay outputs, but, if during normal operation by the protection system or by an operator is captured and logged, it can be considered as a maintenance test. However, periodic maintenance testing may be required via regulation based upon the specifics of applications.

Problems of the different elements of the fault-clearing system can be of two main types: if the system does not operate when it has to, and if it operates when it should not. These two types of problems are detected when the system is in service and an event occurs. The operation needs to be analyzed in order to determine the reason and take some corrective action to prevent future incorrect operation of the system. After the corrective action is implemented, it needs to be tested following acceptance or other types of test described earlier.

Maintenance testing is the main use case for virtual isolation because it includes testing of different functions or function elements in an energized substation. The level of isolation in this case depends on the testing philosophy and may include isolation of a function element, a subfunction, or a function for the complete IED for testing. These options need to be carefully considered as part of the development of a standard testing philosophy. This requires a new approach, due mainly to the fact that IEC 61850 allows the control of the mode and behavior of different functional elements, something that is not possible in conventional IEDs.

14.3 IEC 61850 Testing-Related Features

In order to support the testing of IEC 61850 system components in energized substations, Edition 1 of the standard defined many different features that could be used for testing, such as:

- The possibility to put a function or a functional element (logical nodes or logical devices) in a test mode;
- The possibility to characterize a GOOSE message as a message being sent for test purpose;
- The possibility to characterize a service of the control model as being sent for test purpose;
- The possibility to flag any value sent from a server in the quality as a value for test purpose.

However, Edition 1 was not very specific on how to use these features. As a consequence, they were not supported by all vendors because interoperability could not be guaranteed.

This has been improved with Edition 2 [2–4] of the standard. Besides more detailed specifications on how to use the existing features, additional features have been added. However, because it is possible to have some substations with a mixed edition system where there are Edition 1 devices, the testing needs to be performed after detailed analysis and design of the test plans.

14.3.1 Modes of a Function

A logical node or a logical device can be put in test mode using the data object Mod of the LN or of LLN0. Once a logical node is in test mode, every common data class defined in IEC 61850-7-3 contains a quality attribute that characterizes the quality of the information from the server. Test is one such identifier that may be used to classify a value being a test value that should not be used for operational purposes and it plays a key role in implementing the virtual isolation in digital substations. Every function element in an IEC 61850-based digital substation produces an output depending on the values of the quality attribute and its operating mode.

The behavior of a logical node depends on the combination of the mode of the logical node and the mode of the logical device to which it belongs. Table 14.1 shows all the different combinations and the resulting logical node behavior as defined in IEC 61850-7-4.

When the behavior of a logical node is on, it will process data objects with the Test identifier in the quality attribute FALSE and the Test of its data objects will be FALSE as well. When the behavior of a logical node is Test, it will process data objects with the Test identifier in the quality attribute TRUE and the Test of its data objects will be TRUE as well.

If a logical node can produce a physical output that, for example, can lead to the tripping of a breaker in an energized substation, the Test/Blocked mode is used instead of Test. The behavior is similar to Test, with the only exception being that a physical output is not produced.

Table 14.1 Mode and Behavior of Logical Nodes

LDMode	*LNMode*	*LNBeh*
On	On	On
On-blocked	On	On-blocked
Test	On	Test
Test/blocked	On	Test/blocked
Off	On	Off
On	On-blocked	On-blocked
On-blocked	On-blocked	On-blocked
Test	On-blocked	Test/blocked
Test/blocked	On-blocked	Test/blocked
Off	On-blocked	Off
On	Test	Test
On-blocked	Test	Test/blocked
Test	Test	Test
Test/blocked	Test	Test/blocked
Off	Test	Off
On	Test/blocked	Test/blocked
On-blocked	Test/blocked	Test/blocked
Test	Test/blocked	Test/blocked
Test/blocked	Test/blocked	Test/blocked
Off	Test/blocked	Off
On	Off	Off
On-blocked	Off	Off
Test	Off	Off
Test/blocked	Off	Off
Off	Off	Off

The behavior is explained in Figures 14.2 and 14.3. A command to operate can be either initiated by a control operation or by a GOOSE message that is interpreted by the subscriber as a command. If the command is initiated with the Test flag set to FALSE, it will only be executed if the function (LN or logical device) is on. If the device is set to Test mode, it will not execute the command.

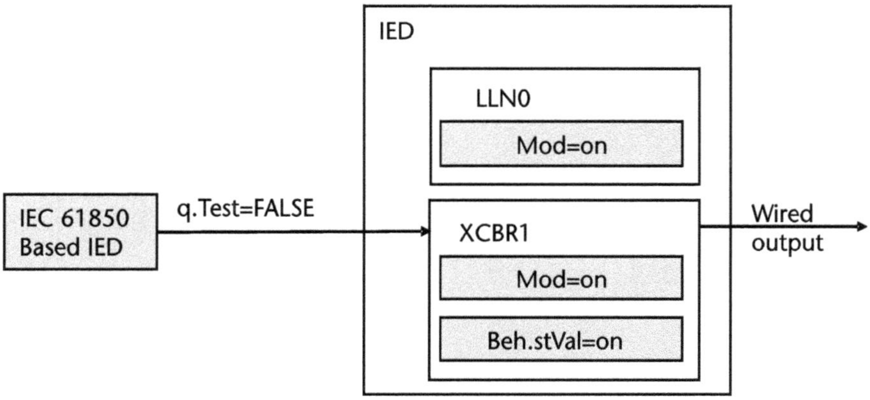

Figure 14.2 GOOSE message in the normal operating mode.

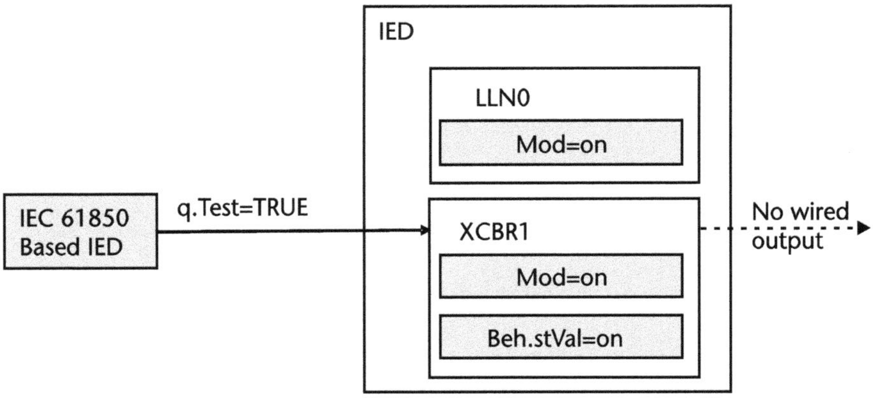

Figure 14.3 GOOSE message in the Test operating mode.

If the command is initiated with the Test flag set to TRUE, it will not be executed if the Function mode is on. If the Function mode is TEST, the command will be executed and a wired output, for example, a trip signal to a breaker will be generated. If the function is set to TEST-BLOCKED, the command will be processed; all the reactions (i.e., sending a command confirmation) will be produced, but no wired output to the process will be activated (Figure 14.3). The mode TEST-BLOCKED is particularly useful while performing tests with a device connected to the process.

14.3.2 Mirroring Control Information

Another feature that has been added is the mirroring of control information. This supports the possibility to test and measure the performance of a control operation while the device is connected to the system.

A control command is applied to a controllable data object. As soon as a command has been received, the device shall activate the data attribute opRcvd. The device shall then process the command. If the command is accepted, the data attribute opOk shall be activated with the same timing as the wired output (Figure 14.4). The data attribute tOpOk shall be the time stamp of the wired output and opOk.

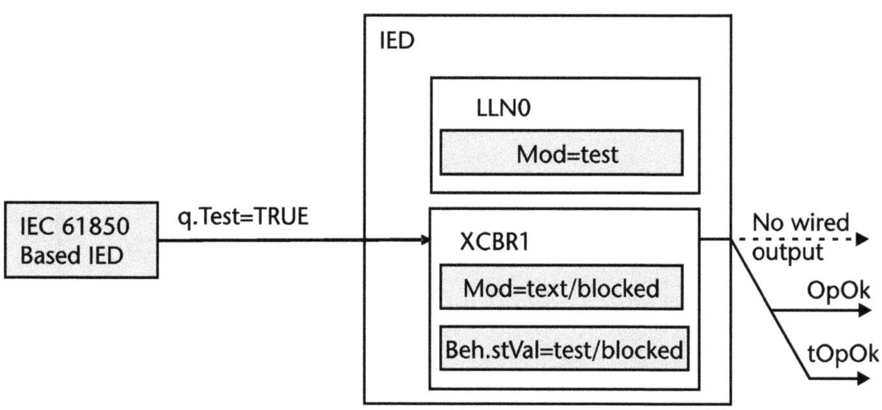

Figure 14.4 Maintenance test with virtual isolation.

These data attributes are produced regardless if the wired output is produced or not; the wired output shall not be produced if the function is in mode Test/Blocked. This allows the testing of the function including its performance without producing a physical output.

Combining the mechanisms described in the previous sections gives us a lot of flexibility that results in improving the reliability of the system because it is possible to perform a maintenance test of different function elements of a device that is connected to an energized substation.

For example, if we want to test the performance of a distance protection without tripping a circuit breaker, the logical device for the protection function shall be set to the mode Test and the logical node XCBR as the interface to the circuit breaker shall be set to the mode Test/Blocked. A test device shall simulate the fault condition required for the test, the distance element will operate sending a message with the quality test set to TRUE.

The breaker controller device containing the XCBR logical node will receive and process that message; however, no output will be generated. The output can be verified through the data attribute XCBR.Pos.opOk and the timing can be measured through the data attribute XCBR.Pos.tOpOk.

14.3.3 Simulation of Messages

Another feature that has been added to Edition 2 is the possibility of subscribing to GOOSE messages or sampled value messages simulated by test equipment. The approach is explained in Figures 14.5 and 14.6. GOOSE or sampled value messages have a flag indicating if the message is the original message or if it is a message produced by a simulation. Also, the IED has in the logical node LPHD (the logical node for the physical device or IED) a data object defining if the IED shall process the original GOOSE or sampled value messages or simulated ones. If the data object Sim is set to FALSE, the IED will process for all GOOSE messages to which it is subscribing the ones from the actual IEDs with the simulation flag set to FALSE. It will ignore the GOOSE messages to which it is subscribing coming from a test device with the simulation flag set to TRUE.

If the data object Sim is set to TRUE, the IED will process from all GOOSE messages it is subscribing to the ones with the simulation flag set to TRUE. If, for

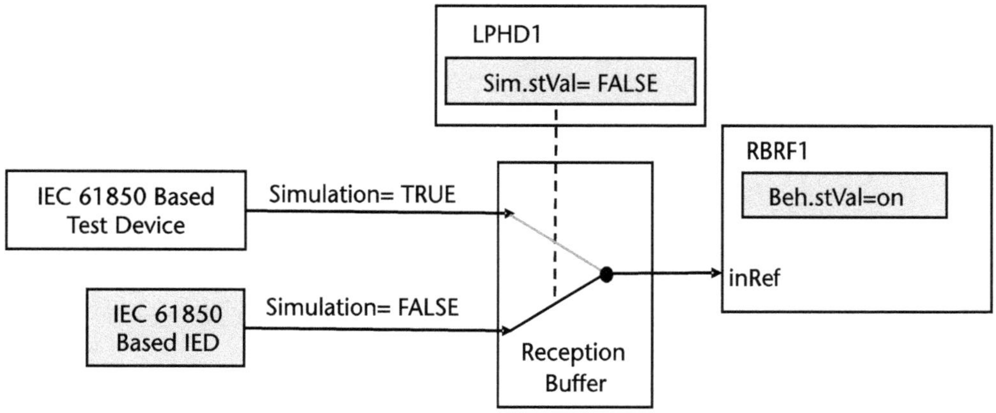

Figure 14.5 Processing of normal IED data.

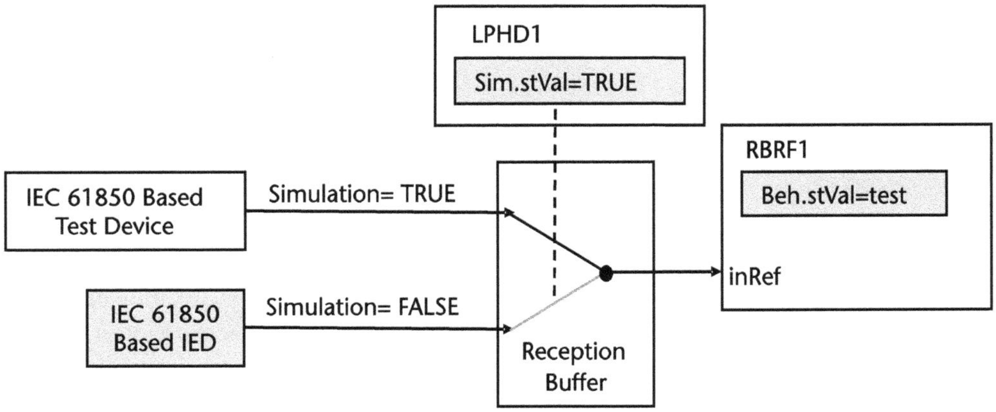

Figure 14.6 Processing of simulated IED data.

a specific GOOSE message, no simulated message exists, it will continue to process the original message. That feature can only be activated for the whole IED, because the IED shall process either the simulated message or the original message.

14.3.4 Advanced Simulation Possibilities

An enhanced simulation feature has been added to the standard that can be used to improve significantly the flexibility and granularity of the functional testing. The concept is explained in Figure 14.7. As described earlier, with Edition 2, the possibility to describe references to the input of a logical node has been added. This is done through multiple instances of data objects InRef of the CDC ORG. That data object has two data attributes providing object references:

- A reference to the object normally used as input;
- A reference to a data object used for testing.

By activating the data attribute tstEna by setting it to true, the function represented in the LN is switched to use the data object referred to by the test reference as input instead of the data object used for normal operation.

With that feature it is, as an example, possible to test a logic function like a breaker failure protection function. Instead of taking the real protection function

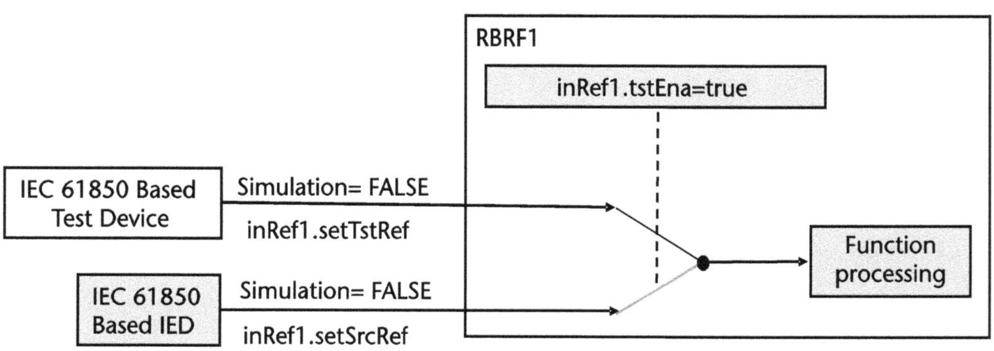

Figure 14.7 Processing of test device data.

operation indications of the different logical nodes as inputs, the node (in that case, RBRF), can be set to use inputs from logical node PTRC in a test device model. A test application can now easily modify the different data objects of the LN PTRC to simulate the test signals that are subject to maintenance testing. That logical node can be external (the data objects being received through GOOSE messages) or it can even be implemented in the IED itself for testing support.

We should keep in mind that, depending on what is the test reference source, although that method allows a detailed functional testing with individually simulated input, it may not necessarily be used for performance testing. This also requires the modeling of test equipment in the substation as such, instead of simulating existing devices that are part of the substation protection and control system. This feature becomes increasingly important with the use of devices with high levels of functional integration and especially in cases of centralized substation protection and control systems.

In this case, the GOOSE or sampled value messages coming from the test equipment do not have to have the simulation bit set to True because they are not simulating other devices but represent data coming from themselves as test equipment.

The above means that the IEDs need to be designed to support both the normal traffic and the test traffic over its communications interface and digital data bus.

14.4 Testing Methods

In order to ensure efficient testing, we need to identify the efficiency criteria (i.e., which resource use should be minimized). The key parameter that we can use is the time that it takes to prepare, execute, analyze, and document the results of the tests.

Functional testing methods can be divided into several categories. They are related to the complexity of the functionality of the individual devices being used in the different levels of the hierarchical system, as well as the types of distributed functions implemented in it. This requires the selection of the right testing method for the specific type of test, as well as the use of testing tools that can automate the testing process.

From this point of view, the testing methods used depend on what we are testing:

- Functional element testing;
- Integration testing;
- Function testing;
- System testing.

A function in this case can be considered as a subsystem with different level of complexity, for example, a system monitoring (SM) function, while the system is the complete redundant protection system.

Regardless of what is being tested, the test object needs to meet the requirement for testability. This is a design characteristic that allows the status (operable, inoperable, or degrade) of a system or any of its subsystems to be confidently determined in a timely fashion. Testability attempts to qualify those attributes of the system design that facilitate the detection and isolation of faults that affect system performance.

From the point of view of testability, a functional element in a protection system is the unit that can be tested, because it is the smallest element that can exist by itself and exchange information with its peers in the protection system.

Another consideration is the purpose of the test and needs to clarify if the tests are performed in relation to acceptance of a new product or function to be used, the engineering and commissioning of a substation component or the complete protection system or its maintenance. From that perspective, different testing methods can be implemented even in the testing of the same functional element or function.

The knowledge of the internal behavior of the test object or, more specifically, the logic or algorithms implemented determine how the tests are being executed. The most commonly used test methods from this point of view are:

- Black box testing;
- White box testing.

An important aspect that needs to be considered during the testing is the availability of redundant devices performing the different protection system functions.

The following sections discuss in more detail the different testing methods listed above.

14.4.1 Black Box Testing

Black box testing is a very commonly used test method where the tester views the test object as a black box (Figure 14.8). This means that we are not interested in the internal behavior and structure of the tested function. Test data are derived only from the specifications without taking advantage of the knowledge of the internal structure of the function.

Black box testing is typically used for:

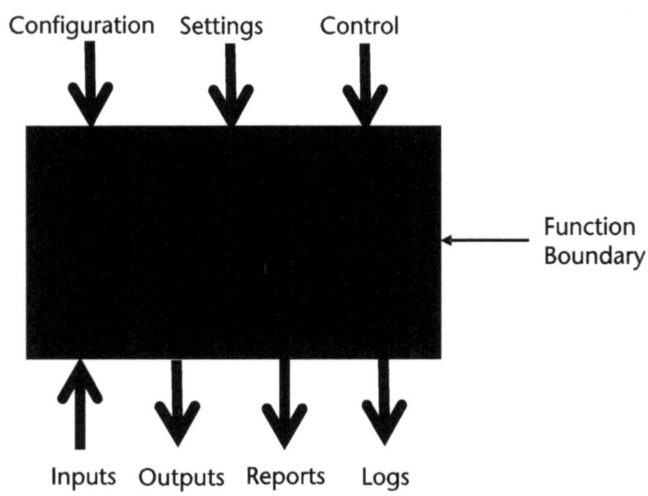

Figure 14.8 Black box testing.

- Functional elements testing;
- Protection system factory testing;
- Protection system site acceptance testing.

As functional elements are defined as units that are the smallest that can exist independently and are testable, it is clear that black box testing is the only method that can be used for their testing.

The response of the test object to the stimuli can be monitored by the test system using the operation of physical outputs, communications messages, or reports.

If the test object is a more complex function that operates based on the interaction of different function elements and we are testing this function using the black box method, everything will be fine if the test is successfully passed. However, if it fails, we have no idea of what was the behavior of the individual elements and which one caused the overall failure of the function.

14.4.2 White Box Testing

White box testing is a method where the test system is not only concerned with the operation of the test object under the test conditions, but also views its internal behavior and structure. In the case of a protection system, it means that it will not only monitor the operation of the system at its function boundary, but also monitor the exchange of signals between different components of the system.

In using this strategy, the test system derives test data from the examination of the test object's logic with consideration of the requirements in the specification. The goal of this test method is to achieve a high level of test coverage by observing the operation of different components of a complex function and the exchange of signals or messages between them under the test conditions.

This method is especially useful when we are testing distributed functions based on different logical interfaces. The observation of the behavior of the subfunctions or functional elements is achieved by monitoring of the exchange of messages between the components of the test object.

However, the test scenarios do not have to be different from the ones used under black box testing.

In IEC 61850-based systems, white box testing is easy to achieve based on the subscription to GOOSE messages whose data sets contain data attributes representing the status of all function elements that are used in the implementation of the tested function (for example, SF_M in Figure 14.9).

14.4.3 Top-Down Testing

Top-down testing is a method that can be widely used for protection systems, especially during site acceptance testing, when we can assume that all the components of the system have already been configured and tested.

Top-down testing can be performed using both a black box and a white box testing method. The testing starts with the complete system, followed by function or subfunction testing and, if necessary, functional element testing.

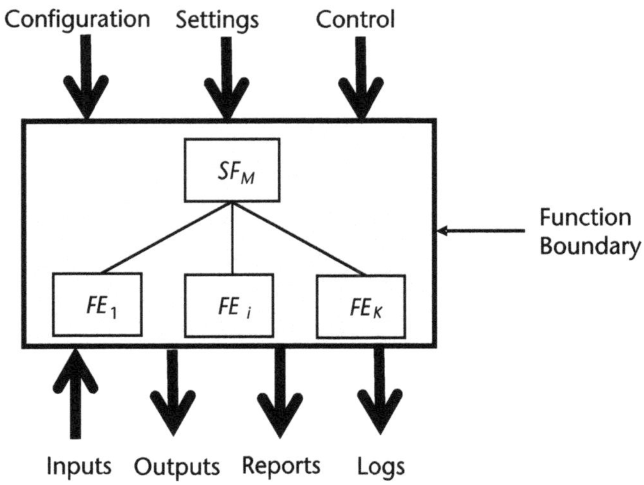

Figure 14.9 White box testing.

In the case of the FAT, when not all components of a system or subsystem are available, it is necessary for the test system to be able to simulate their operation as expected under the test scenario conditions. In this case, the test system creates the stubs for functions or functional elements that are not yet available.

Each functional element is tested according to a functional element test plan, with a top-down strategy.

If we consider a protection system implementation in IEC 61850 for testing using a top-down approach, we will start with the definition of the function boundary.

The testing of the individual components FE_i of a system function (for example, SF_M in Figure 14.9) might be required in the case of failure of a specific test. The function boundary for each of these tests is different and will require a different set of stimuli from the test system, as well as monitoring of the behavior of functional elements using different signals or communications messages.

14.4.4 Bottom-Up Testing

Bottom-up testing is a method that starts with lower-level functions, typically with the functional elements used in the system, for example, PTOC. This method is more suitable for type testing by a manufacturer or acceptance testing by the user.

When testing complex multilevel functions or systems, driver functional elements must be created for the ones not available. The test system must be able to simulate any missing component of the system when performing, for example, FAT.

There are many similarities in the test scenarios used in the bottom-up method, compared to the top-down method. The main difference between the two methods is the order that the tests are performed and the number of tests required.

14.5 Requirements of Testing Tools

It is clear from the previous sections that the testing tools need to support the requirements for all the different types of tests described earlier.

There are two types of tools:

- *Hardware:* The different test devices that generate analog signals or communications messages and monitor the operation of the test objects through hardwired binary signals or communications messages as required by the application.
- *Software:* The different software tools that are used for specific types of test, test configuration, power system conditions simulation, and test assessment and documentation.

The typical testing setup is shown in Figure 14.10 and can be used both for the testing of a multifunctional protection and control device with an analog interface or for the testing of a combination of a merging unit and the protection device with a sampled value interface. In both cases, the test device needs to have the ability to generate current and voltage secondary signals and corresponding simulated streams of sampled values, as well as binary signals and their corresponding GOOSE messages. At the same time, it should have the ability to capture the sampled values coming out of the merging unit and GOOSE messages coming from the protection IED being tested. They are analyzed by the test system to determine if the test object is behaving as expected by the application.

To support the virtual isolation, the test devices should be configurable to operate in a normal operating mode, that is, by sending messages with all test mode-related data objects and attributes set to FALSE. As described earlier, these will be all use cases when there is no need for virtual isolation.

In cases like maintenance testing or commissioning of new bay protection and control schemes in an energized substation, the test equipment should send messages with the simulation bit or test bit set to TRUE, in order to prevent undesired tripping of circuit breakers.

To improve the efficiency of the testing process and reduce the probability for human errors, the testing tools should be configurable by importing the SCD or other SCL files used for the system under test.

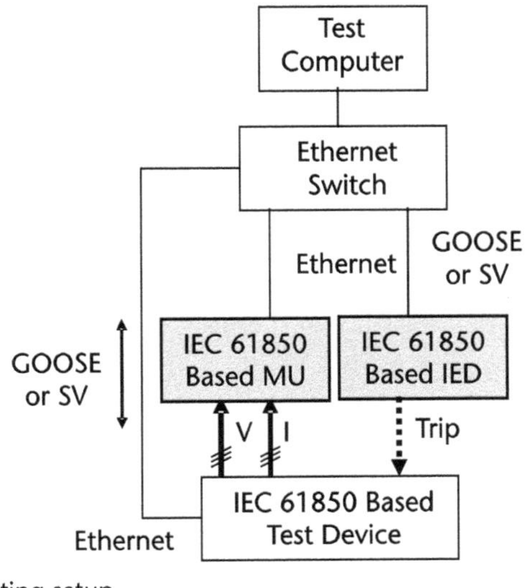

Figure 14.10 Typical testing setup.

Considering the requirements for end-to-end testing of distributed protection schemes that might be using multiple test devices at different physical locations it is necessary for the components of the test system to be synchronized using the PTP as implemented in the system under test.

To cover most of the PAC system tests in any substation environment, the test system should be capable of simulating normal operating conditions as well as different fault scenarios. The simulations should be based preferably on electromagnetic transient simulation methods, but they should also support replaying recorded current and voltage waveforms during actual short-circuit faults or other power system events.

Typically, the analog interface of the test devices should simulate secondary currents and voltages, but, in some cases, during the commissioning of protection and control systems, it may be necessary to inject primary level current and voltages on the instrument transformers in order to verify the proper connection of merging or process interface units in digital substations.

Some more specialized components of the test system should be used to capture communication messages at different locations of the substation network to measure the latency of the different interfaces.

Another requirement related to the improvement of the efficiency of the testing process is the ability to develop standardized test plans as part of the engineering of the system that can be customized for specific applications and configured using different SCL files.

The requirements for the communications interface of the testing system to be used for maintenance of IEC 61850-based digital substations are based on the functions that the test system needs to perform:

- Publish GOOSE messages (Figure 14.11) and sampled values to simulate the specific power system conditions required by the purpose of the test.
- Subscribe to GOOSE messages and streams of sampled values to evaluate the performance of the test object.

Considering that the testing of some centralized functions, such as busbar protection or disturbance recording, may require the use of multiple synchronized test devices, one of the concerns of the protection and control community is cybersecurity. To solve this problem, we can use different methods, such as white listing. This means that all devices that are members of the test system need to be identified and included in the engineering of the system. Dedicated communication ports allocation to test equipment to provide access to the GOOSE or client-server messages from the tested IEDs must be ensured.

The simulation of streams of sampled values by the test system depends on the role that the test devices play in the testing process:

- Acting as specific merging units simulator for the testing of multi-input protection functions, such as transformer or bus differential protection;
- Acting as a generic simulator for testing of a specific function element.

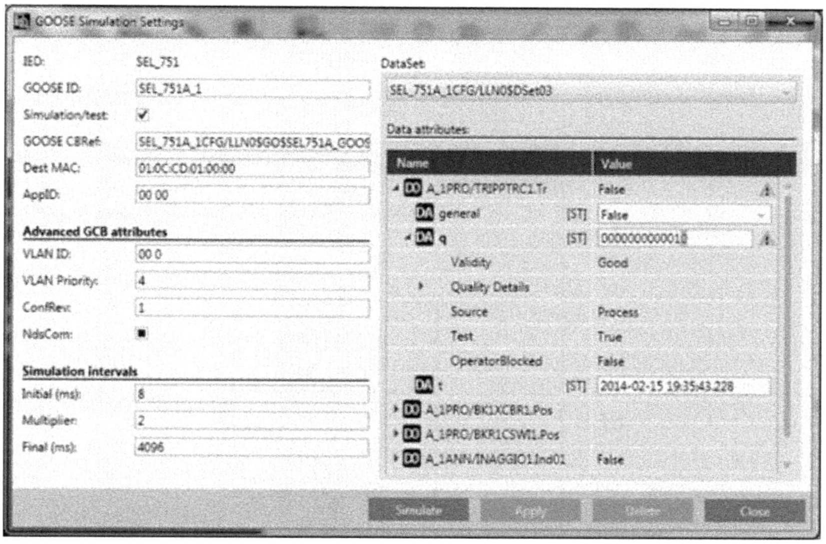

Figure 14.11 Virtual isolation test configuration.

In the first case, the test system will publish messages similar to the ones from the actual process interface devices that it simulates, but with the simulation bit in the message set to TRUE. This will lead to the duplication of the sampled values and GOOSE traffic over segments of the substation communication network.

Setting the test object to process simulated messages by LPHD.Sim=True sometimes creates problems, for example, the operation of a distance protection during the testing of the breaker failure protection in a breaker-and-a-half configuration. This is when the operation of the test system as a generic transient simulator to support the testing of individual protection function elements (for example, RBRF) based on the InRef and TstRef data objects and attributes (see Figure 14.7) in IEC 61850 is a good tool that helps address the abovementioned issue.

14.6 Testing of Low-Power Instrument Transformer-Based Systems

The testing of low-power instrument transformer (LPIT)-based protection applications in principle is not different from the testing of conventional hardwired systems. The main difference is the interface between the output of the instrument transformer and that of the protection and control system. In the conventional case, this is secondary currents and voltages, while in the case of LPITs in the digital substations, it is streams of sampled values. Because the focus of this book is the digitalization of the grid, the testing of LPITs with IEC 61850-9-2 or IEC 61869-9-based interface as shown in Figure 14.12 is described.

An optical current transformer is used as an example. It converts the primary signal into an optical signal that is digitized by embedded proprietary merging units producing as an output a stream of sampled values. Because the user typically does not have the ability to access the optical part of the internal interface to verify that the optical CT is connected to the right phases on the primary side, the primary injection, and monitoring of the stream of sampled values is used.

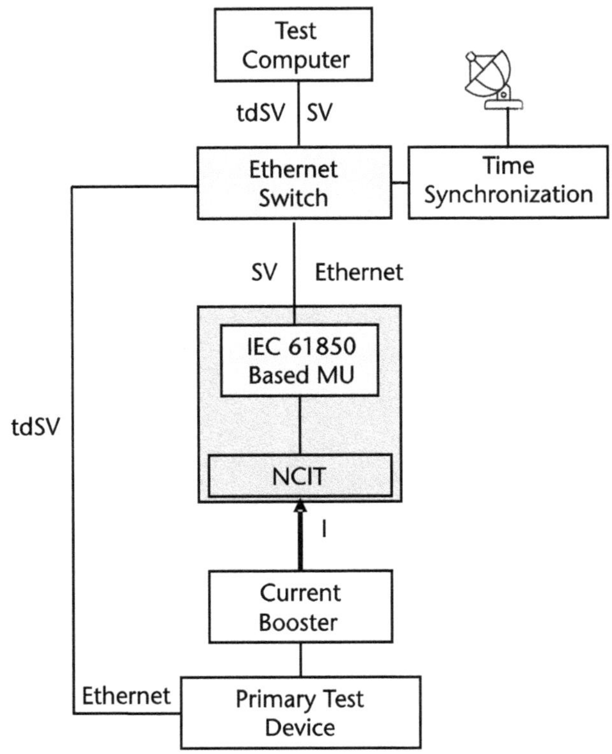

Figure 14.12 LPIT commissioning test setup.

The assumption is that the acceptance testing of the optical sensor is being already performed in a certified high-voltage laboratory and the goal of this test is only to verify that the sensor is not damaged and is properly connected.

Typically, maintenance testing of LPITs is not necessary because, in IEC 61850-based protection and control systems, we can continuously monitor the individual components, which practically eliminates the need for maintenance testing.

If we are testing an LPIT with IEC 61850 9-2 LE output, we need to be able to apply the primary current required by the test and monitor the streaming sampled values in order to determine that the LPIT meets the technical specification and the requirements of the application.

There are different ways to determine the performance of the LPIT, depending on the purpose of the test and the availability of time synchronization between the test device and the test object. One approach is to have the primary test system perform closed-loop testing by injecting a test signal on the primary side of the current sensor. The merging unit converts the optical sensor output into a sampled value stream that is published to the substation network. The primary test system then reads the data back from the network in order to perform the different tests.

To carry out this test, we need to inject sufficiently high current, for example, up to 800A from the primary test set output. If, in some cases, a higher level of up to 2,000A is necessary, it can be achieved by using a current booster connected to the External Booster output to the current transformer's primary side (see Figure 14.12).

The current transformer's Ethernet output is connected to an Ethernet switch, with the test device and test computer connected to the same switch as well.

If the test device and the test object are time-synchronized, a test computer can capture the streaming sampled values and store them for comparison and analysis of the performance of the LPIT.

The same methods described above can be used in case of integration testing of an LPIT with a low-power output and a merging unit with a low-power input.

14.7 Testing of Protection Systems

The testing of modern protection systems can be similar to that of conventional protection systems; the only difference is the interfaces of the test system, which are communications-based instead of hardwired. As a result, the test system will publish GOOSE and sampled value messages and subscribe to GOOSE messages to perform the testing of the different protection functions. Multiple synchronized test devices may be used when necessary as shown in Figure 14.13.

The impact of the simultaneous operation of test equipment used for maintenance testing of energized digital substations with the normal traffic from the stand-alone merging units (SAMU) and the traffic supporting other functions, such as automatic fault records retrieval, needs to be carefully analyzed to ensure that it may not lead to degradation in the performance of the protection functions.

Figure 14.14 shows an example from the testing of the protection of a multifunctional transformer protection IED for a 3-winding transformer connected to a breaker-and-a-half scheme on the high side. This means that, under normal conditions, the IED will subscribe to current and voltage sampled values from four merging units.

If we assume that we will need to perform maintenance testing on the IED and we can use test devices having the ability to simulate up to 3 streams of sampled values, we will need 2 test devices (TD1 and TD2) to perform the test.

Under the test conditions, the transformer protection IED will receive 8 streams of sampled values: 4 from the merging units and 4 from the test devices. This

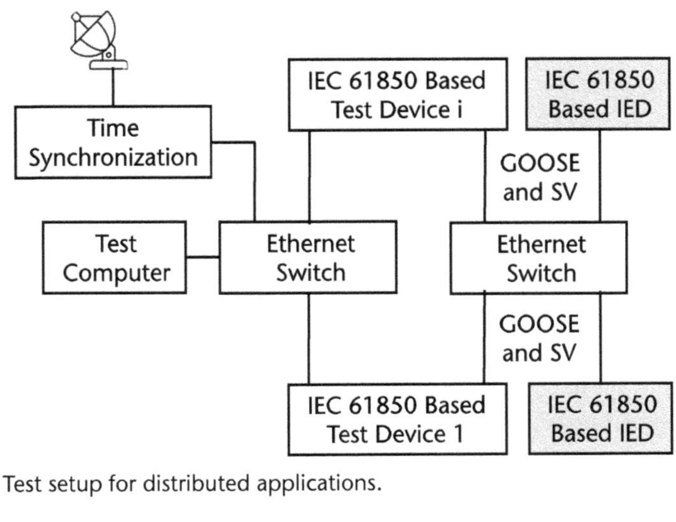

Figure 14.13 Test setup for distributed applications.

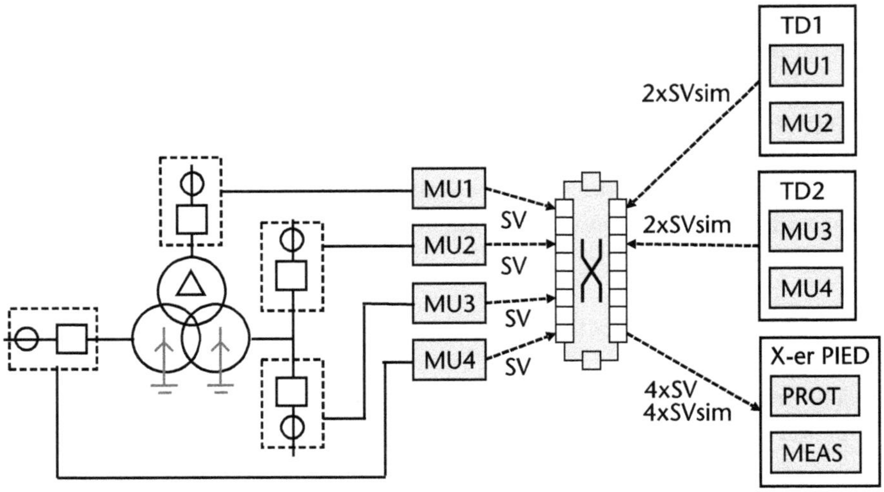

Figure 14.14 Impact of testing on the network traffic.

should be combined with all additional traffic, which can be significant, especially if there are streaming sampled values from sensors for nonelectrical signals such as temperature and pressure.

The above makes it clear that it is very important to include the impact of the testing-related traffic in the analysis for the design of the substation communication system.

14.8 Remote Testing Requirements and Benefits

IEC 61850-based digital substations allow a significant improvement in the efficiency of maintenance testing. This is the result of the availability of testing-related features defined in the standard, which allow the isolation of the test object and testing system from the rest of the live substation without the need for physical switching or connections of equipment in the live substation.

One of the benefits of digital substations is that all devices (PAC IEDs, substation computers, and test devices) are connected to the substation communications network. If there are testing tools that are connected to the network in the substation on a permanent basis, it becomes possible to perform the tests from a remote location. This can be useful in many cases:

- Long distance between the substation and the base of the test staff team;
- Difficult terrain with bad roads;
- Difficult weather conditions;
- Requirements for reduction of outage time because of maintenance.

The remote testing improves the efficiency by eliminating the need to travel to the substation to perform the testing. This leads to the significant reduction in the time spent by the testing team in relation to a specific maintenance test.

Additional savings in time are the result of eliminating the need for connecting the test equipment to the test object.

The ability to isolate only a function element that is being tested improves the efficiency of operation of the electric power system by eliminating the need for an outage during the testing.

In order to be able to perform remote testing the system needs to meet the following requirements:

- Analog and digital interfaces between the process and the PAC system are communications based (IEC 61850 sampled values and GOOSE);
- Support of virtual isolation of test objects;
- Remote secured access to the substation's test system.

The remote testing concept can be implemented as shown in Figure 14.15. The test system in the remote substation includes several components:

- Test computer that runs the testing software supporting IEC 61850 Edition 2 testing features and the required functional testing tools;
- Test devices performing simulation and evaluation of the results from each test.

The interface to the test computer is over a private cloud and requires the use of cybersecurity technology available for remote access from the engineering station by an authorized and authenticated user.

The test engineer or technician accesses the test computer in the remote substation using a remote control tool with advanced cybersecurity features.

Depending on the requirements for the test defined by the type of maintenance testing that needs to be performed, the logical nodes, logical devices, or complete IEDs are set in the required mode in order to ensure their virtual isolation.

14.9 CIGRE Technical Brochure 760

The importance and complexity of the testing of PAC systems in digital substations were the reason for CIGRE Study Committee B5 to establish Working Group B5-53

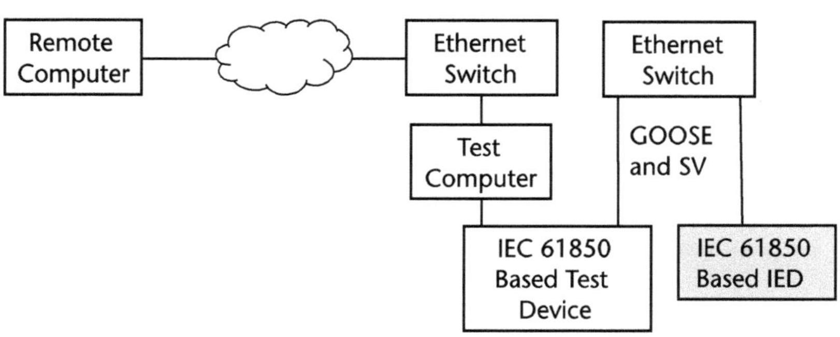

Figure 14.15 Remote test system.

"Test Strategy for Protection, Automation and Control (PAC) Functions in a Fully Digital Substation Based on IEC 61850 Applications." The report is almost 200 pages and covers a wide range of issues related to the testing, as well as the summary of the results from an industry survey. The report was published in March 2019 as CIGRE Technical Brochure 760 [5] and can benefit all PAC specialists involved with IEC 61850-based digital substations.

References

[1] IEC 62351-6:2020 Power Systems Management and Associated Information Exchange—Data and Communications Security—Part 6: Security for IEC 61850, 2020.

[2] IEC 61850-7-2:2020 Communication Networks and Systems for Power Utility Automation—Part 7-2: Basic Information and Communication Structure—Abstract Communication Service Interface (ACSI), Edition 2, Amendment 1, 2020.

[3] IEC 61850-7-3:2020 Communication Networks and Systems for Power Utility Automation—Part 7-3: Basic Communication Structure for Substation and Feeder Equipment—Common Data Classes, Edition 2, Amendment 1, 2020.

[4] IEC 61850-7-4:2020 Communication Networks and Systems for Power Utility Automation—Part 7-4: Basic Communication Structure—Compatible Logical Node Classes and Data Object Classes, Edition 2, Amendment 1, 2020.

[5] CIGRE Technical Brochure 760, Test Strategy for Protection, Automation and Control (PAC) Functions in a Fully Digital Substation Based on IEC 61850 Applications, March 2019.

CHAPTER 15

Digital Substations

15.1 Introduction

Considering that electric power systems have been developing for more than a century, it is easy to understand that there is a huge installed base that will take some time to digitize. However, this process has already started and we are seeing today a wide range of substations with different levels of digitization.

We first need to clearly define what we mean by a digital substation. Based on the global experience, we can divide substations into four main categories:

- Conventional substations are most of the substations around the world that are using only hardwired interfaces between the primary and secondary devices in the substation. All protection and control functions are performed by electromechanical or solid-state devices that do not have communication capabilities, and all interfaces between different relays working in a protection scheme are hardwired.
- Hybrid substations are the growing number of substations all over the world that are using microprocessor-based protection and control devices with some communication capabilities that can be used for data acquisition, event reporting, disturbance recording, and many others. The interfaces with the process and between the devices are still hardwired.
- More advanced versions of hybrid substations are the ones that are using IEC 61850 as the communications protocol and especially GOOSE messages to replace the hardwired signals between the protection and control IEDs.
- Digital substations (Figure 15.1) are the ones where all interfaces between the sensors, IEDs, and other devices performing PAC, measurements, monitoring, and recording are based on digital communications, predominantly using IEC 61850. The only hardwired interfaces are the power supply and the interfaces with the primary equipment, for example, circuit breakers and switches and power and instrument transformers. More advanced versions of digital substations are the ones with nonconventional instrument transformers.

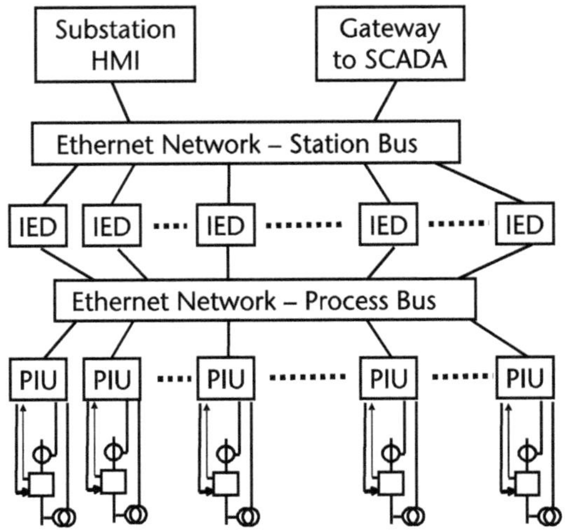

Figure 15.1 Digital substation.

In the following sections of this chapter, we identify the digital substation's components and the different ways of performing the digitization as well as what the benefits are and why the industry should move in this direction.

15.2 IEC 61850-Based Digital Substation

One of the main characteristics of the IEC 61850 standard is that it defines the object models of the different components of a PAC system but does not specify how it should be implemented. In traditional digital substations, their architecture is typically distributed with some devices providing the interfaces to the primary substation equipment and other devices performing PAC, monitoring, and recording functions by exchanging information between the process interface devices and themselves. What is common between a distributed system and a centralized system is that the process interface is identical between the two, while the interactions between the distributed devices over the station bus are replaced by interactions over the digital data bus of the central substation servers.

The process interface functions can be divided into three main categories:

- Switchgear interface;
- Electrical interface;
- Nonelectrical interface.

These functions can be implemented by grouping logical nodes in logical devices. For the logical devices, we use the following naming conventions:

- Switchgear interface unit (SIU) provides a binary status and control interface for circuit breakers and switches.

- Merging unit (MU) converts analog signals (currents and voltages) into time-synchronized streams of sampled values according to IEC 61850-9-2 of IEC 61869.
- Nonelectric interface unit (NEIU) converts analog signals from nonelectric sensors into time-synchronized streams of sampled values according to IEC 61850 9-2 or GOOSE messages according to IEC 61850-8-1.

These logical devices can be placed in individual boxes or can be grouped together in different ways such as:

- Process interface unit (PIU) combines two or more (see Figure 15.2) of the functions listed above.
- Process interface IED (PIIED) combines the functionality of a PIU with local protection, control and/or other noninterface functions.

The different process interface devices communicate with the rest of the substation PAC system using the required IEC 61850 services.

Today most of the existing digital substations have the distributed architecture shown in Figure 15.1 with the functions located in different multifunctional IEDs communicating over the substation station bus using GOOSE messages.

Based on the data from the process interface devices, the IEDs perform different functions such as:

- Protection;
- Measurements;
- Automation;

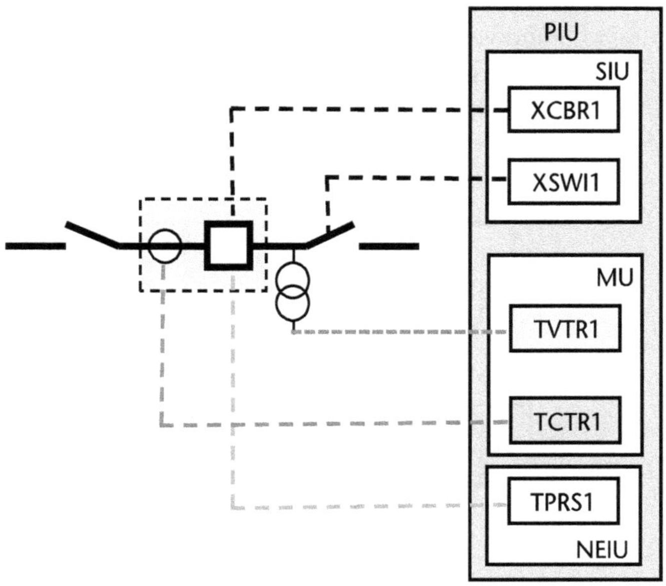

Figure 15.2 PIU.

- Monitoring;
- Recording.

Each of the function elements used in the implementation of the above-listed functions produces a significant amount of information that can be used to improve the performance of different advanced applications. It is very important to note that all the data from the process, as well as the output from the functional elements in the IEDs, is time-stamped based on the accurate time synchronization using IEC 61850 9-3, which supports the precise alignment of the data from all different sources.

It should be noted that, even that most of the decisions of different applications can be made using the locally available data and information, some of them might be misleading if they did not take into consideration the actual topology of the electric power system and the impact of remotely located equipment. That is why it is important also to receive time-stamped information from SCADA or system integrity protection schemes.

15.3 Data in Digital Substation

The IEC 61850-based digital substations use various sensors connected to the primary system equipment. All function elements in such systems are represented in the IEC 61850 model by logical nodes that belong to different groups according to their role in the system.

The sensors in a PAC, monitoring, and recording system belong to the group T. While in Edition 1 of the standard they were only for currents (TCTR) and voltages (TVTR), Edition 2 added many new sensor logical nodes for temperature and vibration that are required by protection, control, or condition monitoring functions of the asset management system.

Figure 15.3 shows an abstract block diagram of the functional decomposition in the IEC 61850 model with the sensor T logical nodes at the bottom performing the digitization and sending sampled analog values (SAV) to the function elements that need them: protection (P group), measurements (M group), and monitoring (S

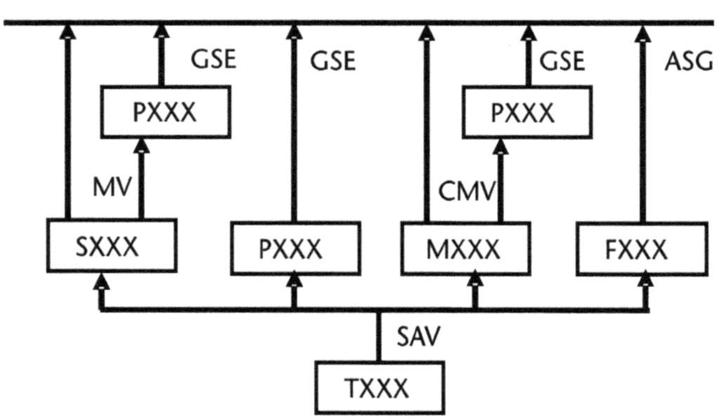

Figure 15.3 Abstract logical nodes interface model.

group). Each logical node represents a function element in the system that is a data source providing data using the services defined in the standard.

All these logical nodes can be combined in logical devices that are integrated in various physical devices publishing the data on the digital substation communications network. Some examples of such devices are:

- Standalone merging units;
- Embedded merging units;
- Switchgear control units;
- Process interface units;
- Nonelectrical sensor units;
- Phasor measurement units;
- Multifunctional IEDs.

The amount of available data in digital substations can be huge, especially when the merging units are publishing streams of current and voltage samples for protection, power quality, and disturbance recording purposes. Today this is typically based on the implementation agreement known as IEC 61850-9-2 LE that defines for protection and control applications a sampling rate of 80 samples/cycle at the nominal system frequency. The digital output publishing rate is 4,000 frames/sec at 50 Hz and 4,800 frames/sec at 60 Hz with one application service data unit (ASDU) per frame, each one representing one sampled value. For power quality monitoring and disturbance recording, the sampling rate is 256 samples/cycle at the nominal system frequency with 8 ASDUs (samples) per frame. It is not difficult to imagine the huge amount of data that this represents.

The switchgear interface units send GOOSE messages any time there is a change of state of the breakers or disconnecting switches.

Because most of today's advanced IEDs also have embedded phasor measurement units (PMUs) that calculate M (for measurement) and P (for protection) class synchrophasors that may be published 4 times per cycle, the amount of data to be processed by the system further increases.

Any operation of a function in an IED also can be configured to send an event report or to log the data, further adding to the data to be processed by the different applications. All of this results in a wide range of different types of data available in digital substations:

- Raw data represented in the model by streaming sample values from electrical or nonelectrical sensors;
- Status data from the switchgear and other primary equipment in the substation;
- Synchrophasor measurements from the P class or M class;
- Status data from multifunctional protection and control IEDs;
- Event report from multifunctional protection and control IEDs;

- Time synchronization data from different clocks;
- Output data from substation-level applications;
- Data received from a remote substation;
- Data received from system integrity protection schemes;
- Data received from the control center.

Depending on the type of function, in some cases, we may need to process sampled values streaming in real time, while in others we will be working with recorded data. It is impossible for all this data to be processed by humans, which is why we need the help of artificial intelligence platforms to solve the big data problem in digital substations.

15.4 Digital Substation Architecture

The digital substation architecture can be considered from two different points of view: logical and physical. The logical architecture has three levels:

- The process level at the bottom performs the digitization of all analog and binary signals from the primary substation equipment and executes the required actions from the substation PAC system.
- The protection and control level digitizes all interfaces between the multifunctional IEDs at the substation bays to support the required functions.
- The station control level interfaces with the protection and control level in order to perform the substation-level functions and also interact with the system level of the grid.

The typical digital substation architecture is shown in Figure 15.1.

The interface between the process level and the protection and control level is vertical and is over the process bus. The interface between the substation IEDs is horizontal and is over the station bus, which is also used for the vertical interfaces with the station control level. This logical architecture can be implemented in many different ways depending on the size and criticality of the substation, as well as the design philosophy. It can be organized on the bay level, voltage level, or substation level.

The communications architecture depends on the specifics of the substation and the requirements for performance, reliability, and security. It can be star, ring, mesh, or hybrid, combining them in order to meet the specific requirements of the substation and the applications.

The process and station bus can be implemented over dedicated physical networks or can share the same physical network.

Early digital substations required a dedicated time synchronization network delivering 1-PPS signals to the merging units. Today the time synchronization is performed over the substation Ethernet network using IEC 61850-9-3.

Redundancy protocols such as PRP and HSR are used to improve the reliability of the system. The price to pay is that the amount of data that is published over the substation local area network is doubled, but monitoring the traffic will allow asset management applications to detect issues with the substation devices and the communications network.

One of the main characteristics of the IEC 61850 standard is that it defines the object models of the different components of a PAC system and the communications services for data exchange. However, it does not specify how it should be implemented.

In traditional digital substations, their architecture is typically distributed with some devices providing the interfaces to the primary substation equipment and other devices performing PAC monitoring and recording functions by exchanging information between the process interface devices and themselves as shown in Figure 15.1.

The new trend of virtualization and digitalization of the electric power grid is leading to the separation between hardware and applications. As a result, we already see digital substations with centralized architecture as shown in Figure 15.4.

The actual IEDs from the distributed system communicating over the substation station bus are replaced by virtual IEDs in the centralized system. What is important to carefully analyze as part of the engineering of such a system is the interface between the process bus and the substation server performing the role of a centralized PAC system that may carry a huge amount of data from the process. It is important also to consider redundant interfaces between the process bus and the server to improve the reliability of the system in case of the loss of a communication link.

15.5 Process Interfaces in Digital Substations

The digitization of the substation starts at the process level with the interfaces between the primary substation equipment and the multifunctional PAC system.

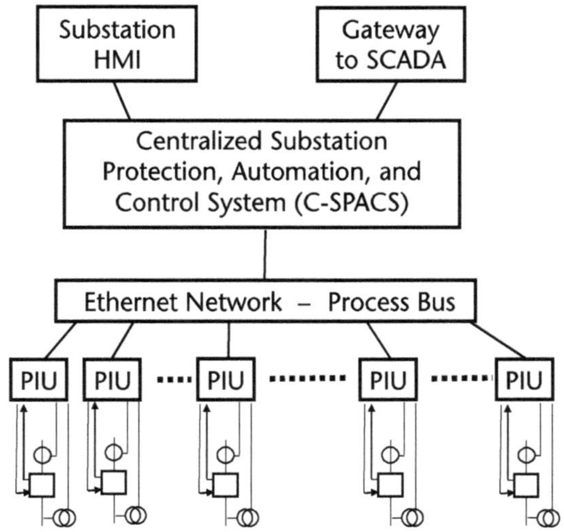

Figure 15.4 Centralized digital substation.

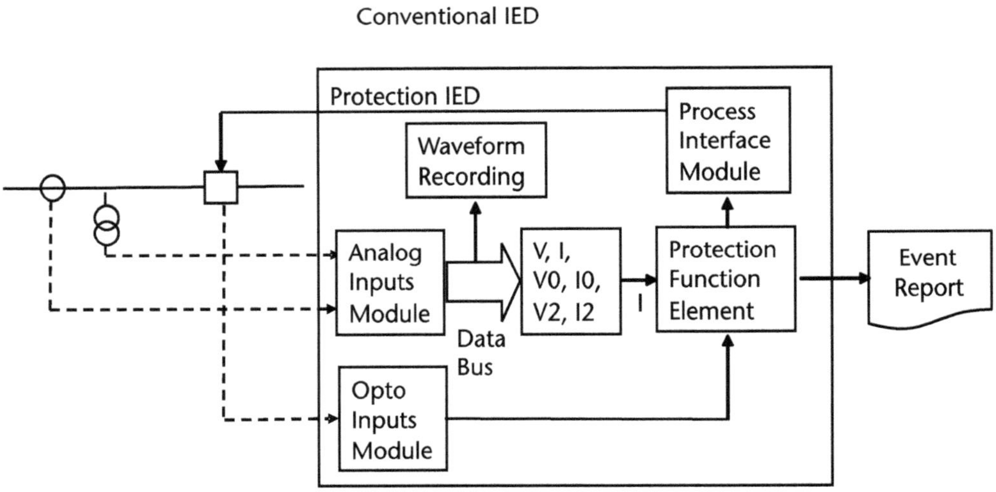

Figure 15.5 Conventional protection interface.

The conventional application shown in Figure 15.5 represents the hybrid substation and connects the secondary current and voltage instrument transformers with copper cables to the analog inputs of the protection and control IEDs located in the substation control house. Copper cables are also used for the binary signals between the switch gear and the control house.

Conventional hardwired protection systems have several deficiencies:

- Susceptible to CT saturation, mainly due to the lead resistance;
- High costs of copper cables;
- Requirements for calibration and testing for a shared CT/VT (e.g., between two different IEDs);
- Requirements for maintenance testing;
- Limited flexibility;
- Requirements for outages in cases of changes in the zone of protection;
- Safety issues related to open CT circuit conditions.

The fact that conventional instrument transformers have been creating safety concerns (explosions) and require separate protection and metering instrument transformers has been pushing utilities to take another look at using the different types of nonconventional instrument transformers available today.

15.5.1 Standalone Merging Unit

Considering the huge number of existing substations with conventional instrument transformers, the digitization of the current and voltage process interface is based on the use of what we call standalone merging units (Figure 15.6). It is preferred that they are connected directly to the secondary conventional instrument transformers because this helps in resolving many of the issues with conventional hardwired systems.

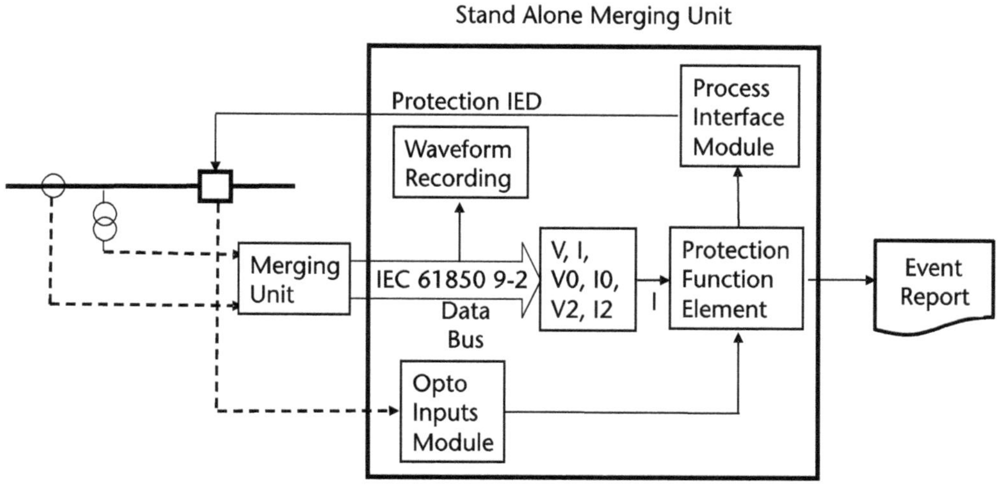

Figure 15.6 Standalone merging unit.

Process bus-based protection applications offer some important advantages over conventional hardwired analog circuits that are especially important in the case of bus protection. The first very important one is the significant reduction in the cost of the system due to the fact that multiple copper cables are replaced with a small number of fiber optic cables.

Using a process bus also results in the significant reduction in the possibility for CT saturation because of the elimination of the current leads resistance when the standalone merging unit is connected directly to the secondary of the instrument transformer.

Process bus-based solutions also improve the safety of the substation by eliminating one of the main safety-related problems, an open current circuit condition. Because the only current circuit is between the secondary of the current transformer and the input of the merging unit that is located right next to it, the probability for an open current circuit condition is very small. It becomes nonexistent if optical current sensors are used.

15.5.2 Low-Power Instrument Transformer Interface

The earlier optical instrument transformers had a low-power analog interface, which was not very useful for protection applications. That is still the case with Rogowski coils.

With the advancement of communications technology and the development of the concept of the merging unit, they are also used for digitizing such low-power sensors. This changed the interface of the protection and other devices in the substation as shown in Figure 15.7.

This interface provides all the benefits that are typical for sampled value-based interfaces with multifunctional protection IEDs. Like all other such interfaces, it is preferred that the merging unit is located as close as possible to the low-power IT output in order to reduce the effect of substation transients on the operation of the protection functions.

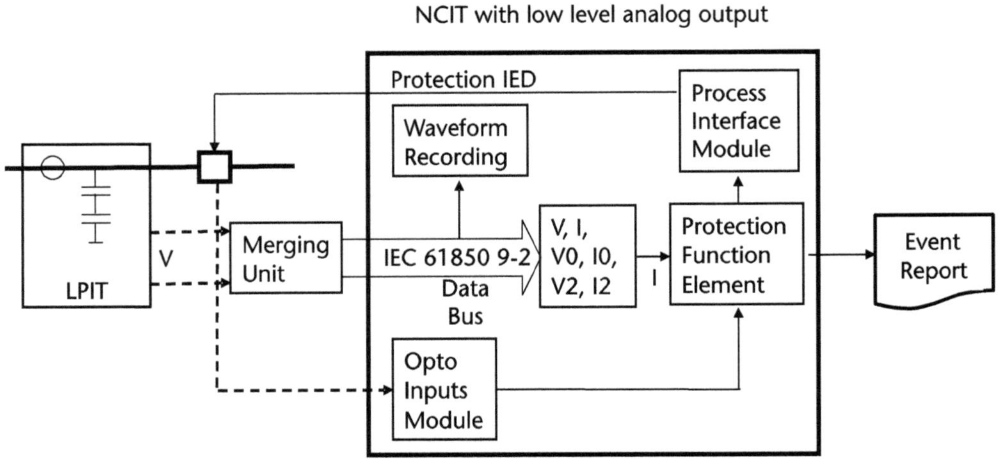

Figure 15.7 Low-power IT interface.

15.5.3 Optical IT with a Direct Sampled Value Interface

Depending on their operating principle (for example, optical CTs), some low-power ITs may have a direct IEC 61850 sampled value interface as shown in Figure 15.8.

Such an interface significantly simplifies the design of the system, because the number of components of the protection and control system is reduced and the number of interfaces is practically limited to fiber optic cables only. The single exception is the hardwired connections that power the individual devices.

One of the challenges with this option is that the testing of the interface with the devices in the system cannot be performed using a secondary injection from the test equipment. It requires a primary injection, which means that the testing of protection functions can be performed only using a sampled value simulation published from the test equipment.

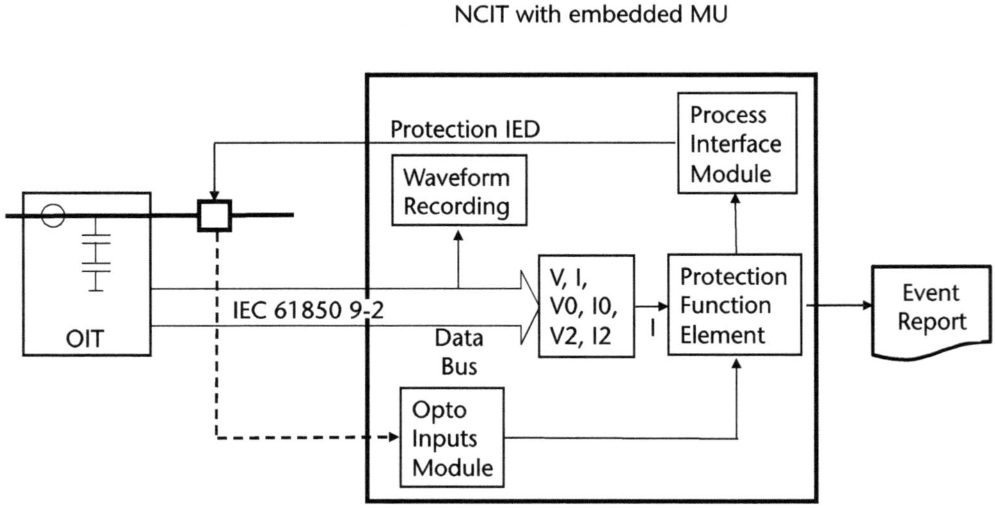

Figure 15.8 Direct IEC 61850 9-2 interface.

15.6 The Business Case for Digital Substations

A business case is an argument, usually documented, that is intended to convince a decision-maker to approve some kind of action, in this case, make a decision to transition from conventional or hybrid to fully digital substations.

The driving force behind moving in this direction is related to the smart grid. The electric power industry is trying to transition into an environment in which we will have a more reliable, secure, and efficient grid. From the point of view of the definition of the business case above, the goal is to convince the decision-makers that the transition from traditionally used hardwired technologies to PAC solutions based on the IEC 61850 standard is the approach that will help to meet the challenges that our industry is facing.

The need for change is driven by the combination of old and new challenges, including:

- High costs of building new substations, including land, equipment, transportation, construction, and installation;
- Labor-intensive process of connecting primary substation equipment and multifunctional IEDs, which leads to increased installation and commissioning costs;
- Labor-intensive mapping of data from multifunctional IEDs to substation human machine interface (HMI), which increases engineering costs and is prone to human errors;
- Limited number of input and output of multifunctional IEDs, which may require the use of additional auxiliary equipment, resulting in increased costs, reduced reliability, and degraded protection scheme performance;
- Proprietary communication interfaces between protection IEDs used in accelerated protection schemes not supporting interoperability and, as a result, reducing the availability and reliability of such schemes;
- Unsupervised hardwired interfaces between relay outputs and opto-inputs in distributed protection schemes not providing indication about the failure of an interface, which may lead to a failure to operate when necessary;
- Labor-intensive maintenance of the interface between multifunctional IEDs and the substation HMI, which requires remapping of IED data to the HMI;
- Safety concerns related to the possibility for open CT circuits, which result in dangerously high voltages that can hurt or even kill employees;
- Labor-intensive maintenance testing requiring traveling to the substation location, which may be under dangerous conditions, raising safety concerns and possibilities for human errors;
- Outages for the testing of protection IEDs or schemes, which are very difficult to obtain and result in reduced electric power system reliability;
- CT saturation and CCVT transients may result in incorrect operation of protection functions;

- Impact of changing substation or power system topology on the performance of protection functions, which has an impact on the sensitivity or selectivity of protection schemes and the fault clearing times;
- High penetration of different types of DERs at all levels of the electric power grid, which requires reduced fault-clearing times at the different levels of the system;
- Electromagnetic pulse (EMP) threats.

The traditionally used technologies have difficulties with resolving the issues as listed above. Taking advantage of the development and implementation of IEC 61850 can help the industry to meet the goals of the smart grid. This has to be further expanded with a description of a vision for the future, including the levels of implementation of the standard at different stages of the migration strategy.

At the same time, while building the business case, we need to also consider the risks and costs of change. This is a very complex issue that needs to take into consideration the short-term and long-term objectives. The analysis should examine the benefits and risks involved with both switching to IEC 61850 based PAC systems and continuing to use the traditional solutions.

It is very important that the goals of the transition to IEC 61850, the benefits, and the risks are presented in a very focused manner without the use of the very technical IEC 61850 terminology. The information included should be understandable by decision-makers without any knowledge of the standard.

Addressing any of the challenges listed above needs to highlight at the end the impact on the bottom line (i.e., how it will improve the efficiency, related to reliability and security).

In the rest of this section, we briefly analyze the benefits and risks of the transition from conventional technologies to IEC 61850-based solutions.

The cost of building new substations or expanding existing substations, especially in densely populated areas can be significantly reduced by using IEC 61850-based digital substations. This cost reduction is most significant when the conventional current and voltage instrument transformers are replaced with nonconventional sensors and the combination of disconnector-breaker-disconnector is replaced by a disconnecting circuit breaker (DCB) available from several major suppliers. Combining them with an optical CT on the same structure can reduce the substation footprint by about 50% of the footprint of a conventional substation. Reducing the number of primary devices also reduces the construction costs due to the smaller number of foundations needed.

Replacing the traditional hardwired analog and binary circuits between the primary substation equipment and instrument transformers (Figures 15.9 and 15.10) by the optical fibers of the IEC 61850 process bus results in a significant reduction in copper cables. When we include also the replacement of the hardwired interfaces between IEDs, it can reach up to 80% in transmission level air insulated substations. This will lead to significant cost savings not only by reducing the cost of the copper cables, but also by limiting the transportation costs. Copper cables remain for the power supply and short connections between primary equipment and marshalling kiosks in the switchyard.

Figure 15.9 Optical CT size.

Figure 15.10 Cables between substation equipment and the control house in a hybrid substation.

The replacement of conventional instrument transformers with nonconventional sensors also reduces the transportation costs because of the significant difference in their weight. For example, a nonconventional current sensor weights less than 15% of the weight of a conventional current transformer and is also smaller in size.

Further construction cost reductions are achieved by the smaller size of the control house due to the reduced number of panels. This is the result of several factors, such as the smaller size of the limited number of IEDs, which do not have terminal blocks for the traditional hardwired interfaces, very limited number of terminal blocks, and high levels of functional integration. In the future, with centralized IEC 61850-based digital substations, the control house as we know it may disappear.

Another set of benefits is related to IED interfaces, installation, and commissioning. The wiring of conventional substations requires a significant amount of time by skilled technicians in order to provide all required interfaces:

- Between the substation equipment in the yard and the panels in the control house;
- Between the panels in the control house;
- Between the panel terminals and the IEDs in the panel;
- Between the IEDs in the panel.

This time is further extended by the preparation time of the copper wires and their labeling. Another issue is the risk for human errors and the requirements for extensive testing in order to ensure the quality of the interfaces.

In the IEC 61850-based digital substations, the thousands of hard wires carrying individual analog and binary signals are replaced by a limited number of fiber optic cables transmitting sampled values, GOOSE messages, or client-server communications (Figures 15.11 and 15.12), which results in significant savings in installation, commissioning, and maintenance costs.

One of the most time-consuming engineering tasks when building a new substation and engineering its PAC system is the development of the substation HMI. In conventional substations using communication protocols, which do not use semantics-based data naming, the mapping of the data from the registers of each IED to the HMI is an extremely time-consuming work that is also prone to human errors.

The use of a standard, semantics-oriented data model in IEC 61850 allows the development of engineering tools that can generate automatically the substation HMI based on extensions to the model that support the visualization of the substation equipment on the HMI in a user-customizable manner. This will lead to significant reduction of the HMI development time, from months to minutes, and will also improve its quality.

The limited number of opto-inputs and relay outputs is another challenge for the development and implementation of some more advanced distributed PAC schemes. In some cases, this problem is solved by using auxiliary relays; however, the price is the reduced reliability of the schemes due to the increased number of devices and the hardwired interfaces between them. The addition of the auxiliary

Figure 15.11 IED panel in a digital substation.

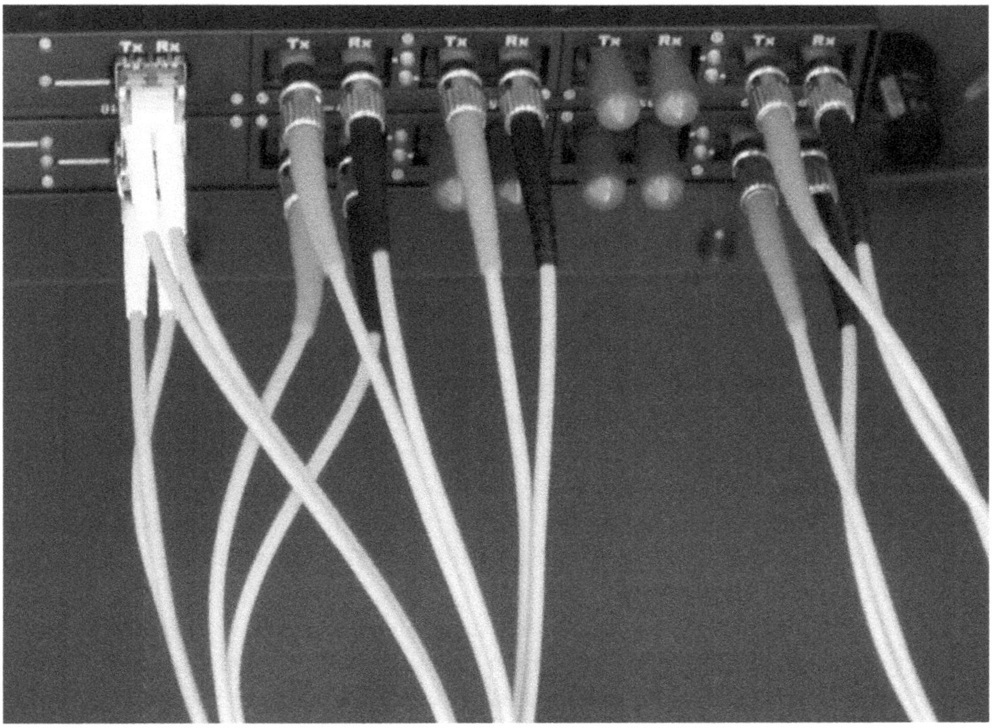

Figure 15.12 Fiber optic cables interface.

relays also increases the operating time of the scheme, which may not be acceptable. This also leads to increased installation, commissioning, and maintenance costs.

Replacing the hardwired interfaces with GOOSE messages eliminates this problem due to the fact that a single fiber connection can carry the equivalent of a practically unlimited number of signals between the IEDs, resulting in significant time and costs savings.

One of the problems with hardwired protection and control schemes is the fact that they are not supervised and, as a result, may represent a hidden failure component. Problems with hardwired interfaces are discovered only due to the failure of a protection scheme to operate as required or during time-based maintenance testing. Maintenance testing is time-consuming and requires outages, which is not acceptable in many cases.

One of the significant benefits of IEC 61850 is the continuous repetition of GOOSE messages, which represent a heartbeat that can be used for interface supervision. As soon as a message is not received, it will indicate a problem with the communications interface or the publishing IED that can be fixed as quickly as possible.

A higher level of monitoring of the communications interfaces is achieved when the IEC 61850 PAC system is using a redundancy protocol such as PRP or HSR. The fact that there are redundant copies of each message allows the detection of communications component failure (if one copy of the message is lost) or failure of the publishing IED (if both copies are not received).

Such monitoring significantly reduces (practically eliminates) the need for time-based maintenance testing, resulting in cost and time savings.

In conventional systems, the failure of an instrument transformer or the wiring between the substation equipment and the protection devices results in the loss of the protection and requires a maintenance crew to go to the substation to fix it. During that time, the reliability of the system is affected.

In an IEC 61850-based digital substation, it is possible to subscribe to a primary and backup data source and automatically switch to the backup in case of the loss of the primary. This maintains the protection and control functions in service during the time that it will take to fix the failed equipment.

The changes in the firmware of IEDs using protocols different than IEC 61850 require remapping of the IED data to the HMI due to changes in the data allocation in the memory of the IED. This work is performed by the utility specialists or the system integrator and requires not only update of the mapping, but also testing to verify that the update is successful.

In the case of IEC 61850-based systems, changes in the firmware do not require remapping. This is because the internal IED data is still available with the same IEC 61850 name as before the update. The mapping of the data from new different registers in the memory to the same IEC 51850 data attributes is the responsibility of the IED manufacturer.

IEC 61850 process bus-based solutions also improve the safety of the substation by eliminating one of the main safety-related problems, an open current circuit condition. Because the only current circuit is between the secondary of the current transformer and the input of the merging unit located right next to it, the probability for an open current circuit condition is very small. It becomes nonexistent if optical current sensors are used.

Maintenance testing is something that needs to be done, regardless of the different challenges that may be faced by the crew:

- Long distance between the substation and the base of the testing team;
- Difficult terrain with bad roads;
- Difficult weather conditions;
- Requirements for the reduction of outage time because of maintenance.

One of the benefits of IEC 61850-based digital substations is that all devices (PAC IEDs, substation computers, and test devices) are connected to the substation communications network. If there are testing tools that are connected to the network in the substation on a permanent basis, it becomes possible to perform the tests from a remote location. This significantly reduces the testing costs and eliminates all the safety travel-related concerns.

Maintenance testing in conventional hardwired schemes requires the isolation of the tested protection device, which is performed typically using a test switch connecting the test device to the test object. In many cases, this requires an outage, due to the unavailability of the protection device. At the distribution feeders, typically there is only a single protection IED, which means that, during the testing, if the feeder is still in service, the protection will be provided by backup functions, for example, in the transformer protection IED. This will result in extended fault-clearing times and may lead to the shutdown of many DERs connected to the distribution system.

The virtual isolation features in IEC 61850 allow the isolation and testing of functions, subfunctions, or even a single function element. In this case, all remaining protection functions are active, which improves the reliability of the system, reduces the fault-clearing time, and eliminates the need for an outage, all very significant benefits compared to the traditional testing practices.

Using an IEC 61850 process bus based on standalone merging units (SAMUs) also results in the practical elimination of CT saturation because of the elimination of the current leads' resistance. CT saturation is eliminated when using nonconventional instrument transformers, because there is no CT circuit.

When a short-circuit fault occurs on a transmission line or distribution feeder connected to a substation with DERs at the distribution level, the voltage drop caused by the fault needs to be considered in the analysis of the performance of the DERs and their ability to ride through the fault.

Using inverse time-delayed overcurrent protection for feeder faults or as a backup protection for busbar faults results in fault clearing in the range of hundreds of milliseconds or even seconds. Such a delayed trip will result in the duration of the voltage sag experienced by a DER in the tripping area of the ride-through characteristic. An accelerated protection scheme can significantly reduce the fault-clearing time and bring it within the stay-connected area of the characteristic.

The challenge for the implementation of protection schemes is that they require multiple hardwired connections between many distribution protection devices that that may not be available based on the equipment used. IEC 61850 GOOSE messages are a technology that can help us to achieve these goals without the need for additional investments.

Based on the analysis of all use cases described above, it is clear that IEC 61850, especially when fully implemented in digital substations with nonconventional instrument transformers and state-of-the-art disconnecting breakers, offers significant savings in time and money. The use of fiber for the interface between the primary substation equipment and the IEDs in the control house significantly reduces the EMP impact on the PAC system.

At the same time, the risks are minimal, and they have to do with the fact that it is a new technology that requires training in order to apply it properly and use it efficiently.

CHAPTER 16

Cybersecurity

16.1 Introduction

If we look at the world around us, it is not very difficult to classify a large electric power system as the biggest machine ever created with multiple interacting parts distributed over whole continents. At the same time, we have reached a point when every aspect of our lives depends on electricity, meaning that the reliable supply of electric power is an essential requirement for every utility.

When we look for a generic definition of security, we can say that there is a consensus that it is freedom from risk or danger. Based on this definition, security is the ability of the power system to withstand imminent disturbances or contingencies.

Implementing the smart grid, Internet of Things, and digital substations based on IEC 61850 over local and wide area communications raises concerns about cybersecurity.

There are many different aspects of the issue related to three core areas known in the computer security industry as CIA:

- *Confidentiality:* Ensuring that information is not accessed by unauthorized persons;
- *Integrity:* Ensuring that information is not altered by unauthorized persons in a way that is not detectable;
- *Authentication:* Ensuring that the users are the persons they claim to be.

However, for the electric power industry, the main goal is to be able to continue operating the grid and delivering power to the consumers, thus making availability the highest priority. This means that, instead of focusing on CIA, we should concentrate on AIC: availability, integrity, and then confidentiality.

Cybersecurity is a huge and extremely important topic that is covered in detail by numerous books. It is impossible to cover everything related to it in a single chapter with a limited space. That is why PAC specialists need to focus on the vulnerabilities of the different components of the PAC systems and think what a hacker can do, how a hack can be detected, and what can be done to deal with it. Like with many other things in life, there is a choice: to get scared and run away from digital technology or face the challenges and deal with them based on a good understanding of the threats and risks and the methods and tools available to use for protection.

16.2 Attack Vectors and Attack Surface

When we think about the digital grid and its importance for economy and the society, it becomes clear that we need to have a very good understanding of its vulnerabilities and what impact an attack on the specific assets may have on it.

There are some terms that are used by cybersecurity professionals when they analyze the vulnerabilities of a specific system that also apply to the PAC systems in the electric power grid. No system is 100% secure and cannot be penetrated by any adversary. That is why it is so important to know the many ways in which the defenses can be breached and identify which are the most vulnerable components of the system and the impact on the security and reliability of the power grid if successfully attacked. This will allow the cybersecurity team to concentrate its efforts on protecting the high-impact targets in order to prevent the most severe cyberattacks.

The collection of all possible cybersecurity attack targets is called the attack surface. When we consider the digitized substations PAC systems and the digitalized engineering process, we can see the huge number of potential cyberattack targets that can have an impact on the reliability of the electric power grid.

An attack vector is a method used by an attacker to illegally access a network or digital devices to exploit system vulnerabilities for different reasons. Hackers use numerous attack vectors to launch attacks that take advantage of weak points in the system, steal login credentials, or cause a breach that allows them to manipulate offline or real-time data. In some cases, they are able to take actions that result in operation of switchgear equipment leading to the weakening of system interconnections or even local and wide area disturbances.

An actual attack may use several different attack vectors. They can be divided into two main categories: passive and active. In many cases, it starts with a passive attack when the hacker penetrates the cyberdefenses and, over a period of time, monitors the system and the behavior of the different actors looking for open ports or other vulnerabilities in order to get detailed information about the target that can be exploited when they decide that is the best time to strike.

Once the hackers have gained enough knowledge about the vulnerabilities of the system, which is the target of the attack, then they can switch from the passive to the active phase and take specific action to achieve their goal. A step in this direction is when an intruder pretends to be a trusted user and steals login credentials that will allow him to gain access privileges to system resources. This can be later used to launch a wider cyberattack against a specific component of the digital grid.

A data breach is any security incident in which sensitive, protected, or confidential data is accessed, stolen, or manipulated by an unauthorized party and they are related to most of the attack vectors against substation PAC systems. Data breaches are the most common and involved in one form or another in cyberattacks, but not all security incidents concern data theft.

Figure 16.1 shows a simplified diagram of the attack surface of a substation PAC system. In the following sections, we briefly discuss the components of the attack surface and their vulnerabilities, followed by an introduction of some attack vectors and what measures can be taken to reduce their impact.

16.2.1 Attack Surface Components

16.2.1.1 Control Center

If an attacker is able to breach the security of the electric power system control center and steal the credentials of a system operator, he or she will gain access to the control of switchgear equipment in critical substations in the system. As was demonstrated by the Christmas cyberattack in Ukraine, such an attack resulted in the tripping of breakers in multiple substations and interruption of the power supply in a significant part of the country. That is why it is so important to use proper authentication, strong passwords, and role-based access control (RBAC). Also, any operator activity should be logged recording what action was taken and who executed it, which will allow detailed analysis following a security breach.

16.2.1.2 Substation HMI

Getting access to the substation HMI in principle is similar to the control center attack but with a significantly lower level of access, only to the switchgear in the

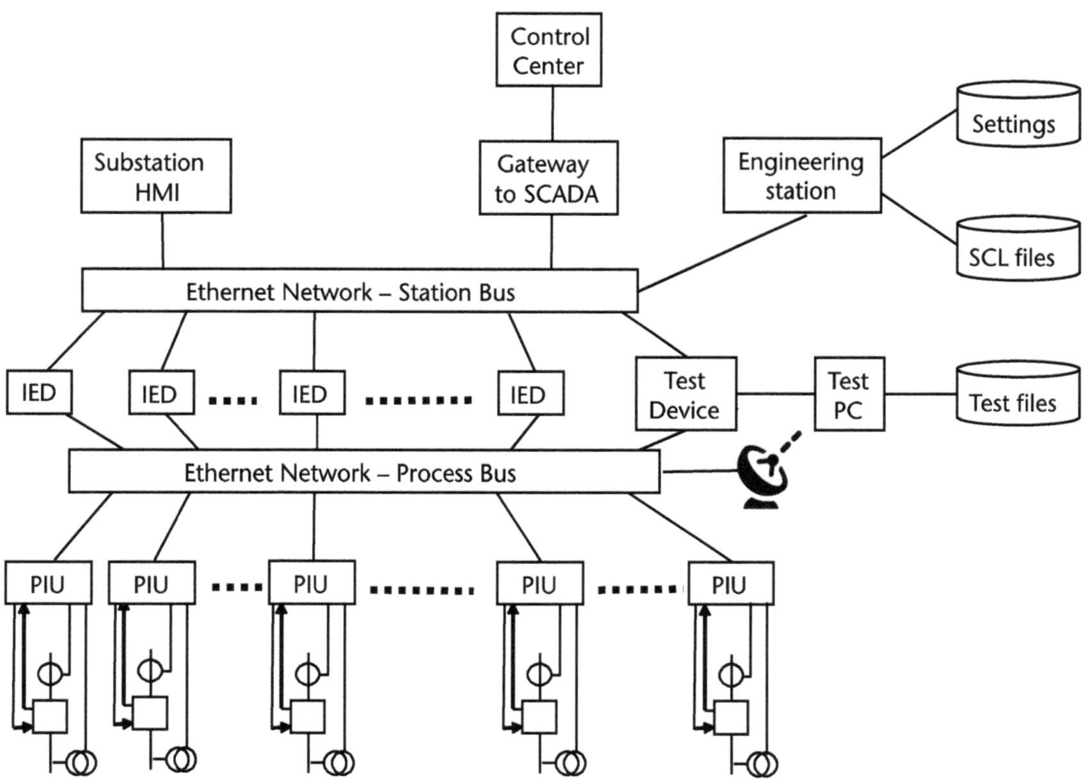

Figure 16.1 Attack surface.

specific substation. Similar measures for protecting credentials and monitoring/logging activities need to be implemented.

16.2.1.3 Engineering Station

Considering the important role that remote communications are playing today in the configuration of multifunctional protection and control devices allowing protection specialists to directly change settings and setting groups or upload or download setting files, it is clear what the impact can be of an attacker getting access to an engineering station and stealing the credentials of employees who are allowed to change the configuration of the devices. By changing the values of the pickup settings of different protection elements, which will make them more sensitive leading to their operation under increased load conditions, the attacker will achieve tripping of the protected line or distribution feeder causing an outage at a peak load moment in time. If the attacker changes the settings to higher values, thus making the protection less sensitive, the protection will fail to operate when a short-circuit fault occurs, which may lead to damaging the protected equipment. That is why it is necessary to implement proper security procedures to prevent stealing the credentials of employees whose roles give them access to changes of settings.

From the engineering station, it is also possible to download modified setting files that may lead to an undesired operation or misoperation when a short-circuit fault or other abnormal condition occurs.

To reduce the impact of successful attacks affecting the engineering station, it is necessary to log any setting change and place an alarm on the substation HMI and as well make it available to the system operator, who can verify that this was an authorized setting change.

16.2.1.4 Setting Files

If a hacker is able to get access to an unprotected setting file, he or she might be able to manipulate them, which may lead to an operation or misoperation as described above. If the changes in settings are not recognized because of a lack of logging procedures indicating what change was made and who made it, when the setting file is downloaded into the protection device, it may lead to future outages or damages of the protected equipment. Considering the importance of the setting files, they should be controlled as Critical Energy Infrastructure Information (CEII) and have to be protected through strong encryption and being manipulated only by authorized specialists using RBAC. Also, the setting files should be subject to proper testing procedures and version control.

16.2.1.5 Substation Configuration Language Files

The substation configuration language files, both ICD and SCD, contain information about the configuration of all protection and control devices in the substation. If an attacker is able to breach the security and get access to these files, the attacker can gain a complete understanding of the GOOSE messages that are being exchanged between the multifunctional IEDs, which will allow the attacker to execute an attack that is very difficult to prevent.

Considering the importance of the SCL files, they also belong to the CEII category and must be protected through strong encryption and only accessed by authorized specialists using RBAC.

16.2.1.6 Station Bus

If the attacker is able to penetrate the substation security parameter and get access to the station bus, he or she will first be able to monitor all the traffic on the network and gain knowledge of the communication interfaces between the different components of the substation PAC system. Once the attacker has a good understanding of how everything works, he or she can launch an attack by publishing modified, nonsecure, GOOSE messages that will result in the tripping of one or more circuit breakers, thus causing outages and loss of power supply to many customers. This kind of attack is described in detail later in this chapter.

Having access to the station bus also provides an opportunity for executing a denial-of-service attack by constantly publishing high-priority GOOSE messages with a broadcast destination address. To detect this kind of attack, it is necessary to have an intrusion detection system continuously monitoring the traffic on the station bus based on the knowledge of legitimate devices as defined in the SCD file.

16.2.1.7 Process Bus

Considering that in some digital substations the station bus and the process bus are different physical networks, if the hacker is able to penetrate the substation defense and gain access to the process bus, he or she can execute an attack by publishing streams of sampled values, pretending that they are coming from process interface devices or merging units in the substation. Because the time interval between the individual sampled value messages is very short, it is quite difficult for such an intrusion to be successful because it will be easily detected by an adequate intrusion detection system in the substation. The replay of a stream of sampled values is also an issue because it may disable a protection IED if it is not designed to handle such a threat.

It is also possible to execute a denial-of-service attack by publishing multiple streams of sampled values with high priority and using a broadcast destination address.

16.2.1.8 IED

If the attacker is on the station bus, he or she may get access to any of the IEDs connected to it. This will allow the attacker to do different things, especially if the access is not properly protected by an adequate access control based on authentication and the use of strong passwords, as well as RBAC. If the attacker is able to overcome all these defense mechanisms, he or she may be able to change protection settings or get information about GOOSE datasets that can be later used for successful attacks simulating GOOSE messages. If the hacker changes the dead band setting of an analog GOOSE message to a minimum, the load is constantly changing, which will cause the device to continuously publish messages that may have the effect of a denial-of-service attack. This kind of attack can be detected based on the methods described earlier in this section.

16.2.1.9 PIU

If the merging units or the process interface units allow client-server communications, it might be possible for the attacker to get access to their configuration and, for example, modify the CT and VT ratios, thus publishing current and voltage samples with values that are higher or lower than the actual ones. This will result in increasing or decreasing the sensitivity of different protection elements and may result in undesired operation or nonoperation when necessary. This kind of attack can be detected using the same approach as for any other setting change described above.

16.2.1.10 Test Computer

Test computers are typically used in a transient fashion and thus need to be managed and secured as a transient cyberasset and operated by qualified personnel under RBAC.

A test computer can be permanently installed in a digital substation to support remote maintenance testing. As it is an authorized device, if a hacker is able to gain access to it as a legitimate user with proper credentials, he or she can execute an attack by making the test device that it controls publish messages without setting the simulation bit to true, which may lead to protection operation and the tripping of breakers. This can be the result of a cyberattack, but also it is possible if the person performing a maintenance test makes some mistake in the configuration of the published messages.

Because the test computer can be, in principle, controlled by a test specialist in a remote location, a hacker gaining access to this remote computer by stealing the credentials of a test specialist authorized to perform remote testing can execute an attack from this remote computer.

16.2.1.11 Test Files

If a hacker is able to gain access to some test files that have been developed for executing automated maintenance tests, these files can be manipulated by a qualified person to publish GOOSE messages and sampled values when the test is executed that will lead to an actual operation and the tripping of breakers in the substation. This can be accomplished by changing the mode in the test file from test/blocked to test, which will produce a physical output and trip the breaker.

To prevent something like that from happening, it is necessary also to have the test files considered as being in the CEII category and to be protected through strong encryption and only accessed by authorized specialists using RBAC.

16.2.1.12 GPS

Considering the dependence of the digitized grid on precise time synchronization for many different applications, we can identify that as another vulnerability that might be explored by hackers or other individuals with malicious intent. Because most of today's time synchronization systems rely on GPS signals, it is obvious that they are a potential target for attacks. Part of the problem is the built-in weaknesses of GPS communications such as poor signal strength and lack of encryption.

One of the methods that can be used to attack the time synchronization system is by using GPS jamming. This is something that anyone can do with relative ease because a jammer is a fairly simple and inexpensive device that confuses the GPS receiver by emitting radio signals at the same frequency as the GPS. This interference disturbs the ability of the GPS device to determine its correct position or time.

GPS spoofing is a more sophisticated attack in which hackers transmit GPS-like signals and code them in a way that tricks receivers in determining their location or what the time is. For the electric power system applications, it is possible to conduct such a spoofing attack by broadcasting genuine signals with the wrong time stamp, thus making the receiver believe that it is in the right place at the wrong time.

One way of reducing the impact of such attacks is by carefully selecting the antennas' location within the substation and protecting the antennas or other related physical infrastructure. On top of that, using GNSS clocks with embedded rubidium oscillators can eliminate the dependence on the GPS signal and allow the detection of the attack.

16.3 GOOSE Attack

One of the most commonly used features of the IEC 61850 standard used all over the world for different protection and control applications is the GOOSE message, which allows high-speed P2P communications used for different substation protection or transmission line protection applications. The standard has defined different parameters that are transmitted with every message that can be used for intrusion detection and in many cases allow us to prevent successful attacks. However, in the hands of skilled attackers, GOOSE becomes a very dangerous tool that may lead to some significant disturbances if the attacker has selected a high-impact target.

Next we provide examples of such an attack within the substation or on a transmission line protection and describe how the attack can be executed and what measures can be used to limit its impact.

In both cases, the attacker is using the knowledge of the nonsecure GOOSE message structure and repetition mechanism that due to the repetition time, which may be the range of a second, allows the attacker to successfully execute it even from a fairly remote location.

16.3.1 Basic Attack

A basic attack by an unsophisticated hacker simulating protection operation is the case when the attacker has been able to:

- Penetrate the substation LAN (station bus);
- Copy the GOOSE message from a protection IED;

- Change the binary value of a protection or protection related logical node's data attribute from FALSE to TRUE;
- Publish the modified GOOSE message to the station bus at a random moment in time during the repetition interval, as shown in Figure 16.2.

As can be seen from the figure, the hacker has not changed the state and sequence numbers (i.e., the header of the GOOSE message is the same as in the last message published by the IED). This should make it fairly easy to detect the attack and prevent the tripping of the breakers if the subscribing IEDs check the state and sequence numbers as part of the GOOSE message processing. The state number is expected to be the same, but the sequence number should have been incremented by 1. This will serve as an intrusion detection and will prevent the operation of the subscribing device based on the received message. As shown in the figure, after the timing of the maximum repetition interval, a regular message will arrive with the same state number and the sequence number equal to $m + 1$.

16.3.2 Sophisticated Attack

A sophisticated attack by a hacker simulating the protection function operation is the one that is the most dangerous because it may lead to the tripping of a critical interconnection or multiple substation breakers that may lead to a local or wide area disturbance in some cases.

To be able to execute such an attack, the hacker needs first to gain access to the substation LAN and using a network traffic analysis tool such as WireShark (Figure 16.3) to monitor, over a period of time, the GOOSE messages on the network.

The attacker has been able to:

- Penetrate the substation LAN (station bus);
- Copy the GOOSE message from an IED;
- Get the state number using a tool such as WireShark (see Figure 16.3);
- Increment the state number stNum by 1;
- Set the sequence number sqNum to 0;
- Change the binary value of a protection operation-related data attribute from FALSE to TRUE;

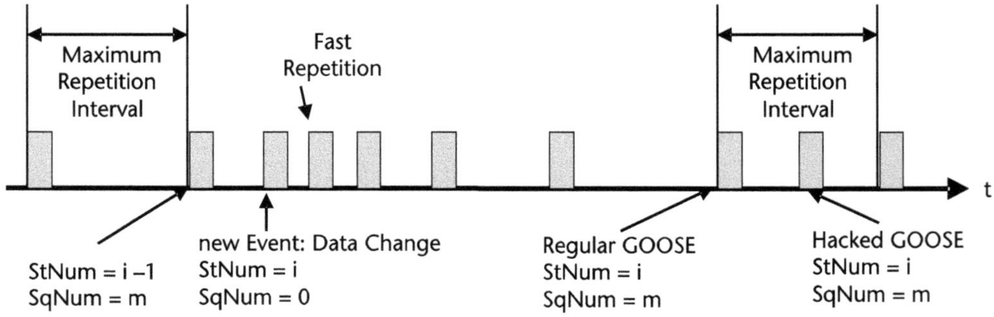

Figure 16.2 Hacked GOOSE message: basic attack.

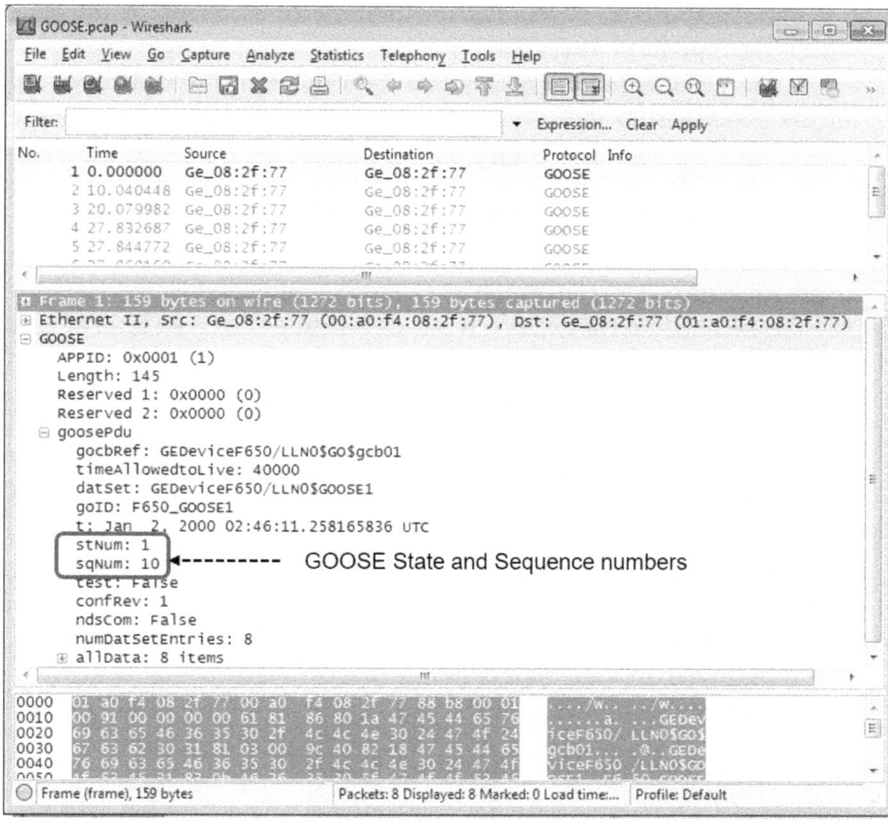

Figure 16.3 Display filter reference: GOOSE.

- Publish the modified GOOSE message to the station bus at a random moment in time during the repetition interval as shown in Figure 16.4.

Because the hacker has changed the state number and the sequence number in the header of the GOOSE message based on the information from the last message published by the IED, it is impossible for a subscribing IED to identify immediately that this is a hacked message and prevent the tripping of the breakers.

This intrusion will be immediately detected when the next message from the publishers in the substation protection and control systems is received; however,

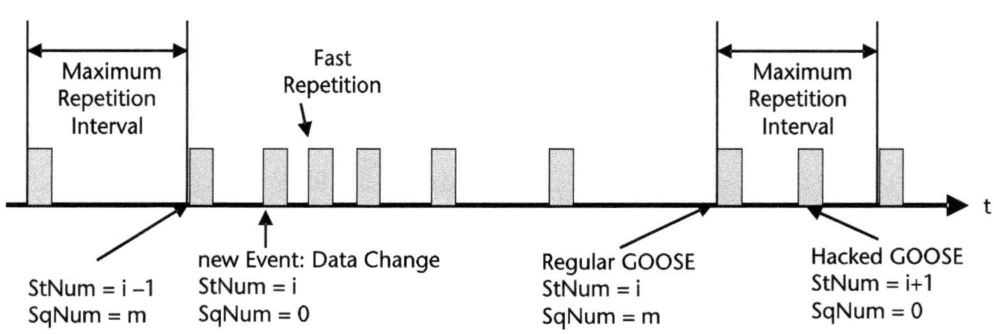

Figure 16.4 Hacked GOOSE message: sophisticated attack.

this may be too late, depending on the time of receiving of the hacked message within the repetition interval.

What might be helpful in this case is the fact that if the intruder is just monitoring the messages on the communications network, he or she may know the data type of the data attributes, but it is not clear what specific piece of data a data attribute in the GOOSE data set represents. So the intruder might have to try changing values of different attributes that he or she expects might lead to a protection operation and that is why it is very important that the intrusion is detected as quickly as possible and proper measures are taken to mitigate it.

The most dangerous case is when the intruder has been able to get access to the IED configuration description information about the content of the GOOSE datasets. If this has happened, then the intruder will know exactly which data attribute value to change in order to cause the maximum damage possible. That is why it is so important to protect all substation language configuration files that contain detailed information of the GOOSE datasets or block unauthorized users to retrieve the information from an IED.

If we assume that this worst-case scenario has actually happened, we still can do something. This is when we need to turn to what we call functional security, relying on our knowledge of the behavior of the electric power system and protection IEDs under specific fault conditions.

16.3.3 Transmission Line Protection Attack

The use of GOOSE messages for transmission line protection schemes already has more than 10 years of history of successful applications, but still many specialists are concerned about the security because of the exposure of the intersubstation messages to cyberattacks. The vulnerability of the protection scheme depends on the selected principle. From a cybersecurity point of view, permissive schemes are better because even if a hacker is able to penetrate the communications link and send a permissive GOOSE by setting:

PDIS2.Str.general = TRUE

indicating that Zone 2 of the distance protection has started, the receiving end is not going to trip if it does not see a fault itself. We can define this as functional security, because the tripping of the local breaker is not activated simply based on the received GOOSE message, but is functionally supervised in the POTT trip logic by the start of the Zone 2 distance element represented by a local PDIS2 logical node.

The case is more complicated for direct transfer trip (DTT) schemes. In principle, they do not require local supervision and, because of that, may lead to a tripping of a breaker when a hacked GOOSE message is received, for example:

RBRF1.OpEx.general = TRUE

indicating failure of the breaker to trip at the remote end of the line. This can be improved if there is redundancy in the data that is included in the GOOSE data set, for example, if the DTT data objects are in the same data set with a different data object indicating the operation of a protection element that initiated the breaker failure protection that resulted in the DTT. It will be more difficult for the attacker to modify both pieces of data that have functional relationship without knowing the specific structure of the data set. This way the subscribing device can check that if both the DTT and the protection element changed states at a reasonable time, it will use that as the functional security criteria to allow or block the tripping of the breaker.

The worst-case scenario is if the intruder knows the structure of the data set so he or she can manipulate related data attributes that may lead to the tripping of a circuit breaker. This is when the functional security should include the same logic that relies on a good understanding of what might be the reason to issue the direct transfer trip. In most cases, this is done for a busbar fault and breaker failure to clear it. The local breaker failure function then sends the DTT signal to the remote end to clear the fault. However, if there is a busbar fault at the remote substation, it should be seen by Zone 2 of the local distance protection. So, when the IED receives the DDT message, it will check if there is a Zone 2 start on that transmission line and, if there is not, this will mean that this may not be a valid DTT message. Any such logic needs to be developed based on a good understanding of the protection schemes at both ends of the line and detailed analysis of the expected behavior during a variety of fault conditions.

16.3.4 Breaker Failure Protection Attack

From the previous example, it is clear that, without proper measures, a malicious attack on a transmission line protection a critical interconnection may be tripped causing some stability problems. It is still tripping of just a single line that under some conditions can be put back in service by auto reclosing.

Probably the worst target of a cyberattack of a critical substation is the breaker failure protection that may result in the tripping of multiple circuit breakers, leading to a significant change in the power system topology that may have impact on the stability of the grid.

If we look at the substation one-line diagram shown in Figure 16.5 and imagine that there is a fault on the transmission line connected to CB6, it will lead to the operation of the protection system that will issue a trip to that breaker. If the breaker fails to trip, the breaker failure protection is going to trip all remaining breakers connected to this section of the bus bar by sending a GOOSE message:

RBRF1.OpEx.general = TRUE

that all protection relays on the corresponding breakers will subscribe to as shown in Figure 16.6.

If a sophisticated attacker has gained access to the substation communications network and sent the GOOSE message indicating the breaker failure operation as described above, all the subscribing IEDs now are going to trip their associated breakers, thus tripping several lines and blocking reclosing, which will result in a weakening of the system and potential local or wide area disturbance. However, if the GOOSE message data set also includes the data object indicating the start of the breaker failure protection RBRF1.Str and there is discrepancy between the values

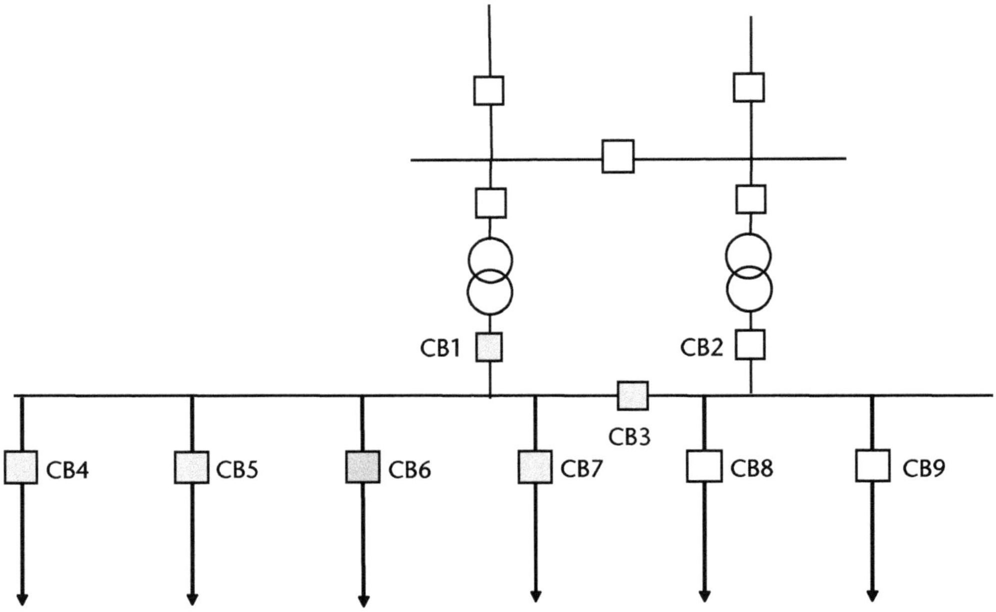

Figure 16.5 Transmission substation.

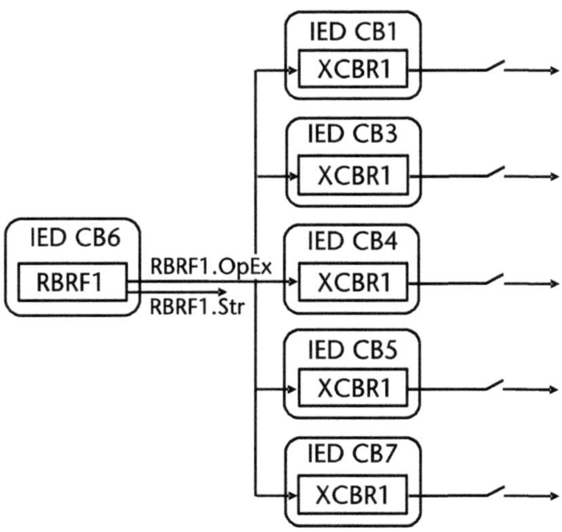

Figure 16.6 Breaker failure operation.

of the start and operate data objects, this can be used as an intrusion detection and to block the tripping of the breakers.

If the attacker has been able to get access to the substation configuration language files containing the definitions of the GOOSE datasets, this attack cannot be prevented unless special measures have been taken in advance considering that possibility. Depending on the substation protection system design, it is possible to identify as part of the engineering process additional fault detection information that might be used to supervise operation when receiving messages indicating breaker failure or other protection operation as part of the functional security.

16.4 Cybersecurity Regulations and Standards

With the digitization and digitalization of the electric power grid, increased use of local and wide area communications, and continuing growth of different cyberthreats, it is becoming more important for the electric power industry to take adequate measures to protect itself and maintain the required levels of reliability and security. That is why different regulating entities, as well as technical and standardization organizations have been working on the development of cybersecurity-related standards and guidelines intended to help the various stakeholders to achieve these goals.

This section lists some of the ongoing developments by the North American Electric Reliability Corporation (NERC), IEC, and IEEE related to cybersecurity and gives a brief description of the focus of each document that will help the readers to identify where to focus their attention. It does not claim that this is all that is out there, but it is a starting point for specialists with cybersecurity concerns.

16.4.1 NERC Critical Infrastructure Protection

The NERC Critical Infrastructure Protection (NERC CIP) is a set of standards developed with the goal to regulate, monitor, and enforce the cybersecurity aspects of the Bulk Electric System (BES) in North America, thus providing a framework to secure critical assets and improve the reliability and efficiency of the electric power grid.

These standards are required by law, which means that all entities under the NERC's regulation must comply with them to ensure that BES and its users are protected from all threats that may affect their normal operation. This requires the identification of critical assets and regular assessment of vulnerabilities that may threaten their secured operation using firewalls and cybersecurity monitoring tools. At the same time, it is necessary to develop and implement contingency plans to respond to unplanned events or cyberattacks. The failure of North American entities to comply with the NERC CIP standards may result in significant monetary fines, sanctions, or other actions.

The NERC CIP is organized into 9 standards containing specific requirements that utilities must follow to identify critical assets, create control mechanisms, enforce the security of their systems, and recover any affected assets following a cybersecurity incident.

NERC CIP-002-5.1a: BES Cyber System Categorization

The goal of this standard is to ensure that identified BES cyberassets are protected from security breaches that could result in faulty operations or system disturbances. It categorizes various BES cybersystems based on the impact of any interruption on the reliability of the electric power supply with a focus on the length of the power outage. Cyberassets are broadly categorized as:

- Electronic access control or monitoring systems;
- Physical access control systems (PACS);

- Protected cyberassets.

NERC CIP-003-8: Security Management Controls

This standard establishes requirements for delegating authority to senior management to develop policies around consistent and sustainable security management controls. This ensures clear accountability related to the handling of emergency situations.

NERC CIP-004-6: Personnel & Training

This standard focuses on reducing the exposure of BES to cyberrisks from personnel by ensuring that the staff and third-party contractors receive proper training consisting of:

- Cybersecurity awareness and training every 15 months;
- Risk and access control management including programs for personnel risk assessment, access management, and revocation or removal of personnel access privileges.

NERC CIP-005-6: Electronic Security Perimeter(s)

This standard requires entities to create electronic security perimeters (ESPs) around cyberassets in order to protect them from misoperation and instability. This establishes a virtual barrier that allows the monitoring of data flows and the control of network access to critical assets. Outside access to the ESP is through a specified and monitored electronic access point. It also requires data encryption and remote access control.

NERC CIP-006-6: Physical Security of BES Cyber Systems

This standard defines the requirements for physical security based on plans for restricted physical access, as well as visitor control considering escorts and detailed visitor logs. It also requires maintenance and biennial testing of the physical security perimeter.

NERC CIP-007-6: System Security Management

The goal of this standard is to secure all technical, operational, and procedural elements within ESPs, including critical and noncritical cyberassets. This includes ports and services, malicious code prevention, and security patches, as well as system access control and security event monitoring.

NERC CIP-008-6: Incidence Reporting and Response Planning

This standard provides guidelines on how to respond to cyberincidents and requires entities to have a proper response plan. It defines a process to identify, classify, report, and respond to incidents, as well as the requirement for the plan's review, update, and testing.

NERC CIP-009-6: Recovery Plans for BES Cyber Systems

This standard defines requirements for recovery plans that can help entities to recover from a cybersecurity incident that has affected the functioning of their cybersecurity systems. It specifies who is responsible and when the plan should be activated, as well as the review, update, and communication process.

NERC CIP-010-3: Configuration Change Management and Vulnerability Assessments

The goal of this standard is to detect and prevent any unauthorized changes to cybersystems through system configuration controls and active vulnerability testing. It defines three compliance areas:

- Configuration change management that defines an authorization process for changes in system components, such as operating systems, software, and ports;
- Configuration monitoring for unauthorized changes, every 35 days;
- Vulnerability assessments, every 15 months.

NERC CIP-011-2: Information Protection

This standard specifies the requirements to identify high-impact information that could influence the functioning of BES if it is compromised and maliciously misused. It also specifies protocols for information protection, reuse, and disposal.

NERC CIP-014: Physical Security

This standard specifies the requirements to identify and protect transmission stations and substations and their associated primary control centers, which, if damaged or becoming inoperable as a result of a physical attack, could lead to a wide area disturbance, uncontrolled separation, or cascading within an interconnection.

16.4.2 IEC 62351 Communication Network and System Security

Considering the fact that the digitization and digitalization of the electric power grid expose it to different attack vectors, Working Group 15 of IEC TC 57 was given the task of developing the IEC 62351 series of standards to address different aspects of information security for power system protection and control operations. The main goal is to ensure the security of the communication protocols defined by IEC TC 57, such as the IEC 60870-5 and IEC 60870-6 series, the IEC 61850 series, and the IEC 61968 and IEC 61970 series, by developing standards or technical reports on end-to-end security. To avoid any misunderstanding, end-to-end security is defined as:

- Safeguarding information in a secure telecommunication system by cryptographic or protected distribution system means from point of origin to point of destination;
- Safeguarding information in an information system from point of origin to point of destination.

To meet this goal, WG 15 has developed and continues working on a series of documents covering various aspects of information security as applied to power system operations, the specification of security standards for the IEC TC 57 communication protocols. The different parts of the standard are listed below with short descriptions of their content.

IEC TS 62351-1 Communication Network and System Security—Introduction to Security Issues

As the title indicates, this technical specification [1] provides an introduction to the remaining parts of the standard in order to familiarize the reader with various aspects of information security as applied to power system operations. This publication is of core relevance for the smart grid.

IEC TS 62351-2 Glossary of Terms

This technical specification [2] includes mostly a collection of cybersecurity terms used in the IEC 62351 series, most of which have been formally defined by other standards organizations. It covers the key terms used in the documents and is not meant to be a definitive list. Most terms used are included with references to where they were originally defined.

IEC 62351-3 Communication Network and System Security—Profiles Including TCP/IP

This part of the standard [3, 4] ensures the cybersecurity of TCP/IP-based transport layer communications for SCADA and telecontrol protocols. It specifies how to provide confidentiality, integrity protection, and message-level authentication between communicating entities at either end of a TCP/IP connection.

IEC 62351-4 Profiles Including MMS and Derivatives

This technical specification [5, 6] provides support at the application layer for authentication during handshake for the MMS-based applications. Amendment 1 provides support for extended integrity and authentication also for the data transfer phase by shared key management and data transfer encryption at the application layer and it provides end-to-end security with zero or more intermediate entities. It also extends the support for application protocols using other protocol stacks and XML encoding.

IEC TS 62351-5 Security for IEC 60870-5 and Derivatives

This technical specification [7] is focused on the security of operation of all protocols based on or derived from IEC 60870-5: Telecontrol Equipment and Systems—Transmission Protocols. It specifies messages, procedures, and algorithms for IEC 60870-5-101, 60870-5-102, 60870-5-103, and 60870-5-104. A new edition adds the capability to change update keys remotely, uses security statistics to aid in detecting attacks, discards unexpected messages, and adds permitted security algorithms.

IEC 62351-6 Security for IEC 61850

This standard [8] is dedicated to providing cybersecurity for IEC 61850. It specifies messages, procedures, and algorithms for securing at least the operation based on IEC 61850-8-1, IEC 61850-8-2, IEC 61850-9-2, and IEC 61850-6, thus covering security for client-server, Layer 2 GOOSE, and sampled values. This document is intended to help the members of the working groups developing and using the protocols. Some of the material can also help people to better understand the requirements for securing IEC 61850.

IEC 62351-7 Network and System Management (NSM) Data Object Models

The goal of this technical specification [9] is to define a set of abstract NSM data object models that are specific to the power system and will support the remote monitoring of the health and condition of components of the smart grid, such as IED and DER systems and other systems that are important to power system operations. The new edition of the standard includes revisions of the NSM object data model with the UML model adopted for NSM objects' description and the SNMP MIBs' translation included as code components that are also available as electronic machine-readable file.

IEC 62351-8 Role-Based Access Control for Power System Management

The goal of this standard [10] is to ensure that access to data is restricted and enable RBAC for power system management. It defines human users, automated systems, and software applications as subjects that are assigned to specified roles that restricts their access to the resources necessary for their roles. The standard defines a set of mandatory roles to be supported, as well as the exchange format for defined specific or custom roles. It applies to subjects' direct wired local access to an object and wireless or dial-up remote access. RBAC for IEC 61850 is an extension of IEC 62351-8 in TR 61850-90-19.

IEC 62351-9:2017 Cyber Security Key Management for Power System Equipment

As the name of this standard [11] indicates, it specifies cryptographic key management to protect digital data and its communication. It discusses the generation, distribution, revocation, and handling of public-key certificates and cryptographic keys, including asymmetric keys and symmetric keys for groups (Group Domain

Interpretation (GDOI)). It assumes that keys and cryptography infrastructures have been selected and focuses on the management techniques that will help to guarantee interoperability among different vendors by specifying or limiting key management options to be used.

IEC TR 62351-10:2012 Security Architecture Guidelines

This technical report [12] introduces security architecture guidelines for power systems based on essential security-related components and functions and their interaction, including the mapping of these security controls to the general system architecture of power systems. It can serve as a guideline to system integrators on applying available standards to securely deploy power generation, transmission, and distribution systems.

IEC 62351-11 Security for XML Documents

Considering the use of XML for different documents defined by IEC TC 57-related standards that play an important role in the secure operation of the grid, this standard [13] specifies a schema, procedures, and algorithms for securing XML documents that are used in different domains. It can be considered as a profile of existing W3C standards for XML document security and provides some additional extensions.

IEC TR 62351-12 Resilience and Security Recommendations for Power Systems with Distributed Energy Resources (DER) Cyber-Physical Systems

Considering the constantly growing dependence of electric power systems on interconnected DERs, this technical report [14] focuses on improving the resilience of the electric power grid by providing cybersecurity recommendations and engineering/operational strategies for the different stakeholders of distributed generation and storage devices. It takes under consideration the hierarchical architecture of cyberphysical DER systems.

IEC TR 62351-13 Guidelines on Security Topics to Be Covered in Standards and Specifications

This technical report [15] provides general guidelines for the working groups developing standards and specifications on what security topics could or should be covered in these documents used in the electric power industry.

16.4.3 IEC 62443 Industrial Communication Networks—Network and System Security

The ISA/IEC 62443 series of standards was developed by the ISA99 committee and adopted by the IEC with a focus on the specific requirements of operational technology. It provides a flexible framework to address and mitigate current and future security vulnerabilities in industrial automation and control systems (IACSs). The series includes four parts:

- Part 1 covers topics that are common to the entire series.
- Part 2 focuses on methods and processes associated with IACS security.
- Part 3 is about requirements at the system level.
- Part 4 provides detailed requirements for IACS products.

The standard takes a risk-based approach to cybersecurity using the defense-in-depth strategy and introduces the concept of security levels that can be applied to zones, conduits, channels, and products.

16.4.4 IEEE Power and Energy Society Standards

The IEEE Power and Energy Society (PES) has also recognized the importance of cybersecurity for the reliable operation of the electric power system and developed several standards addressing different threats in order to prevent the impact of cyberattacks. This section provides a brief overview of such standards and ongoing activities.

IEEE C37.240-2014 Cyber Security Requirements for Substation Automation, Protection and Control Systems

This standard is the result of a joint effort between the IEEE PES Substations and Power Systems Relaying Committees and is focused on the applicability of the NERC CIP and NIST Smart Grid security efforts for substation automation, protection, and control systems. It presents comprehensive engineering practices that can be used to achieve the required levels of cybersecurity of automation, protection, and control systems regardless of the criticality of the cyberassets based on trust and assurance of data in motion, data at rest, and incident response.

The goal of the standard is to achieve a balance between technical and economic feasibility without disturbing legitimate activities, especially during disturbances.

IEEE 1686-2013 Standard for Intelligent Electronic Devices Cyber Security Capabilities

Considering that IEDs are building blocks with a critical role in the digitized grid, they have to support functions enabling the implementation of critical infrastructure protection programs. This standard defines the features to be provided in the IEDs to meet such requirements. The security of IED access, operation, configuration, firmware revision, and data retrieval are covered.

References

[1] IEC TS 62351-1:2007 Power Systems Management and Associated Information Exchange—Data and Communications Security—Part 1: Communication Network and System Security—Introduction to Security Issues, 2007.

[2] IEC TS 62351-2:2008 Power Systems Management and Associated Information Exchange—Data and Communications Security—Part 2: Glossary of Terms, 2008.

[3] IEC 62351-3:2014 Power Systems Management and Associated Information Exchange—Data and Communications Security—Part 3: Communication Network and System Security—Profiles Including TCP/IP, 2014.

[4] IEC 62351-3:2014+AMD1:2018+AMD2:2020 CSV Consolidated Version Power Systems Management and Associated Information Exchange—Data and Communications Security—Part 3: Communication Network and System Security—Profiles Including TCP/IP, 2020.

[5] IEC 62351-4:2018 Power Systems Management and Associated Information Exchange—Data and Communications Security—Part 4: Profiles Including MMS and Derivatives, 2018.

[6] IEC 62351-4:2018+AMD1:2020 CSV Consolidated Version Power Systems Management and Associated Information Exchange—Data and Communications Security—Part 4: Profiles Including MMS and Derivatives, 2020.

[7] IEC TS 62351-5:2013 Power Systems Management and Associated Information Exchange—Data and Communications Security—Part 5: Security for IEC 60870-5 and Derivatives, 2013.

[8] IEC 62351-6:2020 Power Systems Management and Associated Information Exchange—Data and Communications Security—Part 6: Security for IEC 61850, 2020.

[9] IEC 62351-7:2017 Power Systems Management and Associated Information Exchange—Data and Communications Security—Part 7: Network and System Management (NSM) Data Object Models, 2017.

[10] IEC 62351-8:2020 Power Systems Management and Associated Information Exchange—Data and Communications Security—Part 8: Role-Based Access Control for Power System Management, 2020.

[11] IEC 62351-9:2017 Power Systems Management and Associated Information Exchange—Data and Communications Security—Part 9: Cyber Security Key Management for Power System Equipment, 2017.

[12] IEC TR 62351-10:2012 Power Systems Management and Associated Information Exchange—Data and Communications Security—Part 10: Security Architecture Guidelines, 2012.

[13] IEC 62351-11:2016 Power Systems Management and Associated Information Exchange—Data and Communications Security—Part 11: Security for XML Documents, 2016.

[14] IEC TR 62351-12:2016 Power Systems Management and Associated Information Exchange—Data and Communications Security—Part 12: Resilience and Security Recommendations for Power Systems with Distributed Energy Resources (DER) Cyber-Physical Systems, 2016.

[15] IEC TR 62351-13:2016 Power Systems Management and Associated Information Exchange—Data and Communications Security—Part 13: Guidelines on Security Topics to Be Covered in Standards and Specifications, 2016.

CHAPTER 17

DER Integration

17.1 Introduction

There is no doubt that one of the main characteristics of the electric power systems of the twenty-first century is the large penetration of DERs. At the beginning, they were not a problem from the point of view of PAC specialists; they were small and isolated, and there were not that many of them. However, the pressure to solve global warming by replacing fossil fuel-based power production with green power produced by renewable energy resources (Figure 17.1), together with the advancements in technology, created an environment where we now see in service many wind 8–10-MW generators, while some prototypes are getting close to 15 MW with turbine blades reaching lengths of more than 100m (328 feet). These huge machines are amazing compared to the windmills used for centuries or the first electricity-generating wind turbines from 125 years ago. They are also connected to the electric power system in a very different way. While at the end of the nineteenth century and most of the twentieth century wind generators were isolated single machines predominantly used to produce electricity in rural areas without connection to the distribution system, today they can be all sizes, connected to any level of the electric power system. The types of generators used, the methods of connection to the system, and the fact that they are not limited to covering the load at their location, but, in many cases, they deliver power to the system that creates a lot of challenges for the protection and control systems.

Wind is not the only renewable resource. Solar power is another major player with a very rich history as well. While at the beginning and through the centuries concentrating solar rays using mirrors or magnifying lenses was the main method that is still used today to produce steam that can be used to generate electricity, the photovoltaic effect discovered in the twentieth century resulted in the development and widespread use of solar panels to power watches, calculators, and homes or deliver power from a solar farm to the electric power grid.

There are many others (tidal, micro-hydro, fuel cells, fly wheels, biomass, to name a few), all with different characteristics and different behavior during abnormal system conditions. All of them need to be integrated in the electric power system, as well as its protection and control systems. This is one of the big challenges that our industry faces, because, in many cases, the owners of the DERs do not

Figure 17.1 Wind farm.

understand that their single or multiple machines need to behave in a certain way during a fault on the interconnection to the system or during a wide area disturbance. They need to provide information about the transient and dynamic models of the DER that they are connecting to the grid. They need to continuously communicate with the energy management, protection, and control systems. They need to receive commands and execute the signals. Integration means that each individual unit has to become part of the whole and part of the system. This will then help to maintain the stability of the system during wide area disturbances.

The digitization of the electric power system based on the IEC 61850 standard provides the tools to achieve these goals in the most efficient way.

17.2 IEC 61850-7-420

The development of multifunctional IEDs for PAC by some of the leading global manufacturers that started later in the last century leading to the digitization of the electric power grid brought significant benefits in improving its reliability and security. However, it also introduces some challenges, mostly related to the fact that the different suppliers were using their own communication protocols. The lack of interoperability between devices produced by different manufacturers was making difficult the integration of equipment in substation PAC systems. To solve this problem, IEC TC 57 started work on the development of the IEC 61850 standard.

In parallel to these developments, the electric power grid was going through a period of different changes: the introduction of independent power producers and the penetration of DERs of different kinds. While at the beginning their numbers were limited, they have been going through an exponential growth and have been contributing significant amounts of electric power to the grid, making it clear that they have to interface with the existing protection and control systems. The operation of the DERs depends on the algorithms implemented in their controllers, which are also using communication protocols and thus facing similar issues like the protection and control devices. The use of different communication technologies presents significant technical difficulties and also increases integration, installation, and maintenance costs, putting pressure on the industry to have one international standard that defines the communication and control interfaces for all DER devices and their controllers.

The successful development and implementation of the IEC 61850 standard demonstrated that it can be used as a cornerstone technology for the smart grid and it makes sense to integrate the DERs using it as well. That is why IEC TC 57 created Working Group 17 Communication Systems for Distributed Energy Resources (DER) and gave it the task to identify which of the existing logical nodes and communication services defined in the IEC 61850 standard can be reused and what extensions are required to meet the needs of this new domain.

What makes the development of such standard so challenging is that it has to model generating sources that are very different from what has been used in the traditional electric power systems. Instead of having to protect and control a limited number of large synchronous generators, in this case, we have to deal with a large number of relatively small energy resources of several different kinds. Many of them also depend on the changing weather conditions, which makes their power output sometimes difficult to predict and schedule.

Another major difference is that, while in the traditional grid we typically have generators and loads (with the exception of pump storage power plants), in the case of DERs, we may have storage that acts as a generator at one moment in time and as a load in another.

Last but not least, because of the large number of small units that need to be monitored and controlled, it is not reasonable to expect that this will be done directly by the energy management system. What makes a lot more sense is to have a system to aggregate the information from the individual units at the site and to provide that to the control center, and at the same time when it receives a command from the control center, it will make a decision what the behavior of the individual units at the plant will be and control them. The same applies also to fulfilling the electricity market requirements and obligations.

The results of the efforts of the members of Working Group 17 were published as Edition 1 of the standard IEC 61850-7-420:2009 Communication Networks and Systems for Power Utility Automation—Part 7-420: Basic Communication Structure—Distributed Energy Resources Logical Nodes [1]. As the name indicates, it defines the IEC 61850 information models to be used in the exchange of information with DERs of different kinds, such as:

- Wind generators;
- Photovoltaic panels;
- Micro-hydro;
- Fuel cells;
- Combined heat and power;

- Energy storage.

Where possible, the IEC 61850-7-420 information model uses existing IEC 61850-7-4 logical nodes, but also defines DER-specific logical nodes to fill the gaps.

This standard introduces the D group of logical nodes that are specific to the DER domain. They are grouped together in the document according to the systems that they are to be used in:

- Logical nodes for DER management systems;
- Logical nodes for DER generation systems;
- Logical nodes for specific types of DER;
- Logical nodes for auxiliary systems.

While the majority of new logical nodes in this standard belong to the D group, there are also some additions to the Z group logical nodes for further power system equipment (for example, ZINV: inverter for converting direct current to alternating current) and the M group logical nodes for metering and measurement (for example, MFLW: flow characteristics, including air, oxygen, water, hydrogen, and/or other gases or liquids used for fuel and for the fuel cell processes).

Figure 17.2 shows a generic DER interconnection model with some new logical nodes (in black) and existing logical nodes (in gray).

This standard also defined a couple of new DER-related common data classes (CDC), for example, for scheduling.

Since the publication of Edition 1, the amount of generation produced by the different kind of distributed renewable energy resources and especially from wind and solar has grown exponentially, becoming a critical source of power for many utilities around the world. At the same time, in response to the growing concerns of global warming and the exploding electric vehicle industry, we have a completely new type of DER. While everything covered in Edition 1 is at a fixed location and with a fixed interface to the grid, the electrical vehicles can be connected to any charging station at different moments in time, thus introducing completely new modeling requirements.

Microgrids also became available at university campuses or industrial facilities, while integrating DERs have different kinds, usually with inverter-based connections to the grid and with the completely different behavior compared to the synchronous machines. This also required some developments in the standards to cover their modeling for PAC applications.

Considering that many DERs are connected to distribution feeders changing their characteristics, it also became necessary to address the issues related to distribution automation.

All of the above resulted in the expansion of the scope of Working Group 17, which had to change its name to "Power System Intelligent Electronic Device Communication and Associated Data Models for Microgrids, Distributed Energy Resources and Distribution Automation."

At the same time, the development process of the core IEC 61850 standard has evolved as well, becoming strictly object oriented and based on UML. This had a

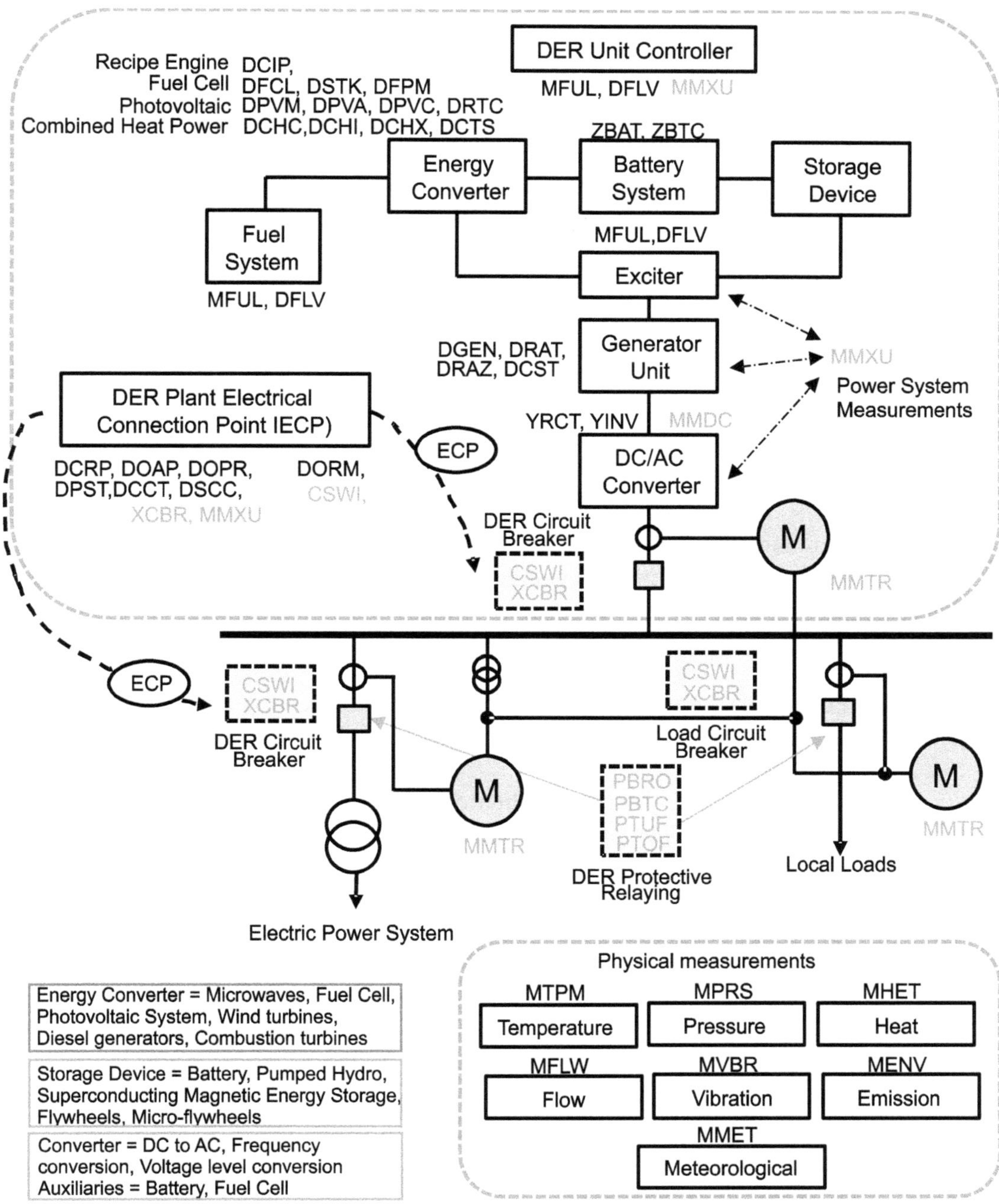

Figure 17.2 Logical nodes for DER systems.

significant impact on the revision of the standard that was published in 2021 as IEC 61850-7-420:2021 Communication Networks and Systems for Power Utility Automation—Part 7-420: Basic Communication Structure—Distributed Energy Resources and Distribution Automation Logical Nodes [2]. This edition is a technical revision and cancels and replaces Edition 1, which was published in 2009.

The focus of this Edition 2 of the standard is expanded to cover not only distribution automation systems but also DER-based systems including:

- Distribution-connected generation systems;
- Energy storage systems;
- Controllable loads;
- DER management systems, including aggregated DER.

It defines specific logical nodes containing the necessary data objects for DER and distribution automation (DA) functions, reflecting the requirements of various countries and regions grid codes for DER interconnection. It also defines a generic modeling approach for different kinds of DERs such as generation, storage, controllable loads, and their physical and virtual aggregation.

An extremely important extension is the modeling support of a wide range of operational functions covering specific grid codes functions as expressed in IEEE 1547 [3] and EN 50549 [4].

Some of the original LNs defined in Edition 1 of IEC 61850-7-420 have first been reassessed within technical reports, such as IEC TR 61850-90-7 [5], IEC TR 61850-90-9 [6], and IEC TR 61850-90-10 [7]. As a result, some have been modified and renamed, while others have been deprecated. The changes in the model compared to Edition 1 are due mainly to changes to the definitions of DER-related operational functions.

17.3 DER Function Modeling Principles for Protection and Control Applications

The modeling of DER systems and their components is very challenging because it includes numerous devices of many different kinds and various modes of operation. Because, in many cases, multiple DERs are components of a virtual power plant, for example, a wind farm, which exports power to the electric power system, they must be integrated into the different PAC systems.

Considering the large number of DERs already installed and being installed in many countries around the world, they have developed standards defining requirements for the behavior of DERs or DER-based plants during varying electric power system conditions in order to ensure the stable operation of the grid.

With IEC 61850 becoming the dominant communications protocol that brings significant benefits to the industry and allows the more efficient integration of devices of different types into systems, it was clear that it should become the protocol of choice for the DER industry.

The modeling of DER systems is further complicated by the many different functions that they need to perform to support the stable and reliable operation of the electric power grid, such as:

- Voltage ride-through;
- Frequency ride-through;
- Voltage and reactive power control (volt-var control);
- Voltage and active power control (volt-watt control);
- Active power-reactive power mode;

- Constant reactive power mode;
- Constant power factor (PF) mode.

One of the main requirements in the development of the standard was to support interoperability and flexibility. To meet these requirements, the standard includes a combination of some mandatory and many optional data objects and data attributes. This results in significant interoperability challenges caused by the differences in the implementation of the functional models by the different manufacturers.

The solution to addressing these challenges is to improve the interoperability by limiting the flexibility of the implementation of the standard. This requires the development of profiles of the standard focused on meeting the requirements of the different countries or, in some cases, even of large electric power systems within a country. Profiles are required to provide a specification for vendors to use in building, so that the resulting products will interoperate, and at the same time to provide a specification that a customer may reference for procurement. This is possible because in a profile:

- The functional model is based on the defined modeling hierarchy.
- All data objects and attributes included in the profile are mandatory, even if they are optional in the core standard.
- The certification testing is based on the profile definitions.

17.4 Ride-Through Modeling Requirements

In this section, we use the ride-through function as an example of the modeling approach that can be used for many other DER-related functions.

Short-circuit faults are part of life for the electric power system and have an impact on its operation and stability. As can be seen from Figure 17.3, during a single-phase-to-ground fault, the voltage drops in the faulted phase and increases in the healthy phases, creating conditions that can damage electric power system equipment and impact the operation of many other devices or systems.

The impact of the fault depends on two characteristics. The first characteristic of a voltage sag, the depth, is a function of the type of fault, the fault location, and the system configuration. It will also be affected by the state of the distributed generator, if it is in service or not. The level of voltage increase is also affected by the grounding of the interface transformer and should also be taken into consideration.

This is something that we cannot control, but we have to study in order to be able to predict or estimate the effects of different faults on the sensitive equipment. The second characteristic of the voltage sag, duration, is the parameter that we can control by properly applying the advanced features of multifunctional protection relays or by using communications-based protection schemes.

According to many country codes, a DER should not trip regardless of the voltage value in the first few cycles after the start of a short-circuit fault or voltage

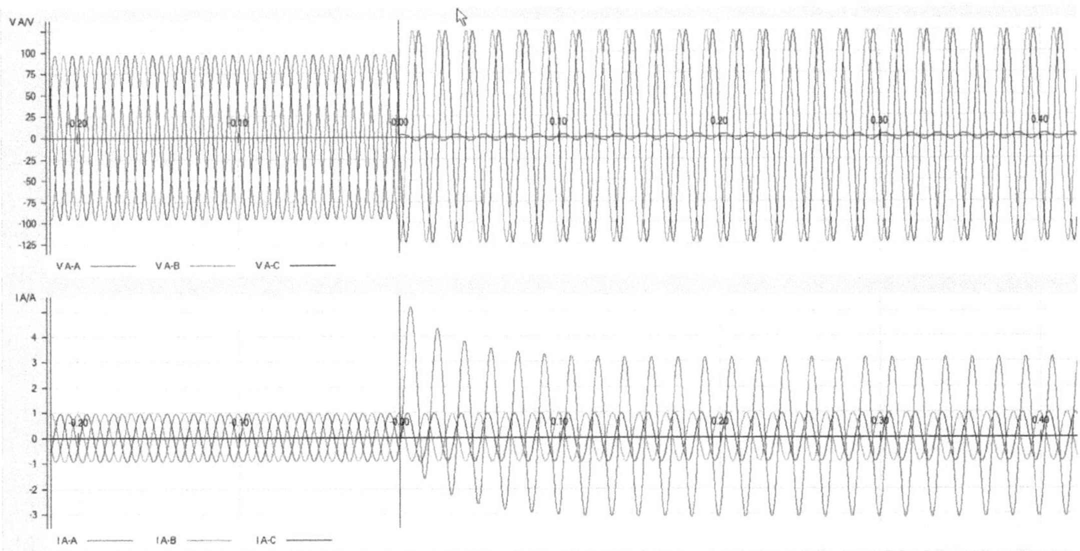

Figure 17.3 Voltage and current waveforms for single-phase-to-ground fault.

disturbance. Trip is defined as inhibition of immediate return to service, which may involve disconnection.

The implementation of this functionality depends on grid code compliance and can be based on the use of one or multiple instances of undervoltage (PTUV) elements to detect in which zone of the characteristic the voltage is, while DLVT (low-voltage-ride-through) will provide the synthesis status needed by the power management function and/or the protection system of the resource to take the action required by the grid code, for example, must trip or cease to energize.

As can be seen from Figure 17.4, for the European low-voltage ride-through characteristics, single undervoltage elements will be sufficient to model it.

The undervoltage characteristic TmVChr and the settings of the PTUV instance should be selected to meet the specific country grid code requirements, as shown in Figure 17.5. When the voltage level during a short-circuit fault is lower than the setting PTUV.StrVal, it will start the undervoltage element and if it is in the operating part of the characteristic PTUV.Op.general will become true. This will be used by logical node DLVT to issue a trip signal through a PTRC.

However, that is not the case for the North American ride-through requirements defined in the IEEE 1547 standard, which specifies high-voltage and low-voltage requirements (Figure 17.6). Two curves are required for the voltage-time profiles, but only the low-voltage "Must Trip" is mandatory defining:

- Low-voltage Momentary Cessation boundary;
- Low-voltage Must Trip boundary.

According to this definition, for a DER to meet the IEEE P1547 requirements, the overall disconnection time from the fault inception needs to be 160 ms.

This time depends on several factors:

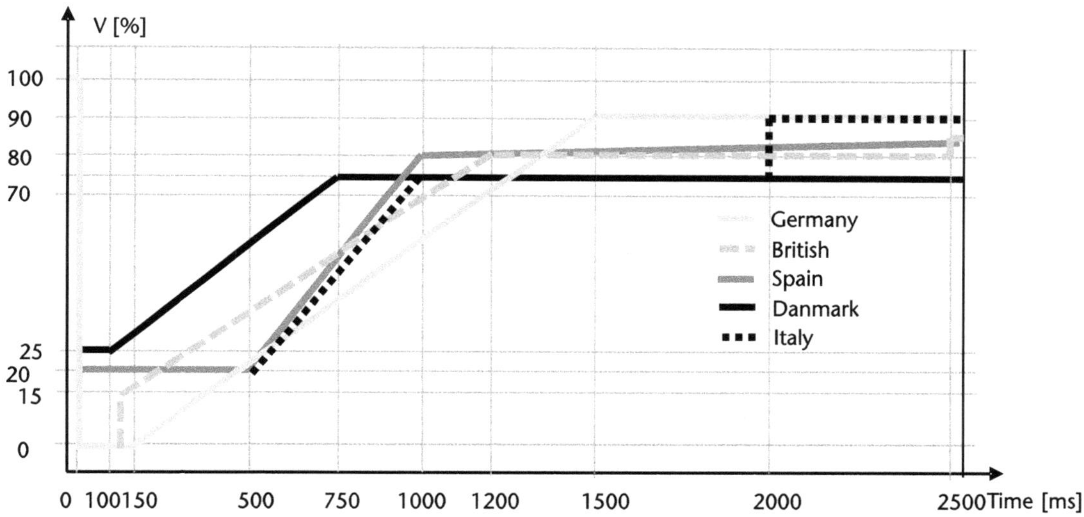

Figure 17.4 Voltage ride-through characteristics of European countries.

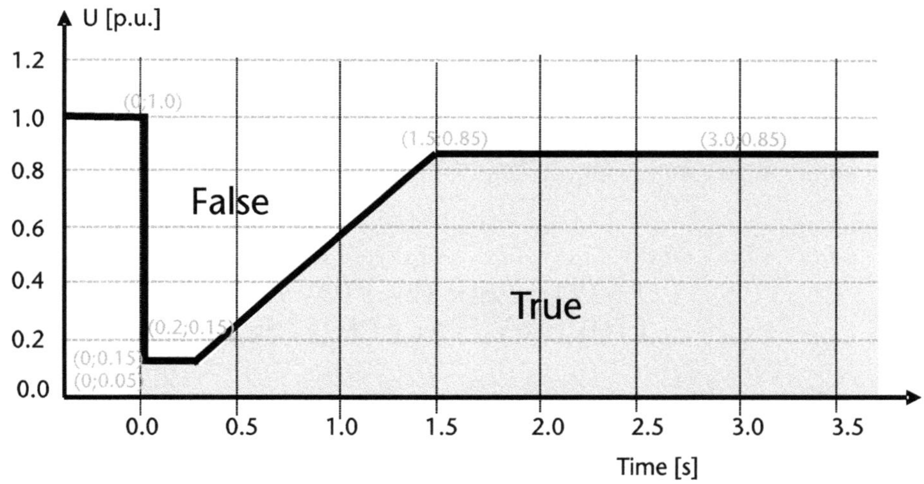

Figure 17.5 PTUV characteristic.

- The time for the PTUV or PTOV to start: for most of today's devices, this is about 20 ms;
- The time delay setting OpDlTmms;
- The breaker tripping time: we can assume about 80 ms for a 5-cycle breaker at 60 Hz.

Based on such analysis, the DER voltage trip function time delay setting will be

$$\text{PTUV2.OpDlTmms} = 60 \text{ ms}$$

We should not forget that during a single phase to ground fault it is possible to have simultaneously undervoltage in the faulted phase and overvoltage in the healthy phases. This is shown in the waveforms in Figure 17.3.

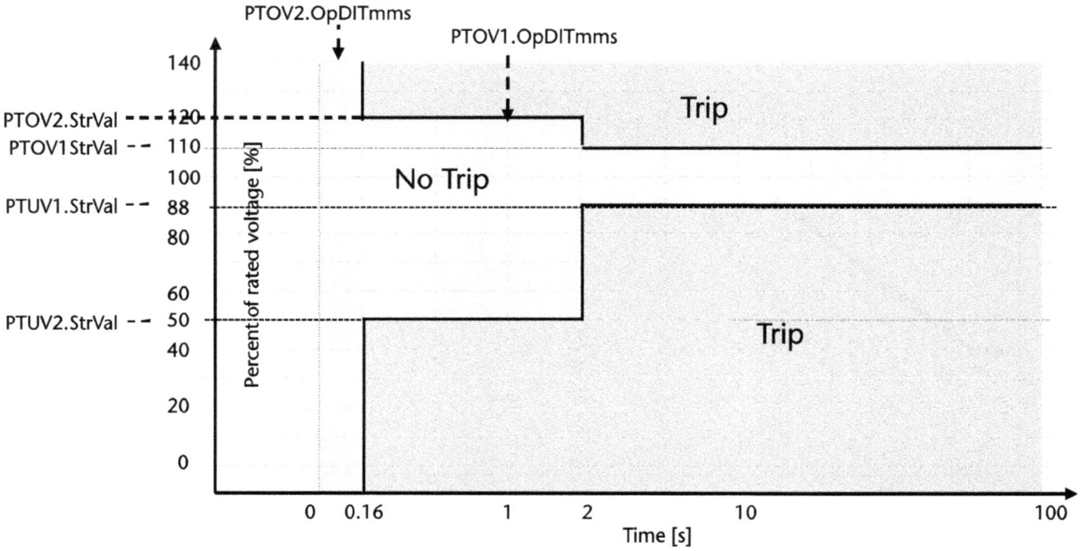

Figure 17.6 IEEE 1547 PTUV and PTOV settings for a voltage disturbance trip.

In this case, the faulted phase is phase A and the healthy phases are B and C. As a result, depending on the values of the rms voltage measurements, we may have:

$$Va < PTUV2.StrVal$$

and

$$PTOV1.StrVal < Vb < PTOV2StrVal$$
$$PTOV1.StrVal < Vc < PTOV2.StrVal$$

This condition will simultaneously start the timers for the corresponding undervoltage and overvoltage disturbance.

From the above analysis, it is clear that the implementation of the PTOV and PTUV functions and the IEC 61850-based DER controller will require the monitoring of the rms values of the measured voltages in all three phases.

17.5 DER Management

The transition of the electric power grid from a system with a limited number of large electric power stations controlled by the energy management system to balance the load and generation, as well as maintain voltage and frequency to a system with constantly growing number of DERs of different kinds and sizes is bringing a lot of very different requirements for the management of the DERs in a way that will maintain the required high levels of system reliability and security.

Considering that we may have a rooftop solar array producing a few kilowatts at the low-voltage level or a wind power plant generating a few hundred megawatts, it is clear that we need to define methods that will allow us to integrate all these different types of resources in a way that will allow them to be managed by the transmission or distribution system operators. It should also be possible to control their behavior by system integrity protection schemes when necessary to prevent wide area disturbances.

The management of transmission and distribution grids with increasingly higher levels of DERs is one of the central issues that many utilities around the world are already facing. It is going to get even more complicated due to the fact that many of these DERs are interfacing with the grid through inverters that operate based on the control algorithms implemented in them by the developers. In the case of areas that might become isolated from the rest of the grid and have only inverter-based DERs supplying the power, frequency is not anymore an indicator about the balance between loads and generation, which will require an energy management system to do this job.

If we think about the huge number of DERs that can be distributed all over the territory of an electric power utility, it becomes clear that it simply does not make sense to have direct interface between individual DERs and the energy management system. The number of communication links and the traffic on the network will be humongous. Therefore, we need to have DER management systems (DERMSs). One of their main functions is to be the intermediary between a DER system and the different system level entities such as:

- Electricity markets;
- Transmission system operator;
- Distribution system operator;
- System integrity protection scheme.

Each one of them will have a specific interface with the DERMS with specific performance and cybersecurity requirements. Depending on the electric power system condition at any moment in time, they will send to the DERMS requests for specific action that needs to be taken by the DER system to maintain the optimal performance of the electric power grid.

One of the critical functions of a DER system is to monitor the state of each one of the DERs that belong to it and, based on this, provide aggregated information about the state of the system to the DERMS. Different parts of this information are then sent to the system operators, the electricity market, or the SIPS.

When any action needs to be taken by the DER system, a command or request will be sent to the DERMS, which will be forwarded down the system hierarchy to the DER controllers for execution.

A DERMS can be defined as an intelligent software platform responsible for the optimal management of medium-scale to large-scale DER systems, or DER groups that are containing large numbers of small-scale DERs aggregated via DER aggregators. Their main task is to make the real-time availability of the distributed resources useful to the grid-level systems and thus support in an optimal way the reliable and secure operation of the electric power system.

From Figure 17.7, we can clearly see the hierarchical architecture of DER management. The number of layers of this hierarchy and the complexity of the architecture depend on the specific location and the number and types of the individual DERs being integrated.

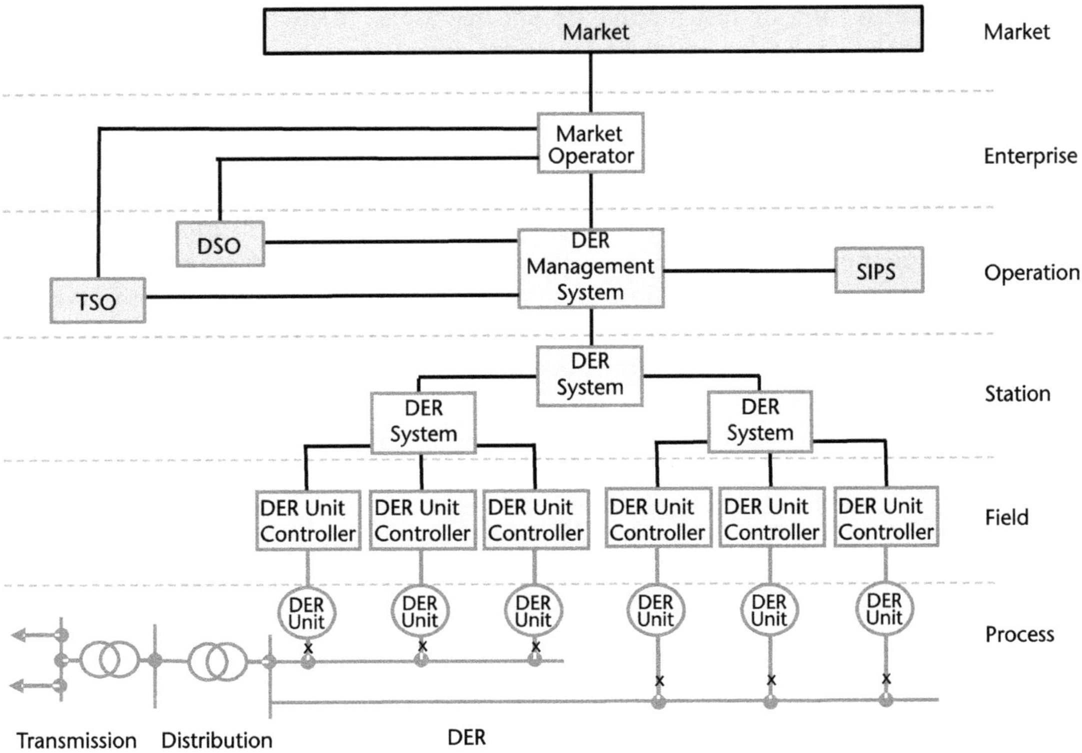

Figure 17.7 DER management system.

In some cases, the DERMS may also need input from load aggregators in order to be able to balance the output from the renewable resources and the load when there is no connection to the grid.

At the bottom of that hierarchy are the DER controllers that receive commands or requests from the upper layers of the DER system to change their mode of operation or output depending on the electric power system conditions. In some cases, intelligent controllers can operate independently based on the electric power system parameters that they monitor.

17.6 DER Controller IEC 61850 Modeling Principles

The modeling of an IEC 61850-based complex multifunctional IED such as DER controller is possible only when there is good understanding of the problem domain. At the same time, we should keep in mind that the models apply only to the communication-visible aspects of the IED.

The functions in relatively simple IED are fairly easy to understand and group together in order to build the object model. That is not the case for the more complex devices like a DER controller IED. For example, the voltage disturbance function has different components that interact with each other in a specific way that needs to be taken into consideration in the model.

The DER IEC 61850 model is based on the principles described in Chapter 9.

The DER controller is modeled as a server containing different functions such as:

- Process interface;
- Measurements;
- Power management;
- Electric power system disturbance operation.

The functions are represented in the model by logical devices contained in the server.

Multiple instances of different logical nodes become components of different protection, control, monitoring, and other functions in an electric power system automation system. They are used to represent individual steps in a function, for example, a different undervoltage level for the trip operation during a voltage disturbance.

The functionality of a complex multifunctional IED such as a DER controller is multilevel and is challenging to model. A simple example is the voltage disturbance function. It contains two subfunctions:

- Undervoltage;
- Overvoltage.

The process interface is represented by the logical nodes from:

- Instrument transformers and sensors group T, for example, TVTR (voltage transformer);
- Switchgear group X, for example, XCBR (circuit breaker).

Based on the sampled values (SAV) from the voltage and current sensors, a measurement logical node like MMXU can calculate the instantaneous values (CMV) of the currents and voltages and make them available to protection (like PTUV) or DER-related (like DLVT) functions.

An example of the use of this modeling principle for voltage tripping is shown in Figure 17.8.

TCTR1 and TVTR1 digitize the current and voltage signals at the point of common coupling (PCC) and provide sampled values to rmsMMXU1. The calculated rms values of the phase voltages are then used by the PTOV and PTUV to determine if there is an overvoltage or undervoltage condition that requires a trip and forward this to PTRC1 to execute the tripping. This information is also provided to DHVT1 and DLVT1.

17.7 Electrical Reference Point Modeling Considerations

According to IEEE P1547, the reference point of applicability (RPA) is the location where the interconnection and interoperability performance requirements specified

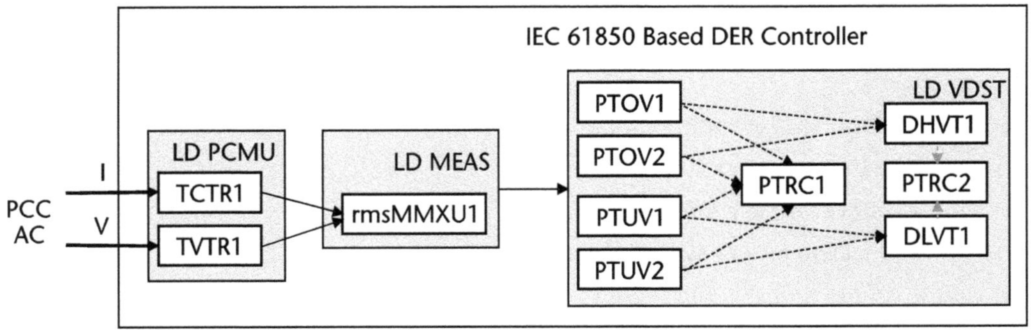

Figure 17.8 Modeling of voltage disturbance tripping.

in the standard shall be met. The electrical quantities referred to in the standard are those at the RPA, unless stated otherwise.

The DER controller has the interfaces from the PCC, which in this case is the electric reference point or RPA.

All interfaces in Figure 17.9 are hardwired to the terminals of the DER controller. The analog to digital conversion of the current and voltage transformers is performed by the analog interface of the IED. The binary signals from the auxiliary contacts of the circuit breaker and the disconnecting switches are wired to the opto inputs of the IED and used to detect the status of the PCC switchgear.

The modeling principle for the electric reference point is shown in Figure 17.10.

TCTR and TVTR digitize the current and voltage signals and provide sampled values to MMXU logical nodes.

PoCMMXU1 calculates the instantaneous measurements of the electrical parameters at the output of the PV system inverter.

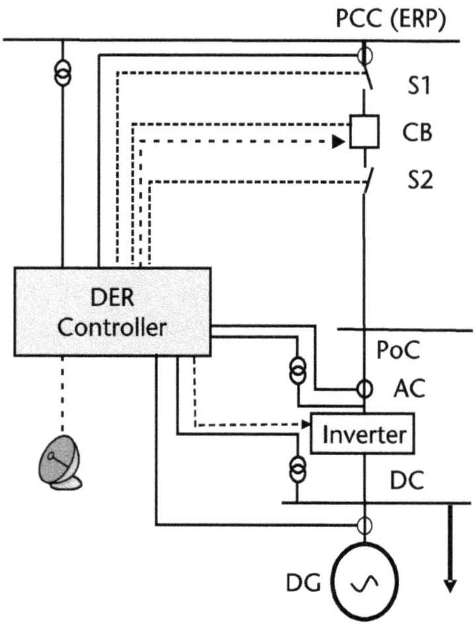

Figure 17.9 One-line diagram for the interface of the DER controller.

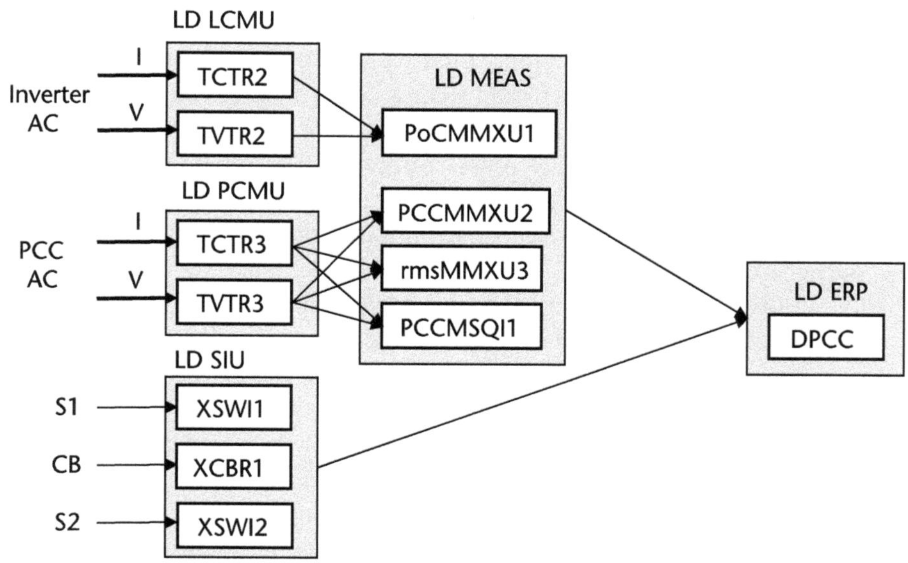

Figure 17.10 Modeling of the electrical reference point.

PCCMMXU2 calculates the instantaneous measurements of the electrical parameters at the PCC.

The voltages' magnitude and phase angles from MMXU1 and MMXU2 can be used in the DPCC logical node to determine the status of the PCC represented by the data object EcpIsldSt.

The calculated rms values of the logical node rmsMMXU3 are required to meet the monitoring requirements defined in IEEE P1547.

The maximum magnitude of phase voltage MaxPhVPhs and the minimum magnitude of phase voltage MinPhVPhs measurements are necessary to monitor the EPS conditions at the PCC and are used as an input to the voltage tripping and ride-through function.

The calculated sequence component values from the logical node PCCMSQI can be used to monitor and detect short-circuit faults or open phase conditions in the electric power system as required by IEEE P1547.

Logical nodes XSWI and XCBR provide to DPCC status information for the disconnecting switches and the circuit breaker that can be used for determining the PCC status.

17.8 DER Controller Interface Requirements

The DER controllers play a critical role in the integration of different type of DERs in the electric power system.

DERs can be connected in many different ways to the electric power system. The DER controller needs to be able to support the electric power grid energy management system based on different modes of operation and configuration settings under varying electric power system conditions and power output from the renewable resource.

Depending on the type of DER and its location, the interface requirements may change. This is especially valid for IEC 61850-based systems in which the hard-wired connections between the process and the DER controller can be replaced with high-speed P2P communications.

Considering the varying power output of photovoltaic DERs that depends on the season and time of day in order to perform its configured functionality, the DER controller needs interfaces to monitor the DC current, voltage, and power output of the PV system.

In order to determine if the inverter is responding correctly to the commands from the DER controller to the inverter and providing required active and reactive power and voltage phase angle, the controller needs to measure the currents and voltages at the point of coupling. That is why it needs to be connected to the current and voltage transformers on the output of the inverter.

To be able to respond to different events in the electric power system, the DER controller also needs to measure the currents and voltages at the PCC to the electric power system. This means that it needs to be connected to the current and voltage transformers at the PCC.

To monitor the status of the switching devices at the PCC, the DER controller needs to be connected to the auxiliary contacts of:

- The circuit breaker CB;
- The disconnecting switches S1 and S2.

Based on this status information, the controller will know if it is possible or not to connect to the electric power system.

In case of a fault on the line where the PCC is, the protection IED at the substation will send a direct transfer trip (DTT) signal to the DER controller. This means that the controller should have:

- An interface to the substation protection IED that will allow it to receive the DTT signal;
- An interface to the circuit breaker trip coil to be able to disconnect from the electric power systems.

The DER controller also should be able to connect the DER or reconnect it after it has been disconnected. This will require an interface with the close circuit of the circuit breaker.

Some functions that the DER controller performs require the monitoring of the voltage angles on the electric power system and DER side. Disturbance and event reporting also require time synchronization. That is why the controller needs to have a time synchronization interface.

Last but not least, the DER controller needs to interface with the DERMS, which will send it the required DER operating mode at any moment in time based on the state of the electric power system and the electricity market.

All the above-described interfaces are shown in Figure 17.11.

The DER interfaces and its functionality, as defined in IEEE P1547, will determine the model based on the IEC 61850 standard.

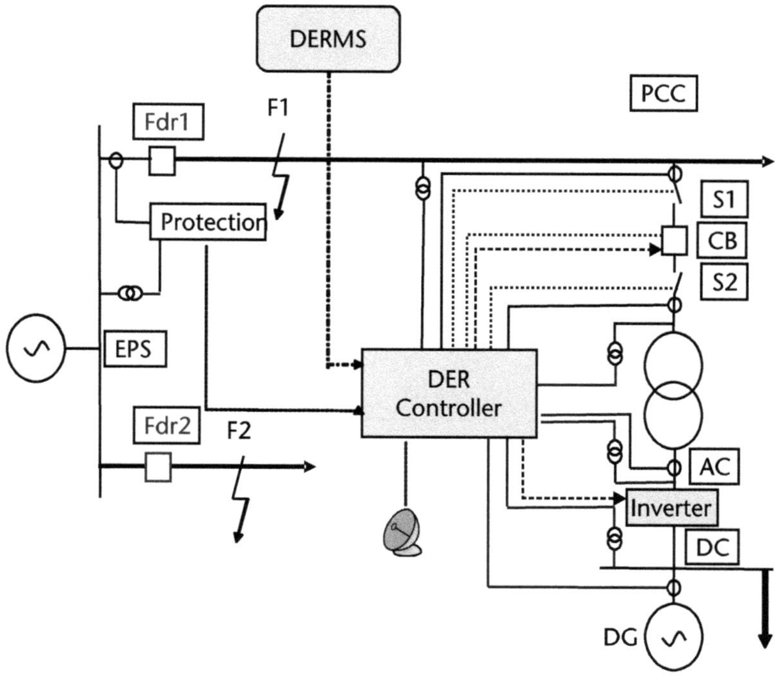

Figure 17.11 DER controller interfaces.

For PV systems with a limited number of PV panels that take a relatively small space with a short distance between the DER controller and the PCC, we expect that all interfaces described above will be implemented using hardwiring.

However, it is well known that hardwiring presents significant challenges, such as:

- CT saturation;
- Safety concerns from high voltage in the case of an open current transformer circuit;
- High construction and installation costs;
- High maintenance costs;
- Lack of condition monitoring of the hardwired interfaces between the IED controller and the circuit breaker at the PCC.

Considering the trend towards the digitalization of the electric power grid based on IEC 61850 and the maturity of the process interface solutions using IEC 61850 GOOSE and sampled value messages, the IEC 61850-based controller should be capable of having the PCC interfaces using P2P and client-server communications.

Figure 17.12 illustrates an implementation of an IEC 61850-based DER controller with a hybrid process interface to the PCC and point of coupling:

- Hardwired analog interface to the DC output from the PV;
- Hardwired analog interface to the AC output from the inverter;
- IEC 61850 sampled value analog interface to the PCC;

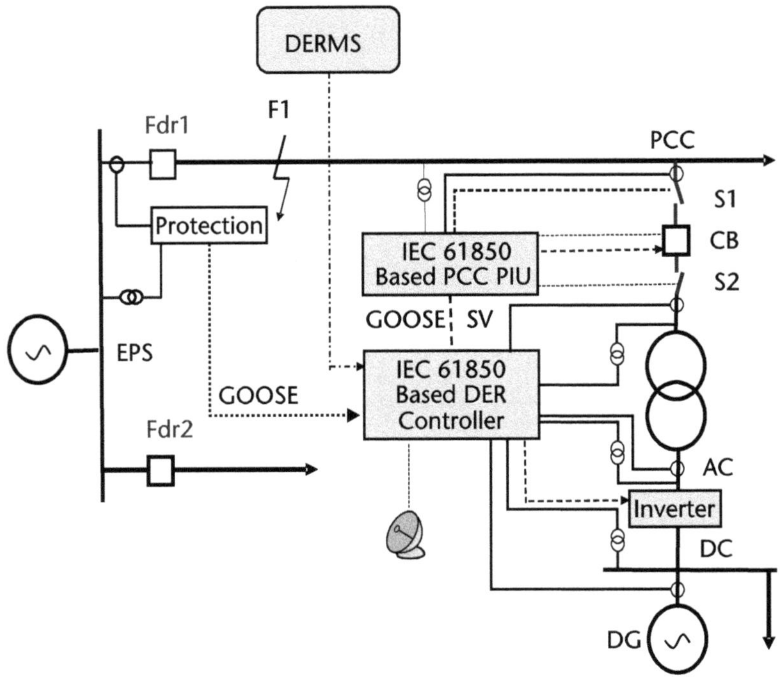

Figure 17.12 Hybrid IEC 61850-based DER controller interface.

- IEC 61850 GOOSE interface to the PCC switchgear.

The IEC 61850-based PCC process interface unit (PIU) is located at the PCC and combines the functionality of two types of interfaces: analog and binary. It is hardwired to:

- The secondary of the instrument transformers;
- The auxiliary contacts of the circuit breaker and disconnecting switches;
- The trip and close circuits of the circuit breaker;
- The open and close circuits of the disconnecting switches;
- The interfaces between the PCC PIU and the DER controller is over a fiber optic or wireless communication link using IEC 61850 GOOSE and sampled value messages.

This implementation of communications-based interface between the PIU and the DER controller offers significant benefits:

- Practically eliminates CT saturation and the safety concerns from high voltage in case of open current transformer circuit;
- Significantly reduces construction and installation costs;
- Significantly reduces maintenance costs;
- Supports condition monitoring of the interfaces between the IED controller and the circuit breaker and switches at the PCC.

17.9 Protection of Systems with a High Penetration of DERs

The protection of systems with a high penetration of DERs with inverter-based interfaces presents two main groups of challenges related to their nature:

- Requirements for ride-through capabilities during short-circuit faults;
- Very low fault contributions.

The first group means that the design of the protection system will result in reduced fault-clearing time in order to keep the duration of the voltage sag caused by the fault outside of the tripping part of the characteristic.

The second group leads to the need for using a different approach compared to the traditional protection solutions that will support fast fault clearing.

In Chapter 10, we introduced several schemes using GOOSE messages to achieve faster fault clearing at the distribution level of the system. When we look at the transmission level, we can achieve the same goal by using line differential protection; however, it requires high-speed communications between the protection devices at both ends of the protected line, which may not be always available or are expensive to install. That is why typically the transmission lines are protected by distance protection. The problem with this solution is that conventional distance protection does not provide instantaneous tripping for all faults on the protected transmission line due to the fact that typically the last 20% of the protected line are covered by the Zone 2 element that typically operates with a time delay of 300 to 400 ms (Figure 17.13).

As we saw from the analysis of the different ride-through characteristics, typically the required fault clearing time is below 200 ms, which means that, if the fault is cleared with the Zone 2 time delay, many of the DERs are going to trip, which in areas with high penetration may lead to a local or wide area disturbance.

Communications-based schemes allow considerable improvement in the overall fault-clearing time for any fault within the zone of protection, while at the same time they do not have the high-speed communication requirements that line differential protection has. This is because, in these schemes, a signaling channel is

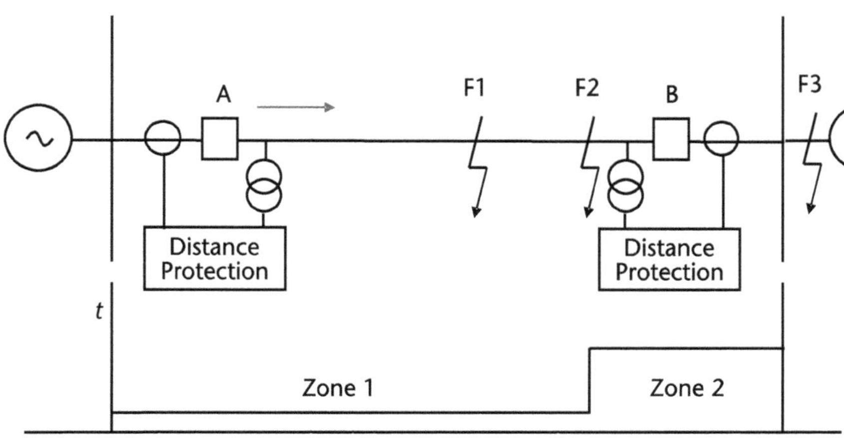

Figure 17.13 Distance protection.

used to transmit simple on/off data (from a local protection device). This provides additional information to the remote end protection device that can be used to accelerate in-zone fault clearance or to prevent operation for external faults.

These teleprotection schemes can be grouped into three main operation modes. In each mode, the decision to send a command is made by a local protective relay operation.

In intertripping (direct or transfer tripping) applications, the command is not supervised at the receiving end by any protection relay and simply causes a breaker trip operation. As no checking of the received signal by another protection device is performed, it is absolutely essential that any noise on the signaling channel is not seen as being a valid signal. In other words, an intertripping channel must be very secure.

In permissive applications, tripping is only permitted when the command coincides with a protection operation at the receiving end. As this applies a second, independent check before tripping, the signaling channel for permissive schemes does not have to be as secure as for intertripping channels.

In blocking applications, tripping is only permitted when no signal is received, but a protection operation has occurred. In other words, when a command is transmitted, the receiving end device is blocked from operating even if a protection operation occurs. Because the signal is used to prevent tripping, it is clear that a signal is received whenever possible and as quickly as possible. In other words, a blocking channel must be fast and dependable.

The protection function that sends the permissive or blocking signal to the remote end determines the type of scheme used. If this is a distance element, we usually talk about permissive underreaching or overreaching schemes or blocking schemes. If a directional element is used to initiate the transmission of a signal to the remote end of the protected line, we have directional comparison schemes. Directional comparison schemes can be permissive or blocking, with directional elements initiating the signal transmission and providing the supervision at the receiving end.

The most commonly used accelerated transmission line protection scheme is the permissive overreaching transfer trip (POTT) that requires the start of the Zone 2 distance elements at both ends of the protected line as shown in Figure 17.14.

For a short-circuit fault at F1, it will be seen by the protection at both ends of the line in Zone 1 and it will be cleared without any significant time delay. For a fault at F2, it will be seen in Zone 1 by the relay at end B and in Zone 2 by the relay in A, thus leading to a delayed tripping of that breaker. If this delayed fault clearing is not acceptable, we need to use an accelerated protection scheme as shown in Figure 17.14. In this case, we can consider that it is a POTT scheme in which each relay will send the permissive signal to the remote end when its Zone 2 distance element starts. If the local Zone 2 element has started and the permissive signal is received from the remote end, the scheme operates and trips the local breaker.

In this case, the fault-clearing time will depend on the Zone 2 starting time and the communication channel delay, which will be much faster than the Zone 2 tripping time.

One of the concerns of the protection community related to the high levels of penetration of inverter-based resources is that, when a fault occurs, some distance elements may not operate and, as a result, expose the DERs in the neighboring

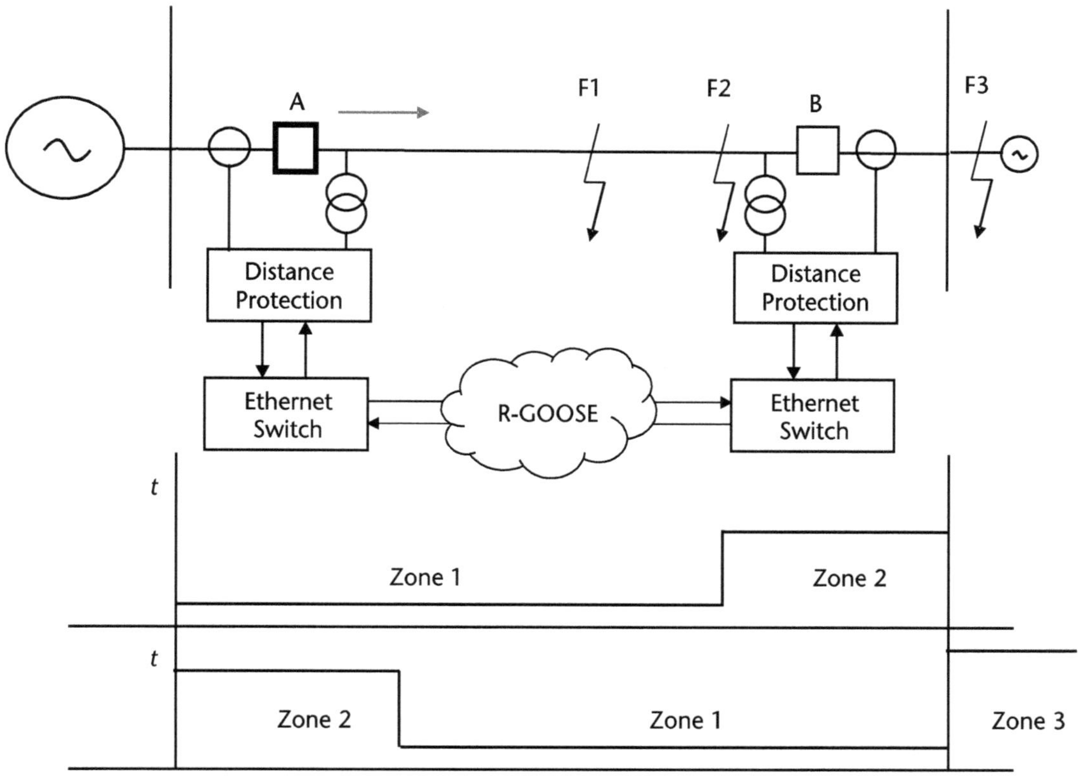

Figure 17.14 Accelerated distance protection scheme with weak infeed logic.

system to lower-voltage levels that may be in the tripping zone of the ride-through characteristic. We can use a POTT scheme to accelerate the tripping, but such a scheme requires the Zone 2 distance element at both ends of the line to start, which results in the sending of a permissive signal to the remote ends as described above.

If one end of the line is connected to a substation in an area with a high penetration of inverter-based DERs when the fault occurs, there might not be enough fault current to start the Zone 2 element (i.e., we have end B with a weak source). This is where the weak infeed logic can help.

For a fault at F2, it will be seen in Zone 2 of the line for the relay at the end with a strong source (A) and it will send the permissive signal to the weak end protection. Because of the weak source at B, the fault will not be seen by Zone 1 or Zone 2 because of the low fault current level, so a traditional POTT scheme will not operate.

If Zone 2 of the protection in A starts for a fault F3 that is not on the protected line but behind the protection in B, it will again send a permissive signal to B but the POTT scheme will not operate again. To solve this problem, we can implement a weak infeed logic, which, when a permissive signal is received from A and a reverse-looking Zone 3 of the protection at B has not started, indicating that the fault is on the protected line, will send back a permissive signal to A for the accelerated tripping and trip breaker B.

In this case, the fault-clearing time will be increased by the communication time from B to A but will be still much faster than the Zone 2 tripping time.

The reverse-looking distance element at B is going to see the F3 fault because of the fault current coming from the strong end of the line, making the weak-infeed logic work.

Considering that typically a Zone 2 time delay is in the range of 300 to 400 ms, the operating time of this logic will be much faster, somewhere about 70 to 80 ms, which will keep the fault-clearing time outside of the tripping zone of the ride-through characteristic.

The implementation of such an accelerated scheme can be achieved in IEC 61850-based protection systems using routable GOOSE message (i.e., without the need for dedicated communication channels).

References

[1] IEC 61850-7-420:2009 Communications Systems for Distributed Energy Resources (DER)—Logical Nodes, 2009.

[2] IEC 61850-7-420:2021 Communication Networks and Systems for Power Utility Automation—Part 7-420: Basic Communication Structure—Distributed Energy Resources and Distribution Automation Logical Nodes, 2021.

[3] IEEE 1547-2018 IEEE Standard for Interconnection and Interoperability of Distributed Energy Resources with Associated Electric Power Systems Interfaces, 2018.

[4] EN 50549-1:2019 Requirements for Generating Plants to Be Connected in Parallel with Distribution Networks—Part 1: Connection to a LV Distribution Network—Generating Plants Up to and Including Type B, 2019.

[5] IEC TR 61850-90-7:2013 Object Models for Photovoltaic, Storage and Other Inverter-Based Applications, 2013.

[6] IEC TR 61850-90-9:2020 Object Models for Electrical Energy Storage, 2020.

[7] IEC TR 61850-90-10:2017 Object Models for Schedules, 2017.

CHAPTER 18

Migration Strategy

18.1 Introduction

In the previous chapters, we looked at what PAC systems are, what the IEC 61850 standard is, and its impact on the evolution of these systems. The benefits of the digital transformation from conventional hardwired systems to communications-based solutions are obvious and numerous, so there is no doubt that this is the future of our industry. The question is not if, but when, this transformation will take place in the different utilities.

As with almost everything in life, the answer to this question starts with: "It depends." It depends on:

- What are the objectives of the transformation?
- What is the existing technology being used in the substations?
- How comfortable is the technical staff with today's state-of-the-art technology?
- How far have standard PAC schemes been implemented?
- Is the transformation going to affect only greenfield installations or will it also apply to existing substations?
- How far is the transformation intended to go?
- What skills does the workforce need to have to succeed?
- Is it necessary to make changes in the organization of the company?

For any utility to successfully migrate from their existing solutions, it is necessary to develop a meaningful and solid migration strategy that will answer all of the questions listed above and many others that will appear in the process.

In this chapter, we look at the evolution of PAC technology, which will help us to see where we are and define where we would like to go. Since IEC 61850 has been developed, keeping in mind the wide range of existing implementations and in order to provide solutions for the integration of legacy devices of different types, we discuss with some examples how this can be done.

18.2 The Evolution of PAC Systems

18.2.1 Electromechanical Systems

Since the beginning of electric power systems, each utility has had one main goal: the secure and reliable operation of the system. That is why from the early days engineers have been trying to find the best possible ways to detect short-circuit faults or other abnormal conditions and isolate the failed equipment as quickly as possible so the service to the affected customers can be restored.

While the goal remained the same through the history of electric power, the tools that are available to achieve it have been changing through the years with the advancements of technology.

From the beginning of the last century, the best available and continuously developing protection and control technology was electromechanical protective relays and measuring and control devices. These were the available solutions for the first 50 years, and many such systems are still operational today.

All the interfaces of these electromechanical systems with the primary equipment in the substation, as well as between them, were hardwired (Figure 18.1). Another characteristic is that the individual relays are single function devices that must be wired together to implement the required protection and control scheme.

Due to their electromechanical design, their performance changes over time, which requires scheduled maintenance to ensure their required performance. Because their interfaces cannot be monitored, their failure can be detected only in case of misoperation or during maintenance testing.

18.2.2 Electromechanical and Solid-State Systems

Around the middle of the last century, with the development of electronics, the next generation of protection equipment was born: static or solid-state relays. The name is related to the fact that these devices do not have moving parts. In many applications, these relays were faster than their electromechanical equivalents and can be set move precisely using the electronic circuits. Also, they did not have the same maintenance requirements.

From the functional and interface points of view, they were quite similar to the electromechanical ones, which made the migration from electromechanical to solid state quite straightforward.

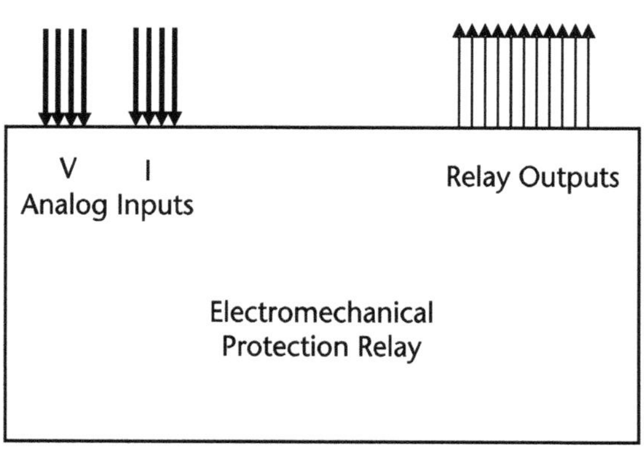

Figure 18.1 Electromechanical protection relay interfaces.

This migration was accepted by many protection specialists because combining electromechanical and solid state relays was a very good solution from a reliability point of view; it provided two completely different operating principles helping to avoid potential common-mode failure issues. That is why what many utilities did at the time was replacing one line of electromechanical relays with their solid-state equivalents and keeping the second one as it is.

18.2.3 Electromechanical and Microprocessor-Based Protection Systems

The development in computer technology and the invention of microprocessors had a significant impact on the continuing evolution of protection devices.

With growing interest in the development of numerical protection devices that started in the early 1970s, the first commercial products started to appear around a decade later. They did not only implement the protection functions as numerical algorithms, but they also introduced some additional nonprotection functions such as measurements, fault location, and event reporting. This delivered some significant benefits compared to the electromechanical and solid-state relays:

- The protection characteristics did not change with time.
- The settings were specified by entering a number.
- There is a record of when a protection function operated and what were the pre-fault, fault, and post-fault currents and voltages helping with the analysis of the operation.
- Commissioning and maintenance were simplified by being able to read the currents and voltages applied to the relay at any moment in time.
- In some cases when there was a communication link available, it was even possible to access the relay remotely for setting changes or event analysis.
- There were some self-monitoring functions providing an alarm in case of detecting some problem with the hardware or firmware paving the way for event-driven maintenance.

All these benefits made the migration to microprocessor-based protection devices easier, especially as most of the interfaces remained hardwired. To ease the process, some manufacturers even offered on the market digital relays that were designed to fit within the case of existing electromechanical devices. This made the transition much more efficient because no wiring had to be touched.

What made the migration also more acceptable was that still each different domain in the utility had their own devices. The initial steps in functional integration were quite limited so every group insisted that they need dedicated devices to meet their requirements, so there were no needs for changes in the company's organization.

One of the biggest challenges that the microprocessor-based relays introduced was the addition of communication capabilities (Figure 18.2). Most protection and control specialists had never dealt with communications, which, on top of everything, were not very reliable and were quite slow. Things were further complicated by the fact that most manufacturers were defining proprietary communication

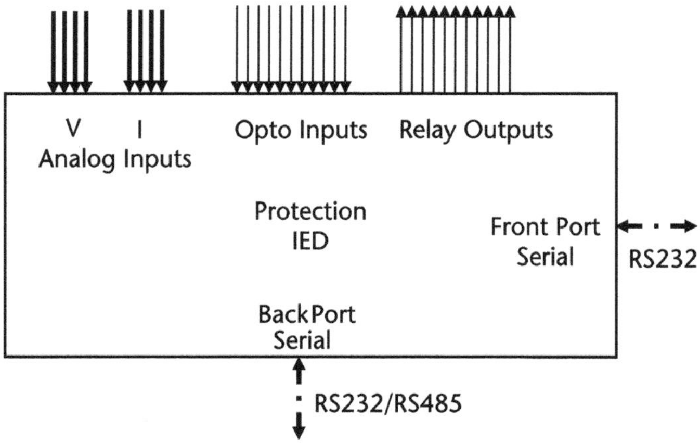

Figure 18.2 Legacy IED interfaces.

protocols with some using RS 232 and others using RS 485 as the communication interface. This made the integration of devices from different manufacturers in a PAC system very challenging and expensive. The communications interface to the IEDs also introduced cybersecurity concerns.

The migration strategy to microprocessor-based relays varied between different utilities, but what was quite common was using them in greenfield substations and replacing failing electromechanical or solid-state relays with microprocessor-based relays in the existing installations.

18.2.4 Hybrid Digital Substations

The continuing evolution of computer and communications technologies, as well as the experience of utilities in the design of substation PAC systems, led to the next revolution in our industry: the development of advanced multifunctional IEDs and the standardized communications and engineering environment based on the IEC 61850 standard.

The process that started later in the last century developed the tools allowing the digitization of all interfaces between the IEDs and the primary substation equipment, as well as between the IEDs themselves.

As we have seen throughout this book, the IEC 61850 standard brings some very significant benefits that support the reliable and secure operation of the electric power grid in the most efficient way. However, it introduces also some significant challenges, because it requires a very good understanding of object-oriented engineering concepts and advanced communications technologies.

Although these technologies are much easier to understand and accept by young protection and control engineers with some computer programming background, they may be quite difficult to accept by experienced protection specialists who had been working with wires throughout their whole career. We should also not forget that the high level of functional integration and the sharing of resources between applications belonging to different domains in the utility require structural reorganization that may not be welcomed by many of the stakeholders.

Another challenge for the migration towards digital substations is the lack of significant global experience with an emerging technology. Because of that, the migration strategies of different utilities have varied significantly to a great extent determined by the people driving the transformation, with some being very excited by the opportunity to improve the functionality and efficiency of the PAC systems, and others being afraid of the change and resisting it because of that.

That is why today the large number of digital substations based on the IEC 61850 standard are hybrid installations with different levels of implementation of functions based on it. Some use only the client-server capabilities, and others go one step further and use GOOSE messages to initiate disturbance recording, but keep hardwired interfaces for all protection-related signals. Some utilities are not afraid to use GOOSE messages for IED-to-IED communications, but still use hardwiring between the protection device and the circuit breaker for tripping. This is still a step in the right direction, but it definitely does not take advantage of the full benefits that the IEC 61850 standard offers.

18.2.5 Fully Digital Substations

Because in the hybrid digital substations the multifunctional IEDs still have hardwired analog circuits and many of the binary signals are also based on wiring between the individual devices, we are still facing many issues common to traditional substations. For example, it is possible to have an open current circuit condition that represents a safety hazard or we may have a loose hardwired connection that will be detected when it fails to operate when necessary or during maintenance testing.

These and many other deficiencies of the more conventional approach that hybrid digital substations represent are eliminated with the transition to fully digital substations. By this term, we understand a substation in which all PAC interfaces to the process and between the different devices are digitized.

A generic example of such an IED is shown in Figure 18.3. In this case, the analog interface is replaced with streaming sampled values, thus eliminating the open current circuit condition hazard and reducing the probability for CT saturation.

All the binary signal exchanges between the switchgear and the IEDs are based on GOOSE messages, which provide high reliability, because they can be easily duplicated and continuously monitored, thus detecting any potential issues as soon as they occur.

Another further improvement available in fully digital substations is the transition from conventional instrument transformers to low-power or optical sensors. They provide a safer working environment and lower transportation and installation costs and lead to the reduction of the substation footprint. The impact of optical connectivity into a substation also lowers the cost of ground isolation.

The migration to such a solution is considered by many utilities that already have experience with hybrid digital substations and are becoming confident in the benefits of the new technology. There are many utilities around the world that are going through pilot projects to become comfortable with the process bus and to verify that all PAC functions operate as expected.

Some utilities have already completed this phase and have standardized on the fully digital solution for all their new substations or for the refurbishment of existing ones. Considering that fully digital substations allow switching to remote

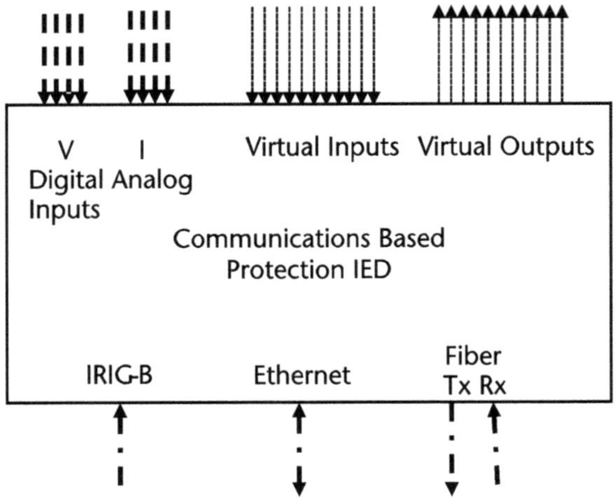

Figure 18.3 IEC 61850-based IED interfaces.

maintenance testing, something that has never been possible before, it is clear that this will require changes in the established testing practices.

18.2.6 Centralized Digital Substations

Digital transformation in all aspects of our everyday lives is becoming an important part of many industries and is also an ongoing process for the electric power systems. The availability of substation-hardened powerful servers, as well as ongoing developments in digital twins, virtualization, and artificial intelligence applications are heading towards the next stage of substation PAC systems, the centralized ones.

This is the case when all the interfaces to the substation process are digitized as in the fully digital substations but the individual multifunctional IEDs communicating over a substation local area network are replaced with virtual devices whose function elements are working together over the digital data bus of a server.

The benefits of such solution are that they improved the reliability of the system as a result of the reduced number of devices used and at the same time the significant reduction in the size of the substation control house due to the replacement of multiple panels with just a few hosting a limited number of servers. A high level of reliability can be achieved by the increased number of redundant servers used.

The migration from the fully digital substation to a centralized version should be quite straightforward, especially if the design and configuration of the centralized solution are using the system configuration language defined by IEC 61850.

Some utilities are already running energize substations with centralized PAC systems based on off-the-shelf technology, thus proving that this is the next stage for the digital transformation of our industry.

18.2.7 Cloud-Based Substations

The ongoing evolution of communication technology is leading to the transfer of many industrial applications to servers in the cloud. This offers some significant advantages because the user does not have to worry anymore about the purchasing, installation, and maintenance of any hardware. It just becomes a matter of using reliable communications and applications that meet the performance requirements.

In the past, the acceptance of such an approach for PAC systems was quite challenging because of the latency of the communications. The development of 5G and other advanced communications technologies supporting high-speed interfaces for distributed applications are making it possible to move the utility applications to the cloud as well.

There is a lot of interest in the industry and there are different working groups addressing the issue of separating the hardware from the software. This means that, like everything else around us, the utilities will be buying specific hardware provided by different suppliers that is designed to act as a cloud-based server and then install and run on it the required PAC applications.

From a practical point of view, this means that in a substation we will be seeing only process interface devices providing sampled values for the analog signals and GOOSE messages for the status information or receiving them to operate the breakers and disconnecting switches.

If the development of cloud-based schemes proves feasible based on extensive testing of their performance and reliability, the migration to such solutions, especially if it happens from a centralized digital substation, should be fairly straightforward because the applications instead of running on the substation server connected over the LAN will be communicating with the server in the cloud.

18.3 Integration of Legacy Devices

18.3.1 Integration in Existing Installations

The existence of thousands of substations with an installed base of electromechanical or solid-state relays that are still meeting the protection requirements of the user presents a challenge to the utilities that are considering the migration to IEC 61850-based solutions. On top of that, many of the substations have been upgraded in the last several years with microprocessor-based relays that do not support IEC 61850. The replacement of all these legacy products will be very time-consuming and costly. The upgrade process requires careful consideration of all available options and the development of a migration path that will allow a smooth transition into the IEC 61850-based protection and substation automation world.

IEC 61850-based protection and control systems typically integrate devices that are designed to support the object models defined in the standard. The concept of a proxy server has been defined during the development of the standard in order to support the integration of legacy devices of different types.

The use of GOOSE messages with legacy protection devices is possible, but some issues regarding the overall performance of such applications under different abnormal conditions need to be analyzed.

Because legacy devices do not support the IEC 61850 protocol and at the same time typically do not have an Ethernet interface, they have to be connected to the substation LAN through a gateway. The gateway device is also called a proxy server. In this case, it provides several functions:

- Conversion of the legacy communications interface (usually RS232 or RS485) to the Ethernet;
- Conversion of the legacy communications protocol (Modbus, DNP 3.0, IEC 870-5-103, other) to IEC 61850;
- Mapping of the data from the legacy device database into the IEC 61850 object models;
- GOOSE interface of the legacy devices with IEC 61850-compliant IEDs on the network.

All of the above allow the legacy devices to look like IEC 61850-based devices to other substation equipment or applications. The main difference between the two types is the performance, as every data exchange between IEC 61850 and legacy devices has to go through the gateway. The performance has to meet the requirements defined in the standard.

The advantages of such architecture include reuse of legacy devices, reduced wiring, improved reliability, and better overall protection and control functionality.

Figure 18.4 shows the client-server relationships of legacy devices and a gateway in the substation protection and control architecture. The legacy IEDs are servers to the IEC 61850 gateway, which interacts with them as a client. The IEC 61850 gateway then interacts as a server with the substation computer client.

18.3.2 Integration of Legacy IEDs

The highest level of complexity of modeling is in the case of integration of advanced legacy protection IEDs similar to the ones discussed above. This also changes the communications architecture in the substation, compared to the purely IEC 61850 communications architecture.

The legacy IEDs in Figure 18.4 are connected to the proxy IED usually through serial communications over RS 232 or RS 485 using Modbus, DNP3, IEC 870-5-103, or another common protocol.

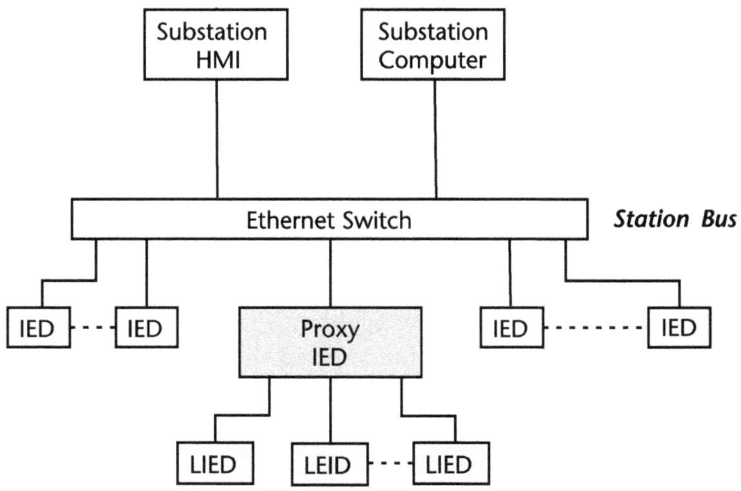

Figure 18.4 Communications architecture with legacy devices.

Because legacy devices are modeled as logical devices in an IEC 61850-based device, it is clear that this will add another layer in the model.

Once all the individual legacy devices object models have been developed, the gateway object model can be completed. It will contain $N + 1$ logical devices, where the logical device LD0 should be reserved for the modeling of the gateway itself.

The legacy physical devices 1 through N are represented in the gateway by logical devices LD1 through LDN, where N is the number of legacy devices interfacing to the substation LAN through the IEC 61850 proxy server (gateway). Figure 18.5 shows a simplified block diagram of the gateway model, where the functions are represented by multiple instances of logical nodes.

It is clear from the above that the logical devices are the IEC 61850 component that enables the building of gateways (proxies) in such a way that logical devices are, from a functional point of view, transparent. Each logical device can be identified independently of its location (in a separate device connected to the network or in a proxy device). By definition, a proxy server is a server that sits between a client application and a real server.

A gateway is a network interconnection device that supports the full stack of the relevant protocol, which it can convert to a non-7-layer protocol for asynchronous transmission over wide area networks.

Logical devices provide information about the physical devices that they use as a host (nameplate and health). They also provide information about external devices that are controlled by the logical device (external equipment nameplate and health). Only those aspects of physical devices that are defined as visible to the network are of interest in IEC 61850.

The logical node zero (LLN0) represents common data of the logical device, and the logical node physical device (LPHD) represents common data of the physical device hosting the logical device.

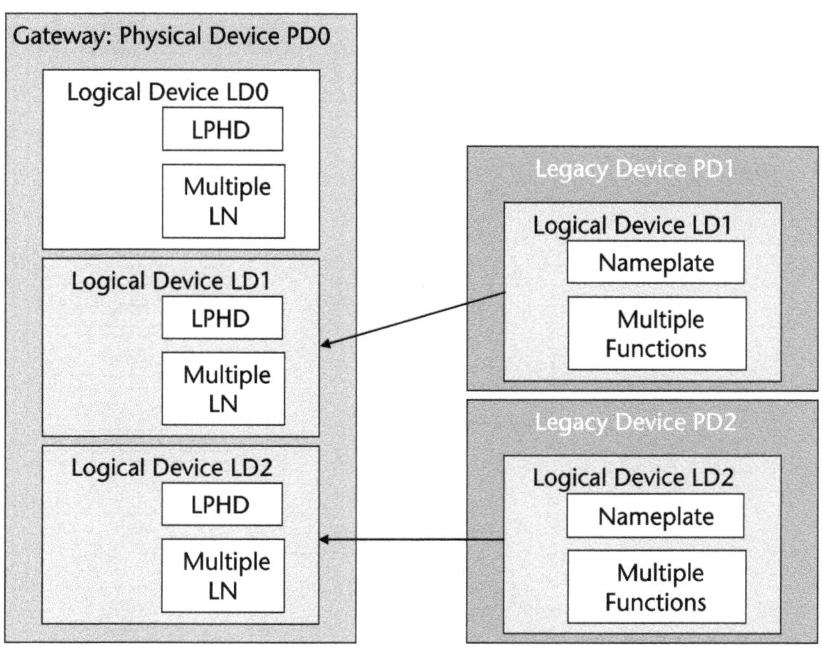

Figure 18.5 Modeling of legacy devices in the gateway.

To represent the information about the proxy/gateway itself, the logical device LD0 shall be implemented in each device that acts as a proxy or gateway. The logical nodes LLN0 and LPHD of LD0 shall represent information about the proxy or gateway device itself.

If a physical device does not provide logical devices that mirror logical devices of other physical devices, then this physical device does not need to provide a LD0. To represent the information about the proxy/gateway itself, the logical device LD0 shall be implemented in each device that acts as a proxy or gateway.

18.3.3 Integration of Electromechanical or Solid-State Relays

The object modeling of the functions in electromechanical or solid-state relays is actually very simple, because they were used to develop the object models of multifunctional protection devices. In the same way that a protection scheme is built from multiple electromechanical or solid-state relays on a panel, it is built from different function elements in a multifunctional microprocessor relay. As the model of any device in IEC 61850-based systems represents communication-visible attributes only, electromechanical and solid-state relays are modeled within physical devices with communication capabilities.

As can be seen from Figure 18.6, the object models of an electromechanical transformer overcurrent and differential protection can be built in different IEC 61850-based devices such as a multifunctional power quality monitoring and recording IED. The electromechanical transformer protection relays are represented by a logical device LD2. An instance of the differential protection logical node PDIF1 in the monitoring IED is used to represent the status of the differential relay:

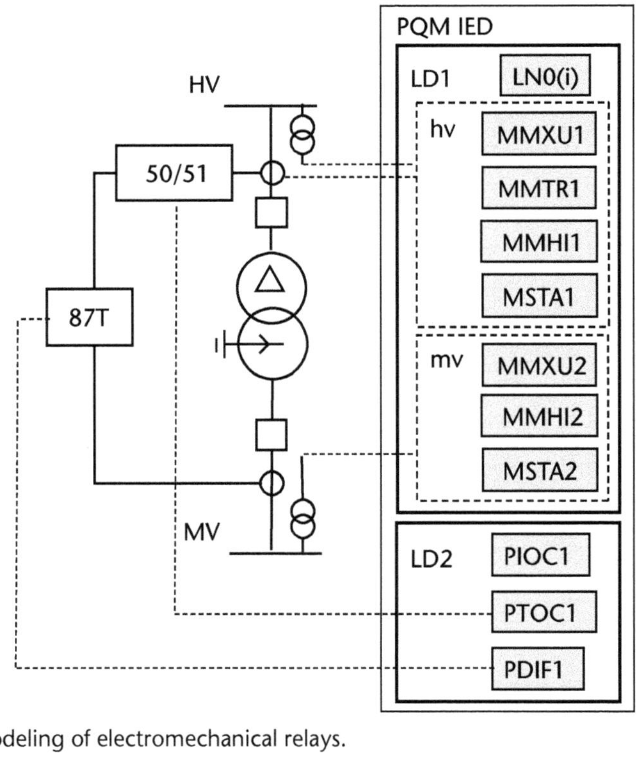

Figure 18.6 Modeling of electromechanical relays.

started (optional) or operated (mandatory). PIOC1 and PTOC1 represent the 50 and 51 overcurrent relays in a similar manner. Nothing else is available from the electromechanical relays. If the monitoring IED can calculate the differential current seen by the relay, it will also be included in PDIF1. This data is then passed to the substation or bay computer that contains the object models of multiple electromechanical or solid-state relays represented as multiple logical devices (LD). These models contain other logical nodes that provide nameplate data. The differential protection logical node PDIF1 in the substation computer will include additional attributes, such as the settings of the differential protection.

The power quality monitoring and recording devices perform several additional functions that satisfy the requirements for the integration of the electromechanical relay. It time-stamps the operation of the relay, thus making it available to the distributed event reporting system. It records waveforms and disturbances for the analysis of the relay performance under different abnormal conditions.

18.3.4 Using GOOSE with Legacy Devices

One of the important differences between IEC 61850 and other communication protocols is the introduction of high-speed P2P communications defined as IEC GOOSE. The most common applications at this stage are for replacement of the hardwired exchange of binary signals: from relay output to opto input.

Because the legacy devices do not support GOOSE messages, this function is performed by the gateway, which will continuously poll the legacy devices for status changes and will form and send the appropriate GOOSE messages to the network. One GOOSE message is sent for each individual logical device in the gateway (i.e., there will be one GOOSE message for each legacy IED).

If a GOOSE message has to be processed by a legacy device, the gateway will subscribe to this message and, after processing it, will send a control signal to the appropriate legacy IED for further action (Figure 18.7). This approach allows the interface of legacy devices with IEC 61850-compliant devices on the substation LAN. However, because the messages between the legacy and the IEC 61850 IED will always go through the gateway, it will be affected by its characteristics and will

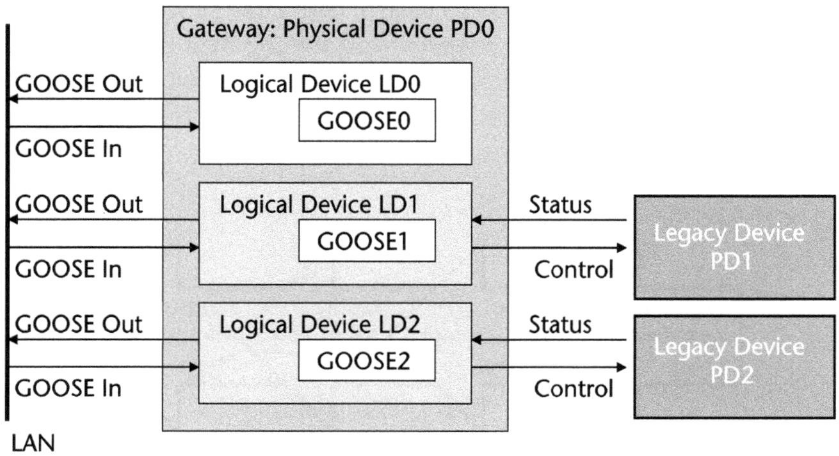

Figure 18.7 Legacy IED GOOSE interface.

always be slower than the pure P2P communications between IEC 61850 IEDs. The control system designer has to evaluate the degradation in performance and determine if this is acceptable. If not, the legacy IEDs must be replaced by IEC 61850-compliant IEDs.

18.4 Migration Process

The digitalization of the electric power industry can be traced through the second half of the last century when large mainframe computers were being used for short-circuit studies and dynamic stability analysis. With the widespread use of computers for many different engineering tasks related to PAC, more of the work that was done by hand and on paper was done by software programs and stored as digital files. The development of microprocessor-based technology revolutionized the protection and control industry by replacing electromechanical and solid-state relays and other devices with multifunctional IEDs. This resulted in what we can consider the baby steps in the digitalization of the substation environment.

Today we are witnessing another revolution: the transition from what we call conventional substations implemented by using multiple specialized devices with hardwired interfaces towards the IEC 61850-based digital substations.

The transition from the conventional grid of the last century into the digital grid of this century is an emerging trend for many utilities across the globe. However, it is not an easy one because it requires leaving the comfort zone of "that's the way we always did it" into a completely different digital environment with which many of the PAC specialists may not be familiar. That is why it requires careful analysis following a strict migration process (Figure 18.8) that can be summarized in five main phases:

- *Phase 1: Digitalization assessment:* In this phase, the users need to assess the impact that digitalization is going to have on all different domains in their electric power system. They have to analyze in detail the state of the art of the technology and the experiences that other utilities have had implementing it.

  However, they have to clearly define the short-term and long-term objectives and make the business case for the digital transformation. Also, it is necessary to perform in this phase risk assessments identifying the challenges that will be faced by implementing digitalization and also the risks of not doing so.

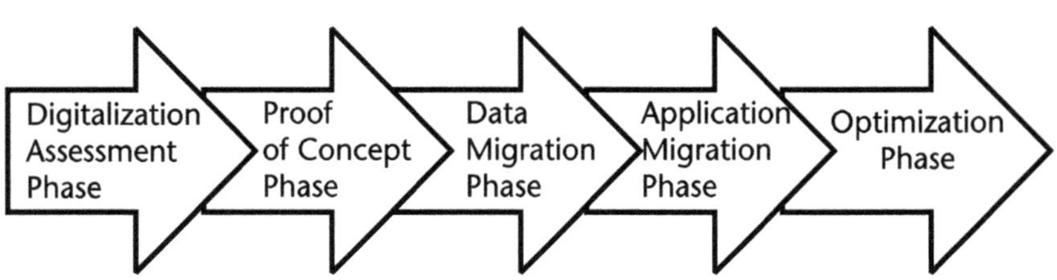

Figure 18.8 Migration process.

It is helpful also to analyze the current state of the PAC systems used within the grid and identify which domains are of higher priority for the transformation. The existing infrastructure, available human resources, and company organizational structure need to also be taken into consideration.

- *Phase 2: Proof of concept:* Considering that one of the main characteristics of the digital transformation of the PAC systems is the high level of functional integration and the sharing of digital resources that improves the efficiency and reliability of the system, it is necessary to assemble a team of domain experts that will identify the required functionality for each individual domain and the overlapping between domains from data management and communications point of view.

  Based on this analysis, the team can design a proof-of-concept project that can be implemented in two phases. The first phase is in a laboratory environment where the different components of the integrated solution can be tested while at the same time the members of the team are getting experience with the new technology and identifying potential issues with the design and the interoperability between the different selected devices used in the project. This also helps to understand how the different applications are interconnected and decide a migration strategy accordingly to meet each use case requirements. This stage also has to show how the new design and selected devices match with the existing installations and if it meets all performance and functional requirements. After all issues have been resolved in the laboratory, the project moves to the second stage in the selected typical substation where the performance of the system can be evaluated in real-life operating conditions. The experience from this project is later used to fine-tune the design of the digitized PAC system.

- *Phase 3: Data migration:* All applications in a digital substation require different kinds of data for their configuration and real-time operation. That is why, in this phase of the migration process, the user needs to perform the data migration. This is a very complex process because it requires the transformation of huge amounts of paper documentation into a digital format that can be processed by the different engineering tools and it has to define how the digitization of all the hardwired interfaces between the PAC devices and the primary substation equipment is going to be implemented. The digitization of all the interfaces between the different IEDs is also addressed at this stage. Because many applications from the different domains might be sharing data, it is very important to identify how exactly this is going to be done, how the data quality is to be maintained, and what kind of test procedures will need to be used.

  Another issue that needs to be addressed is the cybersecurity of the data. This applies to all the different files communicated over the substation LAN as well as over intersubstation communication links. The project is later used to fine-tune the design of the digitized PAC system.

  As it is expected that in many installations the migration process is going to have different stages using combinations of IEC 61850 and legacy protocols, it is necessary to identify the mappings between the data from legacy devices and the IEC 61850 model.

- *Phase 4: Applications migration:* Many of the existing substations have PAC systems based on electromechanical and solid-state relays with each one of them performing a specific protection function based on the utility's philosophy and practice. All this functionality has to be migrated into digital PAC applications defined as function elements within multifunctional IEDs or substation servers. The most efficient way of accomplishing this task is by defining standard schemes that are properly tested to verify that they perform as specified by the designing team.

  Because many of these applications are related to specific substation domains, it is necessary to understand well how they interact with applications belonging to the other domains. This is especially important considering that many of them might be using data from common sources. It is also important to ensure that the implementation of these solutions will be performed in the most efficient way by qualified personnel.

  Another critical part of the applications migration is the development of standardized test plans and procedures to be used when performing evaluation of new products or doing maintenance testing in energized substations. The applications migration strategy has also to include how future upgrades or expansions of the applications or the system will be carried and who will be responsible for doing it.
- *Phase 5: Optimization:* Even with very precise engineering taking into consideration many potential challenges and following extensive testing of the PAC schemes, it is impossible to assume that everything is perfect and operating in the most efficient way. That is why it is very important to keep an eye on the performance of the digital substation and carefully analyze any operation or undesired operation. This will allow to identify potential flows in the design that can be fixed by improving it.

It is also necessary to keep in mind that the technology is constantly evolving and something that was probably the best solution at the time of the initial design of the system may not be the most efficient way of doing things with the new available technology. Also, it is important not to forget that the IEC 61850 standard is continuously evolving based on the global user experience and that some new features might be used to optimize the performance of the PAC system.

18.5 Organizational Changes

Some of the main challenges to the digitalization of the electric power grid are the human factor and the existing organizational structures. If we analyze the conventional substation design from the last century, we see that it is based on the use of panels with dedicated functionality for a specific bay, for example, protection, control, and measurements. Each of these panels contains hardwired electromechanical devices performing different parts required by the function.

The design, selection of equipment, installation, testing, and maintenance for each of these panels is the responsibility of a department or a group within a department. All activities are defined by well-established procedures that in many cases have been used for decades by dedicated teams.

The migration from conventional to the state-of-the-art digital substations introduces some significant differences that require revisiting the organization from the point of view of the engineering, commissioning, and maintenance process.

In the conventional substations of the last century, the utilities and the suppliers had a parallel organizational structure based on the specialization in the main function domains shown in Figure 18.9.

Each of the different domains in the utility or the manufacturer will:

- Define the functional requirements;
- Select the equipment to be used;
- Define the testing requirements;
- Select the testing equipment;
- Design their own system for a specific substation;
- Define the test plans;
- Perform the commissioning;
- Perform the maintenance.

Because digital substations result in high levels of functional integration and sharing of common data sources, they require changes in the organizational structures of both utilities and vendors in order to enable coordination of protection, control, monitoring, and recording activities for development, engineering, and operation. This will require personnel to get out of their comfort zone of a single domain, to learn, maintain the competence, and be efficient in a huge variety of new concepts and tools across different disciplines in order to be able to understand the capabilities and functional interdependencies of the advanced multifunctional IEDs operating in distributed applications over the substation or wide-area network. At the same time, the processes that have been well established and the responsibilities that have been assigned will have to change as well to meet the requirements of the new world.

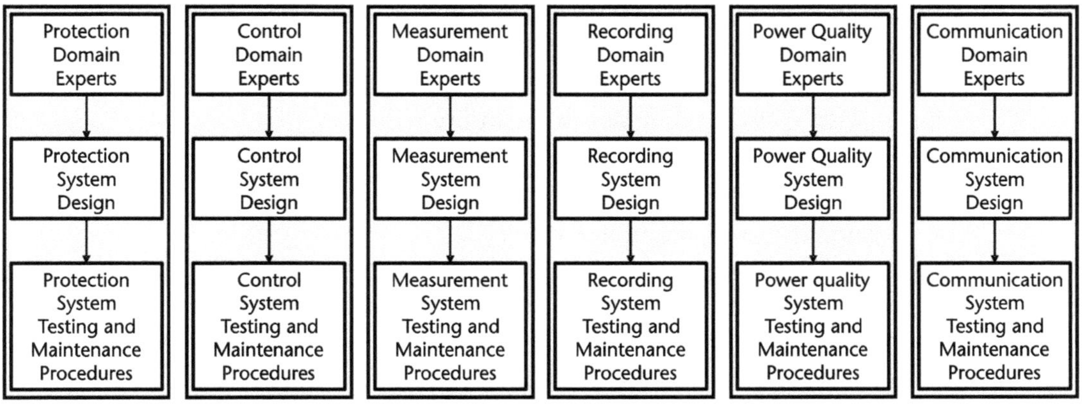

Figure 18.9 Workflow structure for conventional PAC systems.

While there is still a need for an extremely high level of domain expertise by some experts who are responsible to define the functional and testing requirements, at the same time, there is a need for system architects with a broader view of the system requirements and the integration of the different functions in a complete system with high efficiency. Furthermore, this will require ongoing training as new tools and technology are introduced.

Another difficulty introduced by functional integration due to the interdependency between different functions is the need to establish a common process for their enabling or disabling, because it may affect the behavior of the system under certain conditions. This may have an impact on the organization of the type, acceptance, and commissioning testing process.

The users thus must acquire a deeper understanding of all these issues in order to be able to configure and to operate a digital SAS with a high level of functional integration. This represents a major change with respect to the skills required for maintenance and operation of a conventional control system.

The integration of communication functions in the multifunctional substation devices requires a new set of skills in order to be able to configure, commission, maintain, and test such devices and systems. The testing of a system with a high level of functional integration also requires a different set of skills and tools, as well as new methods that will allow the testing of all components and the system as a whole.

Functional integration also potentially has an impact on substation configuration. The advantage is that, in an integrated context, common parameters can be shared allowing the reduction of the work of entering the different values and the checking of the consistency of the parameters of different functions. This may reduce the flexibility of defining different values for corresponding parameters of two integrated functions. In this context, it can be pointed out that particular care has to be taken by the vendors for designing the user interface if many functions are in one IED in order to guarantee a certain user-friendliness.

At the same time, it is important to clearly understand the implementation of specific functions in integrated devices. A good example is the recording of short-circuit faults or other transient events. When making a decision of what devices should be used for transient recording, the designer of the system needs to consider if the recording capabilities of a multifunctional protection IED are sufficient or a dedicated recording device that meets the requirements should be used.

In an integrated functional environment, it is impossible to follow the principles and workflows used for conventional substations as shown in Figure 18.9. A new workflow (see Figure 18.10) will need to be implemented.

Each of the different domains in the utility will:

- Define the functional requirements;
- Define the testing requirements.

Based on these requirements, an integrated team including experts from each domain will:

- Select the multifunctional IEDs to be used;

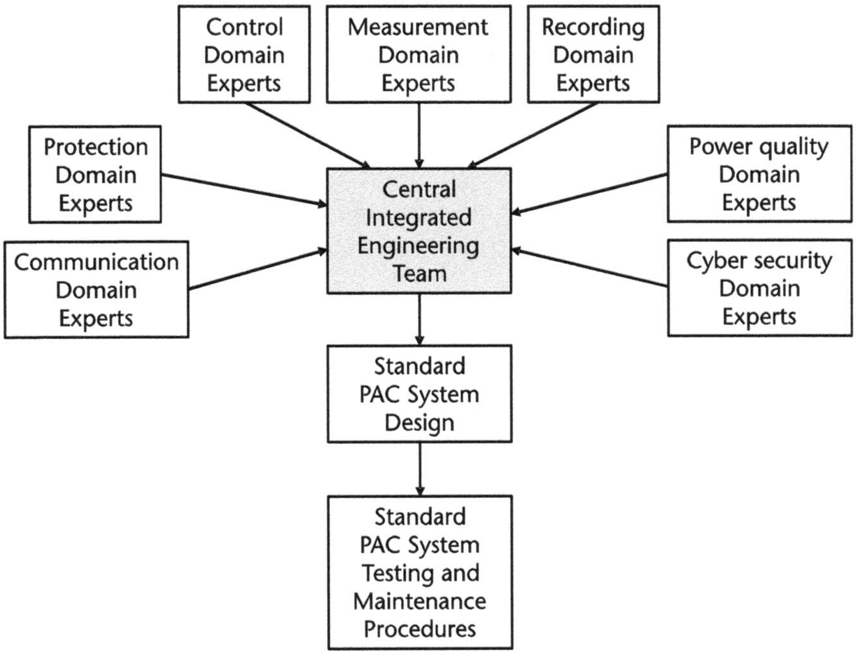

Figure 18.10 Integrated engineering organization.

- Select the testing equipment;
- Design the integrated system for a specific substation (based on standard schemes);
- Define the integrated test plans;
- Perform the commissioning of all functions;
- Perform the maintenance of all functions.

This workflow will require a new approach based on the creation of a hierarchical organization of cross-domain teams that are responsible for the definition of the testing requirements, methods, and tools as part of the system engineering process, including the development of standardized test plans that can support test automation.

Under such a new organizational hierarchy, domain experts for protection, control, automation, monitoring, recording, and power quality will form centralized engineering teams that will define the functional requirements and, based on them, will develop standard PAC schemes.

Following a procurement process, the central integrated team will select the multifunctional IEDs to be used in the PAC schemes and will define the test procedures.

Test procedures should be defined for the different types of tests, such as:

- IED acceptance testing;
- Scheme acceptance testing;
- Factory acceptance testing;

- Commissioning testing;
- Site acceptance testing;
- Maintenance testing.

After extensive testing based on these test procedures and gaining real-life experience through some pilot projects, the PAC schemes are ready for deployment, thus completing the digital transformation.

About the Author

Alexander Apostolov received an MS in electrical engineering, an MS in applied mathematics, and a PhD from the Technical University in Sofia, Bulgaria. He has 49 years of experience in power systems protection, automation, control, and communications.

Dr. Apostolov is presently a principal engineer for OMICRON electronics in Los Angeles, California. He is an IEEE Life Fellow and a member of the IEEE PES Power Systems Relaying and Control (PSRC) Committee. He is a past chairman of the Relay Communications Subcommittee and serves on many IEEE PES working groups. He received the IEEE PES Distinguished Service Award in 2007 and is an IEEE Distinguished Lecturer.

Dr. Apostolov is a member of the IEC TC57 working groups 10, 17, and 19. He has been involved in the development of UCA 2.0 and IEC 61850 for more than 25 years. He is the Convener of CIGRE WG B5.69 "Experience Gained and Recommendations for Implementation of Process Bus in Protection, Automation and Control Systems (PACS)" and is a member of several other CIGRE B5 working groups. He received the 2007 CIGRE Technical Committee Award, the CIGRE 2014 Distinguished Member Award, the 2014 CIGRE USNC Attwood Associate Award, and the 2017 CIGRE Study Committee B5 Distinguished Service Award. He holds four patents and has authored and presented more than 600 technical papers. He is the editor in chief of *PAC World* and the chairman of the PAC World Conference.

Index

D

E

F

G

L

N

O

P

R

S

Artech House Power Engineering Library

Andres Carvallo, Series Editor

Advanced Technology for Smart Buildings, James Sinopoli

The Advanced Smart Grid: Edge Power Driving Sustainability, Second Edition, Andres Carvallo and John Cooper

Applications of Energy Harvesting Technologies in Buildings, Joseph W. Matiko and Stephen Beeby

Battery Management Systems, Volume I: Battery Modelings, Gregory L. Plett

Battery Management Systems, Volume II: Equivalent-Circuit Methods, Gregory L. Plett

Battery Management Systems for Large Lithium Ion Battery Packs, Davide Andrea

Battery Power Management for Portable Devices, Yevgen Barsukov and Jinrong Qian

Big Data Analytics for Connected Vehicles and Smart Cities, Bob McQueen

Design and Analysis of Large Lithium-Ion Battery Systems, Shriram Santhanagopalan, Kandler Smith, Jeremy Neubauer, Gi-Heon Kim, Matthew Keyser, and Ahmad Pesaran

Designing Control Loops for Linear and Switching Power Supplies: A Tutorial Guide, Christophe Basso

Electric Power System Fundamentals, Salvador Acha Daza

Electric Systems Operations: Evolving to the Modern Grid, Second Edition, Mani Vadari

Energy Harvesting for Autonomous Systems, Stephen Beeby and Neil White

Energy Storage Technologies and Applications, C. Michael Hoff

GIS for Enhanced Electric Utility Performance, Bill Meehan

IEC 61850 Demystified, Herbert Falk

IEC 61850: Digitizing the Electric Power Grid, Alexander Apostolov

Introduction to Power Electronics, Paul H. Chappell

Introduction to Power Utility Communications, Harvey Lehpamer

IoT Technical Challenges and Solutions, Arpan Pal and Balamuralidhar Purushothaman

Lithium-Ion Batteries and Applications: A Practical and Comprehensive Guide to Lithium-Ion Batteries and Arrays, from Toys to Towns, Volume 1, Batteries, Davide Andrea

Lithium-Ion Batteries and Applications: A Practical and Comprehensive Guide to Lithium-Ion Batteries and Arrays, from Toys to Towns, Volume 2, Applications, Davide Andrea

Lithium-Ion Battery Failures in Consumer Electronics, Ashish Arora, Sneha Arun Lele, Noshirwan Medora, and Shukri Souri

Microgrid Design and Operation: Toward Smart Energy in Cities, Federico Delfino, Renato Procopio, Mansueto Rossi, Stefano Bracco, Massimo Brignone, and Michela Robba

Plug-in Electric Vehicle Grid Integration, Islam Safak Bayram and Ali Tajer

Power Grid Resiliency for Adverse Conditions, Nicholas Abi-Samra

Power Line Communications in Practice, Xavier Carcelle

Power System State Estimation, Mukhtar Ahmad

Renewable Energy Technologies and Resources, Nader Anani

Signal Processing for RF Circuit Impairment Mitigation in Wireless Communications, Xinping Huang, Zhiwen Zhu, and Henry Leung

The Smart Grid as An Application Development Platform, George Koutitas and Stan McClellan

Smart Grid Redefined: Transformation of the Electric Utility, Mani Vadari

Sustainable Power, Autonomous Ships, and Cleaner Energy for Shipping, John Erik Hagen

Synergies for Sustainable Energy, Elvin Yüzügüllü

A Systems Approach to Lithium-Ion Battery Management, Phil Weicker

Telecommunication Networks for the Smart Grid, Alberto Sendin, Miguel A. Sanchez-Fornie, Iñigo Berganza, Javier Simon, and Iker Urrutia

A Whole-System Approach to High-Performance Green Buildings, David Strong and Victoria Burrows

For further information on these and other Artech House titles, including previously considered out-of-print books now available through our In-Print-Forever® (IPF®) program, contact:

Artech House
685 Canton Street
Norwood, MA 02062
Phone: 781-769-9750
Fax: 781-769-6334
e-mail: artech@artechhouse.com

Artech House
16 Sussex Street
London SW1V 4RW UK
Phone: +44 (0)20 7596-8750
Fax: +44 (0)20 7630-0166
e-mail: artech-uk@artechhouse.com

Find us on the World Wide Web at: www.artechhouse.com

www.ingramcontent.com/pod-product-compliance
Ingram Content Group UK Ltd.
Pitfield, Milton Keynes, MK11 3LW, UK
UKHW050947290926
15470UKWH00005B/1691